UNIVERSITY LIBRARY
UW-STEVENS POINT
P9-CBT-192

ADVANCES IN CHEMICAL PHYSICS

VOLUME 140

EDITORIAL BOARD

Bruce J. Berne, Department of Chemistry, Columbia University, New York, New York, U.S.A.

Kurt Binder, Institut für Physik, Johannes Gutenberg-Universität Mainz, Mainz, Germany

A. Welford Castleman, Jr., Department of Chemistry, The Pennsylvania State University, University Park, Pennsylvania, U.S.A.

David Chandler, Department of Chemistry, University of California, Berkeley, California, U.S.A.

M. S. Child, Department of Theoretical Chemistry, University of Oxford, Oxford, U.K.

William T. Coffey, Department of Microelectronics and Electrical Engineering, Trinity College, University of Dublin, Dublin, Ireland

F. Fleming Crim, Department of Chemistry, University of Wisconsin, Madison, Wisconsin, U.S.A.

Ernest R. Davidson, Department of Chemistry, Indiana Univeristy, Bloomington, Indiana, U.S.A.

Graham R. Fleming, Department of Chemistry, The University of California, Berkeley, California, U.S.A.

Karl F. Freed, The James Franck Institute, The University of Chicago, Chicago, Illinois, U.S.A.

Pierre Gaspard, Center for Nonlinear Phenomena and Complex Systems, Université Libre de Bruxelles, Brussels, Belgium

Eric J. Heller, Department of Chemistry, Harvard-Smithsonian Center for Astrophysics, Cambridge, Massachusetts, U.S.A.

Robin M. Hochstrasser, Department of Chemistry, The University of Pennsylvania, Philadelphia, Pennsylvania, U.S.A.

R. Kosloff, The Fritz Haber Research Center for Molecular Dynamics and Department of Physical Chemistry, The Hebrew University of Jerusalem, Jerusalem, Israel

Rudolph A. Marcus, Department of Chemistry, California Institute of Technology, Pasadena, California, U.S.A.

G. Nicolis, Center for Nonlinear Phenomena and Complex Systems, Université Libre de Bruxelles, Brussels, Belgium

Thomas P. Russell, Department of Polymer Science, University of Massachusetts, Amherst, Massachusetts, U.S.A.

Donald G. Truhlar, Department of Chemistry, University of Minnesota, Minneapolis, Minnesota, U.S.A.

John D. Weeks, Institute for Physical Science and Technology and Department of Chemistry, University of Maryland, College Park, Maryland, U.S.A.

Peter G. Wolynes, Department of Chemistry, University of California, San Diego, California, U.S.A.

ADVANCES IN CHEMICAL PHYSICS

VOLUME 140

Series Editor

STUART A. RICE

Department of Chemistry
and
The James Franck Institute
The University of Chicago
Chicago, Illinois

WILEY

A JOHN WILEY & SONS, INC. PUBLICATION

Copyright © 2008 by John Wiley & Sons, Inc. All rights reserved

Published by John Wiley & Sons, Inc., Hoboken, New Jersey
Published simultaneously in Canada

No part of this publication may be reproduced, stored in a retrieval system, or transmitted in any form or by any means, electronic, mechanical, photocopying, recording, scanning, or otherwise, except as permitted under Section 107 or 108 of the 1976 United States Copyright Act, without either the prior written permission of the Publisher, or authorization through payment of the appropriate per-copy fee to the Copyright Clearance Center, Inc., 222 Rosewood Drive, Danvers, MA 01923, (978) 750-8400, fax (978) 750-4470, or on the web at www.copyright.com. Requests to the Publisher for permission should be addressed to the Permissions Department, John Wiley & Sons, Inc., 111 River Street, Hoboken, NJ 07030, (201) 748-6011, fax (201) 748-6008, or online at http://www.wiley.com/go/permission.

Limit of Liability/Disclaimer of Warranty: While the publisher and author have used their best efforts in preparing this book, they make no representations or warranties with respect to the accuracy or completeness of the contents of this book and specifically disclaim any implied warranties of merchantability or fitness for a particular purpose. No warranty may be created or extended by sales representatives or written sales materials. The advice and strategies contained herein may not be suitable for your situation. You should consult with a professional where appropriate. Neither the publisher nor author shall be liable for any loss of profit or any other commercial damages, including but not limited to special, incidental, consequential, or other damages.

For general information on our other products and services or for technical support, please contact our Customer Care Department within the United States at (800) 762-2974, outside the United States at (317) 572-3993 or fax (317) 572-4002.

Wiley also publishes its books in a variety of electronic formats. Some content that appears in print may not be available in electronic formats. For more information about Wiley products, visit our web site at www.wiley.com.

Library of Congress Catalog Number: 58-9935

ISBN: 978-0-470-22688-9

Printed in the United States of America

10 9 8 7 6 5 4 3 2 1

QD
453
.A27
v.140

CONTRIBUTORS TO VOLUME 140

PHIL ATTARD, School of Chemistry F11, University of Sydney, NSW 2006 Australia

THOMAS BARTSCH, Department of Mathematical Sciences, Loughborough University, Loughborough LE11 3TU, United Kingdom

RAJAT K. CHAUDHURI, Indian Institute of Astrophysics, Bangalore 560034, India

ROBERT J. GORDON, Department of Chemistry, University of Illinois at Chicago, Chicago, Illinois 60607 USA

RIGOBERTO HERNANDEZ, Center for Computational Molecular Science and Technology, Georgia Institute of Technology, Atlanta, Georgia 30332 USA

RAYMOND KAPRAL, University of Toronto, Toronto, Canada

SHINNOSUKE KAWAI, Molecule and Life Nonlinear Sciences Laboratory, Research Institute for Electronic Science (RIES), Hokkaido University, Sapporo 060-0812 Japan

JEREMY M. MOIX, Center for Computational Molecular Science and Technology, Georgia Institute of Technology, Atlanta, Georgia 30332 USA

MALAYA K. NAYAK, Indian Institute of Astrophysics, Bangalore 560034, India

TAMAR SEIDEMAN, Department of Chemistry, Northwestern University, Evanston, Illinois 60208 USA

T. UZER, Center for Nonlinear Science, Georgia Institute of Technology, Atlanta, Georgia 30332 USA

INTRODUCTION

Few of us can any longer keep up with the flood of scientific literature, even in specialized subfields. Any attempt to do more and be broadly educated with respect to a large domain of science has the appearance of tilting at windmills. Yet the synthesis of ideas drawn from different subjects into new, powerful, general concepts is as valuable as ever, and the desire to remain educated persists in all scientists. This series, *Advances in Chemical Physics*, is devoted to helping the reader obtain general information about a wide variety of topics in chemical physics, a field that we interpret very broadly. Our intent is to have experts present comprehensive analyses of subjects of interest and to encourage the expression of individual points of view. We hope that this approach to the presentation of an overview of a subject will both stimulate new research and serve as a personalized learning text for beginners in a field.

STUART A. RICE

CONTENTS

THE SECOND LAW OF NONEQUILIBRIUM THERMODYNAMICS: HOW FAST TIME FLIES

PHIL ATTARD

School of Chemistry F11, University of Sydney, NSW 2006 Australia

CONTENTS

Advances in Chemical Physics, Volume 140, edited by Stuart A. Rice
Copyright © 2008 John Wiley & Sons, Inc.

I. INTRODUCTION

The Second Law of Equilibrium Thermodynamics may be stated:

The entropy increases during spontaneous changes
in the structure of the total system (1)

This is a law about the equilibrium state, when macroscopic change has ceased; it is the state, according to the law, of maximum entropy. It is not really a law about nonequilibrium *per se*, not in any quantitative sense, although the law does introduce the notion of a nonequilibrium state constrained with respect to structure. By implication, entropy is perfectly well defined in such a nonequilibrium macrostate (otherwise, how could it increase?), and this constrained entropy is less than the equilibrium entropy. Entropy itself is left undefined by the Second Law, and it was only later that Boltzmann provided the physical interpretation of entropy as the number of molecular configurations in a macrostate. This gave birth to his probability distribution and hence to equilibrium statistical mechanics.

The reason that the Second Law has no quantitative relevance to nonequilibrium states is that it gives the direction of change, not the rate of change. So although it allows the calculation of the thermodynamic force that drives the system toward equilibrium, it does not provide a basis for calculating the all important rate at which the system evolves. A full theory for the nonequilibrium state cannot be based solely on the Second Law or the entropy it invokes.

This begs the question of whether a comparable law exists for nonequilibrium systems. This chapter presents a theory for nonequilibrium thermodynamics and statistical mechanics based on such a law written in a form analogous to the equilibrium version:

The second entropy increases during spontaneous changes in the dynamic structure of the total system (2)

Here dynamic structure gives a macroscopic flux or rate; it is a transition between macrostates in a specified time. The law invokes the notion of constrained fluxes and the notion that fluxes cease to change in the optimum state, which, in common parlance, is the steady state. In other words, the principle governing nonequilibrium systems is that in the transient regime fluxes develop and evolve to increase the second entropy, and that in the steady state the macroscopic fluxes no longer change and the second entropy is maximal. The second entropy could also be called the transition entropy, and as the reader has probably already guessed, it is the number of molecular configurations associated with a transition between macrostates in a specified time.

This nonequilibrium Second Law provides a basis for a theory for nonequilibrium thermodynamics. The physical identification of the second entropy in terms of molecular configurations allows the development of the nonequilibrium probability distribution, which in turn is the centerpiece for nonequilibrium statistical mechanics. The two theories span the very large and the very small. The aim of this chapter is to present a coherent and self-contained account of these theories, which have been developed by the author and presented in a series of papers [1–7]. The theory up to the fifth paper has been reviewed previously [8], and the present chapter consolidates some of this material and adds the more recent developments.

Because the focus is on a single, albeit rather general, theory, only a limited historical review of the nonequilibrium field is given (see Section IA). That is not to say that other work is not mentioned in context in other parts of this chapter. An effort has been made to identify where results of the present theory have been obtained by others, and in these cases some discussion of the similarities and differences is made, using the nomenclature and perspective of the present author. In particular, the notion and notation of constraints and exchange with a reservoir that form the basis of the author's approach to equilibrium thermodynamics and statistical mechanics [9] are used as well for the present nonequilibrium theory.

A. Review and Preview

The present theory can be placed in some sort of perspective by dividing the nonequilibrium field into thermodynamics and statistical mechanics. As will become clearer later, the division between the two is fuzzy, but for the present purposes nonequilibrium thermodynamics will be considered that phenomenological theory that takes the existence of the transport coefficients and laws as axiomatic. Nonequilibrium statistical mechanics will be taken to be that field that deals with molecular-level (i.e., phase space) quantities such as probabilities and time correlation functions. The probability, fluctuations, and evolution of macrostates belong to the overlap of the two fields.

Perhaps the best starting point in a review of the nonequilibrium field, and certainly the work that most directly influenced the present theory, is Onsager's celebrated 1931 paper on the reciprocal relations [10]. This showed that the symmetry of the linear hydrodynamic transport matrix was a consequence of the time reversibility of Hamilton's equations of motion. This is an early example of the overlap between macroscopic thermodynamics and microscopic statistical mechanics. The consequences of time reversibility play an essential role in the present nonequilibrium theory, and in various fluctuation and work theorems to be discussed shortly.

Moving upward to the macroscopic level, the most elementary phenomenological theories for nonequilibrium thermodynamics are basically hydrodynamics plus localized equilibrium thermodynamics [11, 12]. In the so-called soft sciences, including, as examples, biological, environmental, geological, planetary, atmospheric, climatological, and paleontological sciences, the study of evolution and rates of change is all important. This has necessarily stimulated much discussion of nonequilibrium principles and approaches, which are generally related to the phenomenological theories just described [13–18]. More advanced phenomenological theories for nonequilibrium thermodynamics in its own right have been pursued [19–23]. The phenomenological theories generally assert the existence of a nonequilibrium potential that is a function of the fluxes, and whose derivatives are consistent with the transport laws and other symmetry requirements.

In view of the opening discussion of the two second laws of thermodynamics, in analyzing all theories, phenomenological and otherwise, it is important to ask two questions:

$$\text{Does the theory invoke the second entropy,} \atop \text{or only the first entropy and its rate of change?} \tag{3}$$

and

$$\text{Is the relationship being invoked true in general,} \atop \text{or is it only true in the optimum or steady state?} \tag{4}$$

If the approach does not go beyond the ordinary entropy, or if it applies an optimized result to a constrained state, then one can immediately conclude that a quantitative theory for the nonequilibrium state is unlikely to emerge. Regrettably, for phenomenological theories of the type just discussed, the answer to both questions is usually negative. The contribution of Prigogine, in particular, will be critically assessed from these twin perspectives (see Section IIE).

Moving downward to the molecular level, a number of lines of research flowed from Onsager's seminal work on the reciprocal relations. The symmetry rule was extended to cases of mixed parity by Casimir [24], and to nonlinear transport by Grabert et al. [25] Onsager, in his second paper [10], expressed the linear transport coefficient as an equilibrium average of the product of the present and future macrostates. Nowadays, this is called a time correlation function, and the expression is called Green–Kubo theory [26–30].

The transport coefficient gives the ratio of the flux (or future macrostate velocity) to the conjugate driving force (mechanical or thermodynamic). It governs the dissipative force during the stochastic and driven motion of a macrostate, and it is related to the strength of the fluctuations by the fluctuation-dissipation theorem [31]. Onsager and Machlup [32] recognized that the transport theory gave rise to a related stochastic differential equation for the evolution of a macrostate that is called the Langevin equation (or the Smoluchowski equation in the overdamped case). Applied to the evolution of a probability distribution it is the Fokker–Planck equation [33]. In the opinion of the present author, stochastic differential equations such as these result from a fundamental, molecular-level nonequilibrium theory, but in themselves are not fundamental and they do not provide a basis for constructing a full nonequilibrium theory.

Onsager and Machlup [32] gave expressions for the probability of a path of macrostates and, in particular, for the probability of a transition between two macrostates. The former may be regarded as the solution of a stochastic differential equation. It is technically a Gaussian Markov process, also known as an Ornstein–Uhlenbeck process. More general stochastic processes include, for example, the effects of spatial curvature and nonlinear transport [33–35]. These have been accounted for by generalizing the Onsager–Machlup functional to give the so-called thermodynamic Lagrangian [35–42]. Other thermodynamic Lagrangians have been given [43–46]. The minimization of this functional gives the most probable evolution in time of the macrostate, and hence one might expect the thermodynamic Lagrangian to be related (by a minus sign) to the second entropy that is the basis of the present theory. However, the Onsager–Machlup functional [32] (and those generalizations of it) [35–42] fails both questions posed above: **(1)** it invokes solely the rate of production of first entropy, and **(2)** both expressions that it invokes for this are only valid in the steady state, not in the constrained states that are the subject of the optimization procedure (see Section IIE). The Onsager–Machlup functional (in two-state transition form) is tested against computer simulation data for the thermal conductivity time correlation function in Fig. 8.

On a related point, there have been other variational principles enunciated as a basis for nonequilibrium thermodynamics. Hashitsume [47], Gyarmati [48, 49], and Bochkov and Kuzovlev [50] all assert that in the steady state the rate of first entropy production is an extremum, and all invoke a function identical to that underlying the Onsager–Machlup functional [32]. As mentioned earlier, Prigogine [11] (and workers in the broader sciences) [13–18] variously asserts that the rate of first entropy production is a maximum or a minimum and invokes the same two functions for the optimum rate of first entropy production that were used by Onsager and Machlup [32] (see Section IIE).

Evans and Baranyai [51, 52] have explored what they describe as a nonlinear generalization of Prigogine's principle of minimum entropy production. In their theory the rate of (first) entropy production is equated to the rate of phase space compression. Since phase space is incompressible under Hamilton's equations of motion, which all real systems obey, the compression of phase space that occurs in nonequilibrium molecular dynamics (NEMD) simulations is purely an artifact of the non-Hamiltonian equations of motion that arise in implementing the Evans–Hoover thermostat [53, 54]. (See Section VIIIC for a critical discussion of the NEMD method.) While the NEMD method is a valid simulation approach in the linear regime, the phase space compression induced by the thermostat awaits physical interpretation; even if it does turn out to be related to the rate of first entropy production, then the hurdle posed by Question (3) remains to be surmounted.

In recent years there has been an awakening of interest in fundamental molecular-level theorems in nonequilibrium statistical mechanics. This spurt of theoretical and experimental activity was kindled by the work theorem published by Jarzynski in 1997 [55]. The work theorem is in fact a trivial consequence of the fluctuation theorem published by Evans, Cohen, and Morriss in 1993, [56, 57] and both theorems were explicitly given earlier by Bochkov and Kuzovlev in 1977 [58–60]. As mentioned earlier, since Onsager's work in 1931 [10], time reversibility has played an essential role in nonequilibrium theory. Bochkov and Kuzovlev [60], and subsequent authors including the present one [4], have found it exceedingly fruitful to consider the ratio of the probability of a forward trajectory to that of the reversed trajectory. Using time reversibility, this ratio can be related to the first entropy produced on the forward trajectory, and it has come to be called the fluctuation theorem [56, 57]. An alternative derivation assuming Markovian behavior of the macrostate path probability has been given [61, 62], and it has been demonstrated experimentally [63]. From this ratio one can show that the average of the exponential of the negative of the entropy produced (minus work divided by temperature) equals the exponential of the difference in initial and final

Helmholtz free energies divided by temperature, which is the work theorem [55]. For a cyclic process, the latter difference is zero, and hence the average is unity, as shown by Bochkov and Kuzovlev [58–60]. The work theorem has been rederived in different fashions [57, 64, 65] and verified experimentally [66]. What is remarkable about the work theorem is that it holds for arbitrary rates of nonequilibrium work, and there is little restriction beyond the assumption of equilibration at the beginning and end of the work and sufficiently long time interval to neglect end effects. (See Sections IVC4 and VB for details and generalizations.)

With the exception of the present theory, derivations of the fluctuation and work theorems are generally for a system that is isolated during the performance of the work (adiabatic trajectories), and the effects of a thermal or other reservoir on the true nonequilibrium probability distribution or transition probability are neglected. The existence and form for the nonequilibrium probability distribution, both in the steady state and more generally, may be said to be the holy grail of nonequilibrium statistical mechanics. The Boltzmann distribution is the summit of the equilibrium field [67], and so there have been many attempts to formulate its analogue in a nonequilibrium context. The most well known is the Yamada–Kawasaki distribution [68, 69]. It must be stressed that this distribution is an adiabatic distribution, which is to say that it assumes that the system was in thermal equilibrium in the past, and that it was subsequently isolated from the thermal reservoirs while the work was being performed so that no heat was exchanged during the time-dependent process. This is obviously a very restrictive assumption. Attempts have been made to formulate a thermostatted form of the Yamada–Kawasaki distribution, but this has been found to be computationally intractable [53, 70, 71]. As pointed out earlier, most derivations of the fluctuation and work theorems are predicated on the adiabatic assumption, and a number of authors invoke or derive the Yamada–Kawasaki distribution, apparently unaware of its prior publication and of its restricted applicability.

An alternative approximation to the adiabatic probability is to invoke an instantaneous equilibrium-like probability. In the context of the work theorem, Hatano and Sasa [72] analyzed a nonequilibrium probability distribution that had no memory, and others have also invoked a nonequilibrium probability distribution that is essentially a Boltzmann factor of the instantaneous value of the time-dependent potential [73, 74].

In Sections IVA, VA, and VI the nonequilibrium probability distribution is given in phase space for steady-state thermodynamic flows, mechanical work, and quantum systems, respectively. (The second entropy derived in Section II gives the probability of fluctuations in macrostates, and as such it represents the nonequilibrium analogue of thermodynamic fluctuation theory.) The present phase space distribution differs from the Yamada–Kawasaki distribution in that it correctly takes into account heat exchange with a reservoir during the mechanical work or thermodynamic flux. The probability distribution is the product of a Boltzmann-like term, which is reversible in time, and a new term, which is odd in time, and which once more emphasizes Onsager's [10] foresight in identifying time reversibility as the key to nonequilibrium behavior. In Section IVB this phase space probability is used to derive the Green–Kubo relations, in Section VIIIB it is used to develop a nonequilibrium Monte Carlo algorithm, and in Fig. 7 it is shown that the algorithm is computationally feasible and that it gives a thermal conductivity in full agreement with conventional NEMD results.

In addition to these nonequilibrium probability densities, the present theory also gives expressions for the transition probability and for the probability of a phase space trajectory, in both equilibrium and nonequilibrium contexts, (Sections IVC and VB). These sections contain the derivations and generalizations of the fluctuation and work theorems alluded to earlier. As for the probability density, one has to be aware that some work in the literature is based on adiabatic transitions, whereas the present approach includes the effect of heat flow on the transition. One also has to distinguish works that deal with macrostate transitions, from the present approach based in phase space, which of course includes macrostate transition by integration over the microstates. The second entropy, which is the basis for the nonequilibrium second law advocated earlier, determines such transitions pairwise, and for an interval divided into segments of intermediate length, it determines a macrostate path by a Markov procedure (Sections IIC and IIIC). The phase space trajectory probability contains an adiabatic term and a stochastic term. The latter contains in essence half the difference between the target and initial reservoir entropies. This term may be seen to be essentially the one that is invoked in Glauber or Kawasaki dynamics [75–78]. This form for the conditional stochastic transition probability satisfies detailed balance for an equilibrium Boltzmann distribution, and it has been used successfully in hybrid equilibrium molecular dynamics algorithms [79–81]. Using the term on its own without the adiabatic development, as in Glauber or Kawasaki dynamics, corresponds to neglecting the coupling inherent in the second entropy, and to losing the speed of time.

II. LINEAR THERMODYNAMICS

A. Formalities

Consider an isolated system containing N molecules, and let $\mathbf{\Gamma} \equiv \{\mathbf{q}^N, \mathbf{p}^N\}$ be a point in phase space, where the ith molecule has position $\mathbf{q}_i$ and momentum $\mathbf{p}_i$. In developing the nonequilibrium theory, it will be important to discuss the behavior of the system under time reversal. Accordingly, define the conjugate point in phase space as that point with all the velocities reversed, $\mathbf{\Gamma}^\dagger \equiv \{\mathbf{q}^N, (-\mathbf{p})^N\}$. If $\mathbf{\Gamma}_2 = \mathbf{\Gamma}_0(t|\mathbf{\Gamma}_1)$ is the position of the isolated system at time t given that it was at $\mathbf{\Gamma}_1$ at time $t = 0$, then $\mathbf{\Gamma}_1^\dagger = \mathbf{\Gamma}_0(t|\mathbf{\Gamma}_2^\dagger)$, as follows from the reversibility of Hamilton's equations of motion. One also has, by definition of the trajectory, that $\mathbf{\Gamma}_1 = \mathbf{\Gamma}_0(-t|\mathbf{\Gamma}_2)$.

Macrostates are collections of microstates [9], which is to say that they are volumes of phase space on which certain phase functions have specified values. The current macrostate of the system gives its structure. Examples are the position or velocity of a Brownian particle, the moments of energy or density, their rates of change, the progress of a chemical reaction, a reaction rate, and so on. Let $\mathbf{x}$ label the macrostates of interest, and let $\hat{\mathbf{x}}(\mathbf{\Gamma})$ be the associated phase function. The first entropy of the macrostate is

$$S^{(1)}(\mathbf{x}|\mathbf{E}) = k_\mathrm{B} \ln \int d\mathbf{\Gamma}\, \delta(\mathcal{H}(\mathbf{\Gamma}) - E)\, \delta(\hat{\mathbf{x}}(\mathbf{\Gamma}) - \mathbf{x}) \tag{5}$$

neglecting an arbitrary constant. This is the ordinary entropy; here it is called the first entropy, to distinguish it from the second or transition entropy that is introduced later. Here the Hamiltonian appears, and all microstates of the isolated system with energy E are taken to be equally likely [9]. This is the constrained entropy, since the system is constrained to be in a particular macrostate. By definition, the probability of the macrostate is proportional to the exponential of the entropy,

$$\wp(\mathbf{x}|E) = \frac{1}{W(E)} \exp S^{(1)}(\mathbf{x}|E)/k_\mathrm{B} \tag{6}$$

The normalizing factor is related to the unconstrained entropy by

$$\begin{aligned} S^{(1)}(E) \equiv k_\mathrm{B} \ln W(E) &= k_\mathrm{B} \ln \int d\mathbf{x}\, \exp S(\mathbf{x}|E)/k_\mathrm{B} \\ &= k_\mathrm{B} \ln \int d\mathbf{\Gamma}\, \delta(\mathcal{H}(\mathbf{\Gamma}) - E) \end{aligned} \tag{7}$$

The equilibrium state, which is denoted $\overline{\mathbf{x}}$, is by definition both the most likely state, $\wp(\overline{\mathbf{x}}|E) \geq \wp(\mathbf{x}|E)$, and the state of maximum constrained entropy, $S^{(1)}(\overline{\mathbf{x}}|E) \geq S^{(1)}(\mathbf{x}|E)$. This is the statistical mechanical justification for much of the import of the Second Law of Equilibrium Thermodynamics. The unconstrained entropy, as a sum of positive terms, is strictly greater than the maximal constrained entropy, which is the largest term, $S^{(1)}(E) > S^{(1)}(\overline{\mathbf{x}}|E)$. However, in the thermodynamic limit when fluctuations are relatively negligible, these may be equated with relatively little error, $S^{(1)}(E) \approx S^{(1)}(\overline{\mathbf{x}}|E)$.

The macrostates can have either even or odd parity, which refers to their behavior under time reversal or conjugation. Let $\epsilon_i = \pm 1$ denote the parity of the ith microstate, so that $\hat{x}_i(\mathbf{\Gamma}^\dagger) = \epsilon_i \hat{x}_i(\mathbf{\Gamma})$. (It is assumed that each state is purely even or odd; any state of mixed parity can be written as the sum of two states of pure parity.) Loosely speaking, variables with even parity may be called position variables, and variables with odd parity may be called velocity variables. One can form the diagonal matrix $\underline{\underline{\epsilon}}$, with elements $\epsilon_i \delta_{ij}$, so that $\hat{\mathbf{x}}(\mathbf{\Gamma}^\dagger) = \underline{\underline{\epsilon}}\hat{\mathbf{x}}(\mathbf{\Gamma})$. The parity matrix is its own inverse, $\underline{\underline{\epsilon}}\,\underline{\underline{\epsilon}} = \underline{\underline{I}}$.

The Hamiltonian is insensitive to the direction of time, $\mathcal{H}(\mathbf{\Gamma}) = \mathcal{H}(\mathbf{\Gamma}^\dagger)$, since it is a quadratic function of the molecular velocities. (Since external Lorentz or Coriolis forces arise from currents or velocities, they automatically reverse direction under time reversal.) Hence both $\mathbf{\Gamma}$ and $\mathbf{\Gamma}^\dagger$ have equal weight. From this it is easily shown that $S^{(1)}(\mathbf{x}|E) = S^{(1)}(\underline{\underline{\epsilon}}\mathbf{x}|E)$.

The unconditional transition probability between macrostates in time τ for the isolated system satisfies

$$\begin{aligned}
\wp(\mathbf{x}' \leftarrow \mathbf{x}|\tau, E) &= \Lambda(\mathbf{x}'|\mathbf{x}, \tau, E)\wp(\mathbf{x}|E) \\
&= W_E^{-1}\int \mathrm{d}\mathbf{\Gamma}_1\, \mathrm{d}\mathbf{\Gamma}_2\, \delta(\mathbf{x}' - \hat{\mathbf{x}}(\mathbf{\Gamma}_2))\delta(\mathbf{x} - \hat{\mathbf{x}}(\mathbf{\Gamma}_1))\, \delta(\mathbf{\Gamma}_2 - \mathbf{\Gamma}_0(\tau|\mathbf{\Gamma}_1))\, \delta(\mathcal{H}(\mathbf{\Gamma}_1) - E) \\
&= W_E^{-1}\int \mathrm{d}\mathbf{\Gamma}_1^\dagger\, \mathrm{d}\mathbf{\Gamma}_2^\dagger\, \delta(\mathbf{x}' - \underline{\underline{\epsilon}}\hat{\mathbf{x}}(\mathbf{\Gamma}_2^\dagger))\delta(\mathbf{x} - \underline{\underline{\epsilon}}\hat{\mathbf{x}}(\mathbf{\Gamma}_1^\dagger))\, \delta(\mathbf{\Gamma}_1^\dagger - \mathbf{\Gamma}_0(\tau|\mathbf{\Gamma}_2^\dagger))\delta(\mathcal{H}(\mathbf{\Gamma}_1^\dagger) - E) \\
&= \wp(\underline{\underline{\epsilon}}\mathbf{x} \leftarrow \underline{\underline{\epsilon}}\mathbf{x}'|\tau, E)
\end{aligned} \tag{8}$$

This uses the fact that $\mathrm{d}\mathbf{\Gamma} = \mathrm{d}\mathbf{\Gamma}^\dagger$. For macrostates all of even parity, this says that for an isolated system the forward transition $\mathbf{x} \to \mathbf{x}'$ will be observed as frequently as the reverse $\mathbf{x}' \to \mathbf{x}$. This is what Onsager meant by the principle of dynamical reversibility, which he stated as "in the end every type of motion is just as likely to occur as its reverse" [10, p. 412]. Note that for velocity-type variables, the sign is reversed for the reverse transition.

The second or transition entropy is the weight of molecular configurations associated with a transition occurring in time τ,

$$S^{(2)}(\mathbf{x}', \mathbf{x}|\tau, E) = k_\mathrm{B} \ln \int \mathrm{d}\mathbf{\Gamma}_1\; \delta(\hat{\mathbf{x}}(\mathbf{\Gamma}_0(\tau|\mathbf{\Gamma}_1)) - \mathbf{x}')\, \delta(\hat{\mathbf{x}}(\mathbf{\Gamma}_1) - \mathbf{x})\, \delta(\mathcal{H}(\mathbf{\Gamma}_1) - E) \tag{9}$$

up to an arbitrary constant. The unconditional transition probability for $\mathbf{x} \to \mathbf{x}'$ in time τ is related to the second entropy by [2, 8]

$$\wp(\mathbf{x}', \mathbf{x}|\tau, E) = \frac{1}{W(E)} \exp S^{(2)}(\mathbf{x}', \mathbf{x}|\tau, E)/k_\mathrm{B} \tag{10}$$

Henceforth the dependence on the energy is not shown explicitly. The second entropy reduces to the first entropy upon integration

$$S^{(1)}(\mathbf{x}) = \text{const.} + k_{\text{B}} \ln \int d\mathbf{x}' \, \exp S^{(2)}(\mathbf{x}', \mathbf{x}|\tau)/k_{\text{B}} \tag{11}$$

It will prove important to impose this reduction condition on the approximate expansions given later.

The second entropy obeys the symmetry rules

$$S^{(2)}(\mathbf{x}', \mathbf{x}|\tau) = S^{(2)}(\mathbf{x}, \mathbf{x}'| - \tau) = S^{(2)}(\underline{\underline{\epsilon}}\mathbf{x}, \underline{\underline{\epsilon}}\mathbf{x}'|\tau) \tag{12}$$

The first equality follows from time homogeneity: the probability that $\mathbf{x}' = \mathbf{x}(t+\tau)$ and $\mathbf{x} = \mathbf{x}(t)$ are the same as the probability that $\mathbf{x} = \mathbf{x}(t-\tau)$ and $\mathbf{x}' = \mathbf{x}(t)$. The second equality follows from microscopic reversibility: if the molecular velocities are reversed the system retraces its trajectory in phase space. Again, it will prove important to impose these symmetry conditions on the following expansions.

In the formulation of the nonequilibrium second law, Eq. (2), dynamic structure was said to be equivalent to a rate or flux. This may be seen more clearly from the present definition of the second entropy, since the coarse velocity can be defined as

$$\overset{\circ}{\mathbf{x}} \equiv \frac{\mathbf{x}' - \mathbf{x}}{\tau} \tag{13}$$

Maximizing the second entropy with respect to $\mathbf{x}'$ for fixed $\mathbf{x}$ yields the most likely terminal position $\overline{\mathbf{x}}(\mathbf{x}, \tau) \equiv \overline{\mathbf{x}'}$, and hence the most likely coarse velocity $\overset{\circ}{\overline{\mathbf{x}}}(\mathbf{x}, \tau)$. Alternatively, differentiating the most likely terminal position with respect to τ yields the most likely terminal velocity, $\dot{\overline{\mathbf{x}}}(\mathbf{x}, \tau)$. So constraining the system to be in the macrostate $\mathbf{x}'$ at a time τ after it was in the state $\mathbf{x}$ is the same as constraining the coarse velocity.

B. Quadratic Expansion

For simplicity, it is assumed that the equilibrium value of the macrostate is zero, $\overline{\mathbf{x}} = \mathbf{0}$. This means that henceforth $\mathbf{x}$ measures the departure of the macrostate from its equilibrium value. In the linear regime, (small fluctuations), the first entropy may be expanded about its equilibrium value, and to quadratic order it is

$$S^{(1)}(\mathbf{x}) = \tfrac{1}{2}\underline{\underline{S}} : \mathbf{x}^2 \tag{14}$$

a constant having been neglected. Here and throughout a colon or centered dot is used to denote scalar multiplication, and squared or juxtaposed vectors to denote a dyad. Hence the scalar could equally be written $\underline{\underline{S}} : \mathbf{x}^2 \equiv \mathbf{x} \cdot \underline{\underline{S}}\mathbf{x}$. The thermodynamic force is defined as

$$\mathbf{X}(\mathbf{x}) \equiv \frac{\partial S^{(1)}(\mathbf{x})}{\partial \mathbf{x}} = \underline{\underline{S}}\mathbf{x} \tag{15}$$

Evidently in the linear regime the probability is Gaussian, and the correlation matrix is therefore given by

$$\underline{\underline{S}}^{-1} = -\langle \mathbf{x}\mathbf{x} \rangle_0 / k_\mathrm{B} \tag{16}$$

The parity matrix commutes with the first entropy matrix, $\underline{\underline{\epsilon}}\,\underline{\underline{S}} = \underline{\underline{S}}\,\underline{\underline{\epsilon}}$, because there is no coupling between variables of opposite parity at equilibrium, $\langle x_i x_j \rangle_0 = 0$ if $\epsilon_i \epsilon_j = -1$. If variables of the same parity are grouped together, the first entropy matrix is block diagonal.

This last point may be seen more clearly by defining a time-correlation matrix related to the inverse of this,

$$\underline{\underline{Q}}(\tau) \equiv k_\mathrm{B}^{-1} \langle \mathbf{x}(t+\tau)\mathbf{x}(t) \rangle_0 \tag{17}$$

From the time-reversible nature of the equations of motion, Eq. (12), it is readily shown that the matrix is "block-asymmetric":

$$\underline{\underline{Q}}(\tau) = \underline{\underline{\epsilon}}\,\underline{\underline{Q}}(\tau)^\mathrm{T}\underline{\underline{\epsilon}} = \underline{\underline{\epsilon}}\,\underline{\underline{Q}}(-\tau)\underline{\underline{\epsilon}} \tag{18}$$

Since $\underline{\underline{S}}$ is a symmetric matrix equal to $-\underline{\underline{Q}}(0)^{-1}$, these equalities show that the off-diagonal blocks must vanish at $\tau = 0$, and hence that there is no instantaneous coupling between variables of opposite parity. The symmetry or asymmetry of the block matrices in the grouped representation is a convenient way of visualizing the parity results that follow.

The most general quadratic form for the second entropy is [2]

$$S^{(2)}(\mathbf{x}', \mathbf{x}|\tau) = \tfrac{1}{2}\underline{\underline{A}}(\tau) : \mathbf{x}^2 + \mathbf{x} \cdot \underline{\underline{B}}(\tau)\mathbf{x}' + \tfrac{1}{2}\underline{\underline{A}}'(\tau) : \mathbf{x}'^2 \tag{19}$$

Since $\langle \mathbf{x} \rangle_0 = \mathbf{0}$, linear terms must vanish. A constant has also been neglected here. In view of Eq. (12), the matrices must satisfy

$$\underline{\underline{\epsilon}}\,\underline{\underline{A}}(\tau)\underline{\underline{\epsilon}} = \underline{\underline{A}}'(\tau) = \underline{\underline{A}}(-\tau) \tag{20}$$

and

$$\underline{\underline{\epsilon}}\,\underline{\underline{B}}(\tau)\underline{\underline{\epsilon}} = \underline{\underline{B}}(\tau)^\mathrm{T} = \underline{\underline{B}}(-\tau) \tag{21}$$

These show that in the grouped representation, the even temporal part of the matrices is block-diagonal, and the odd temporal part is block-adiagonal, (i.e., the diagonal blocks are zero). Also, as matrices of second derivatives with respect to the same variable, $\underline{\underline{A}}(\tau)$ and $\underline{\underline{A}}'(\tau)$ are symmetric. The even temporal part of $\underline{\underline{B}}$ is symmetric, and the odd part is antisymmetric.

Defining the symmetric matrix $\underline{\underline{\tilde{B}}}(\tau) \equiv \underline{\underline{B}}(\tau)\underline{\underline{\epsilon}} = \underline{\underline{\epsilon}}\,\underline{\underline{B}}(\tau)^{\mathrm{T}}$, the second entropy may be written

$$\begin{aligned} S^{(2)}(\mathbf{x}', \mathbf{x}|\tau) &= \tfrac{1}{2}\underline{\underline{\epsilon}}\,\underline{\underline{A}}(\tau)\underline{\underline{\epsilon}} : \mathbf{x}'^2 + \mathbf{x} \cdot \underline{\underline{\tilde{B}}}(\tau)\underline{\underline{\epsilon}}\mathbf{x}' + \tfrac{1}{2}\underline{\underline{A}}(\tau) : \mathbf{x}^2 \\ &= \tfrac{1}{2}\underline{\underline{A}}(\tau) : [\underline{\underline{\epsilon}}\mathbf{x}' + \underline{\underline{A}}(\tau)^{-1}\underline{\underline{\tilde{B}}}(\tau)\mathbf{x}]^2 + \tfrac{1}{2}\underline{\underline{A}}(\tau) : \mathbf{x}^2 - \tfrac{1}{2}\mathbf{x} \cdot \underline{\underline{\tilde{B}}}(\tau)\underline{\underline{A}}(\tau)^{-1}\underline{\underline{\tilde{B}}}(\tau)\mathbf{x} \\ &= \tfrac{1}{2}\underline{\underline{\epsilon}}\,\underline{\underline{A}}(\tau)\underline{\underline{\epsilon}} : [\mathbf{x}' + \underline{\underline{\epsilon}}\,\underline{\underline{A}}(\tau)^{-1}\underline{\underline{B}}(\tau)\underline{\underline{\epsilon}}\mathbf{x}]^2 + S^{(1)}(\mathbf{x}) \end{aligned} \tag{22}$$

The final equality results from the reduction condition, which evidently is explicitly [2, 7]

$$\underline{\underline{S}} = \underline{\underline{A}}(\tau) - \underline{\underline{\tilde{B}}}(\tau)\underline{\underline{A}}(\tau)^{-1}\underline{\underline{\tilde{B}}}(\tau) \tag{23}$$

This essentially reduces the two transport matrices to one.

The last two results are rather similar to the quadratic forms given by Fox and Uhlenbeck for the transition probability for a stationary Gaussian–Markov process, their Eqs. (20) and (22) [82]. Although they did not identify the parity relationships of the matrices or obtain their time dependence explicitly, the Langevin equation that emerges from their analysis and the Doob formula, their Eq. (25), is essentially equivalent to the most likely terminal position in the intermediate regime obtained next.

The most likely position at the end of the interval is

$$\overline{\mathbf{x}}(\mathbf{x}, \tau) \equiv \overline{\mathbf{x}}' = -\underline{\underline{\epsilon}}\,\underline{\underline{A}}(\tau)^{-1}\underline{\underline{B}}(\tau)\underline{\underline{\epsilon}}\mathbf{x} \tag{24}$$

If it can be shown that the prefactor is the identity matrix plus a matrix linear in τ, then this is, in essence, Onsager's regression hypothesis [10] and the basis for linear transport theory.

C. Time Scaling

Consider the sequential transition $\mathbf{x}_1 \xrightarrow{\tau} \mathbf{x}_2 \xrightarrow{\tau} \mathbf{x}_3$. One can assume Markovian behavior and add the second entropy separately for the two transitions. In view of the previous results this may be written

$$\begin{aligned} S^{(2)}(\mathbf{x}_3, \mathbf{x}_2, \mathbf{x}_1|\tau, \tau) &= S^{(2)}(\mathbf{x}_3, \mathbf{x}_2|\tau) + S^{(2)}(\mathbf{x}_2, \mathbf{x}_1|\tau) - S^{(1)}(\mathbf{x}_2) \\ &= \tfrac{1}{2}\underline{\underline{A}}'(\tau) : \mathbf{x}_3^2 + \mathbf{x}_2 \cdot \underline{\underline{B}}(\tau)\mathbf{x}_3 + \tfrac{1}{2}\underline{\underline{A}}(\tau) : \mathbf{x}_2^2 \\ &\quad + \tfrac{1}{2}\underline{\underline{A}}'(\tau) : \mathbf{x}_2^2 + \mathbf{x}_1 \cdot \underline{\underline{B}}(\tau)\mathbf{x}_2 + \tfrac{1}{2}\underline{\underline{A}}(\tau) : \mathbf{x}_1^2 - \tfrac{1}{2}\underline{\underline{S}} : \mathbf{x}_2^2 \end{aligned} \tag{25}$$

This ansatz is only expected to be valid for large enough τ such that the two intervals may be regarded as independent. This restricts the following results to the intermediate time regime.

The second entropy for the transition $\mathbf{x}_1 \xrightarrow{2\tau} \mathbf{x}_3$ is equal to the maximum value of that for the sequential transition,

$$S^{(2)}(\mathbf{x}_3, \mathbf{x}_1|2\tau) = S^{(2)}(\mathbf{x}_3, \overline{\mathbf{x}}_2, \mathbf{x}_1|\tau, \tau) \tag{26}$$

This result holds in so far as fluctuations about the most probable trajectory are relatively negligible. The optimum point is that which maximizes the second entropy,

$$\left.\frac{\partial S^{(2)}(\mathbf{x}_3, \mathbf{x}_2, \mathbf{x}_1|\tau, \tau)}{\partial \mathbf{x}_2}\right|_{\mathbf{x}_2=\overline{\mathbf{x}}_2} = 0 \tag{27}$$

The midpoint of the trajectory is $\tilde{\mathbf{x}}_2 \equiv [\mathbf{x}_3 + \mathbf{x}_1]/2$. It can be shown that the difference between the optimum point and the midpoint is order τ, and it does not contribute to the leading order results that are obtained here.

The left-hand side of Eq. (26) is

$$S^{(2)}(\mathbf{x}_3, \mathbf{x}_1|2\tau) = \tfrac{1}{2}\underline{\underline{A}}'(2\tau) : \mathbf{x}_3^2 + \mathbf{x}_1 \cdot \underline{\underline{B}}(2\tau)\mathbf{x}_3 + \tfrac{1}{2}\underline{\underline{A}}(2\tau) : \mathbf{x}_1^2 \tag{28}$$

The right-hand side of Eq. (25) evaluated at the midpoint is

$$\begin{aligned} S^{(2)}(\mathbf{x}_3, \tilde{\mathbf{x}}_2, \mathbf{x}_1|\tau, \tau) = {} & \tfrac{1}{8}[5\underline{\underline{A}}'(\tau) + \underline{\underline{A}}(\tau) + 2\underline{\underline{B}}(\tau) + 2\underline{\underline{B}}^{\mathrm{T}}(\tau) - \underline{\underline{S}}] : \mathbf{x}_3^2 \\ & + \tfrac{1}{8}[5\underline{\underline{A}}(\tau) + \underline{\underline{A}}'(\tau) + 2\underline{\underline{B}}(\tau) + 2\underline{\underline{B}}^{\mathrm{T}}(\tau) - \underline{\underline{S}}] : \mathbf{x}_1^2 \\ & + \tfrac{1}{4}\mathbf{x}_1 \cdot [\underline{\underline{A}}(\tau) + \underline{\underline{A}}'(\tau) + 4\underline{\underline{B}}(\tau) - \underline{\underline{S}}] : \mathbf{x}_3 \end{aligned} \tag{29}$$

By equating the individual terms, the dependence on the time interval of the coefficients in the quadratic expansion of the second entropy may be obtained.

Consider the expansions

$$\underline{\underline{A}}(\tau) = \frac{1}{\tau}\underline{\underline{a}}_1 + \frac{1}{|\tau|}\underline{\underline{a}}_2 + \hat{\tau}\underline{\underline{a}}_3 + \underline{\underline{a}}_4 + \mathcal{O}(\tau) \tag{30}$$

$$\underline{\underline{A}}'(\tau) = \frac{-1}{\tau}\underline{\underline{a}}_1 + \frac{1}{|\tau|}\underline{\underline{a}}_2 - \hat{\tau}\underline{\underline{a}}_3 + \underline{\underline{a}}_4 + \mathcal{O}(\tau) \tag{31}$$

and

$$\underline{\underline{B}}(\tau) = \frac{1}{\tau}\underline{\underline{b}}_1 + \frac{1}{|\tau|}\underline{\underline{b}}_2 + \hat{\tau}\underline{\underline{b}}_3 + \underline{\underline{b}}_4 + \mathcal{O}(\tau) \tag{32}$$

Here and throughout, $\hat{\tau} \equiv \text{sign}(\tau)$. These are small-time expansions, but they are not Taylor expansions, as the appearance of nonanalytic terms indicates. From the parity and symmetry rules, the odd coefficients are block-adiagonal, and the even coefficients are block-diagonal in the grouped representation. Equating the coefficients of $\mathbf{x}_3^2/|\tau|$ in Eqs. (28) and (29), it follows that

$$\frac{1}{2}\left[\frac{1}{2}\underline{\underline{a}}_2 - \hat{\tau}\frac{1}{2}\underline{\underline{a}}_1\right] = \frac{1}{8}\left[6\underline{\underline{a}}_2 - 4\hat{\tau}\underline{\underline{a}}_1 + 4\underline{\underline{b}}_2\right] \tag{33}$$

This has solution

$$\underline{\underline{a}}_1 = \underline{\underline{0}} \quad \text{and} \quad \underline{\underline{a}}_2 = -\underline{\underline{b}}_2 \tag{34}$$

Comparing the coefficient of $\mathbf{x}_1\mathbf{x}_3/|\tau|$ in Eq. (29) with that in Eq. (28) confirms this result, and in addition yields

$$\underline{\underline{b}}_1 = \underline{\underline{0}} \tag{35}$$

No further information can be extracted from these equations at this stage because it is not possible to go beyond the leading order due to the approximation $\overline{\mathbf{x}}_2 \approx \tilde{\mathbf{x}}_2$. However, the reduction condition Eq. (23) may be written

$$\begin{aligned} \underline{\underline{S}} &= \underline{\underline{A}}(\tau) + \underline{\underline{B}}(\tau) - \underline{\underline{B}}(\tau)\underline{\underline{\epsilon}}\,\underline{\underline{A}}(\tau)^{-1}[\underline{\underline{A}}(\tau) + \underline{\underline{B}}(\tau)]\underline{\underline{\epsilon}} \\ &\sim \underline{\underline{a}}_4 + \underline{\underline{b}}_4 + \hat{\tau}[\underline{\underline{a}}_3 + \underline{\underline{b}}_3] + \frac{1}{|\tau|}\underline{\underline{a}}_2\underline{\underline{\epsilon}}|\tau|\underline{\underline{a}}_2^{-1}(\underline{\underline{a}}_4 + \underline{\underline{b}}_4 + \hat{\tau}[\underline{\underline{a}}_3 + \underline{\underline{b}}_3])\underline{\underline{\epsilon}} + \mathcal{O}\tau \end{aligned} \tag{36}$$

The odd expansion coefficients are block-adiagonal and hence $\underline{\underline{\epsilon}}[\underline{\underline{a}}_3 + \underline{\underline{b}}_3]\underline{\underline{\epsilon}} + [\underline{\underline{a}}_3 + \underline{\underline{b}}_3] = \underline{\underline{0}}$. This means that the coefficient of $\hat{\tau}$ on the right hand side is identically zero. (Later it will be shown that $\underline{\underline{a}}_3 = \underline{\underline{0}}$ and that $\underline{\underline{b}}_3$ could be nonzero.) Since the parity matrix commutes with the block-diagonal even coefficients, the reduction condition gives

$$\underline{\underline{S}} = 2[\underline{\underline{a}}_4 + \underline{\underline{b}}_4] + \mathcal{O}\tau \tag{37}$$

1. *Optimum Point*

To find the optimum intermediate point, differentiate the second entropy,

$$\frac{\partial S^{(2)}(\mathbf{x}_3, \mathbf{x}_2, \mathbf{x}_1|\tau, \tau)}{\partial \mathbf{x}_2} = \underline{\underline{B}}(\tau)\mathbf{x}_3 + \underline{\underline{A}}(\tau)\mathbf{x}_2 + \underline{\underline{A}}'(\tau)\mathbf{x}_2 + \underline{\underline{B}}^{\mathrm{T}}(\tau)\mathbf{x}_1 - \underline{\underline{S}}\mathbf{x}_2 \tag{38}$$

Setting this to zero it follows that

$$\begin{aligned}\overline{\mathbf{x}}_2 &= -[\underline{\underline{A}}(\tau) + \underline{\underline{A}}'(\tau) - \underline{\underline{S}}]^{-1}[\underline{\underline{B}}(\tau)\mathbf{x}_3 + \underline{\underline{B}}^{\mathrm{T}}(\tau)\mathbf{x}_1] \sim -\left[\frac{2}{|\tau|}\underline{\underline{a}}_2 + 2\underline{\underline{a}}_4 - \underline{\underline{S}} + \mathcal{O}\tau\right]^{-1} \\ &\quad \times \left[\left(\frac{-1}{|\tau|}\underline{\underline{a}}_2 + \hat{\tau}\underline{\underline{b}}_3 + \underline{\underline{b}}_4\right)\mathbf{x}_3 + \left(\frac{-1}{|\tau|}\underline{\underline{a}}_2 - \hat{\tau}\underline{\underline{b}}_3 + \underline{\underline{b}}_4\right)\mathbf{x}_1 + \mathcal{O}\tau\right] \sim \tfrac{1}{2}[\mathbf{x}_3 + \mathbf{x}_1] \\ &\quad - \tfrac{1}{4}|\tau|\underline{\underline{a}}_2^{-1}[2\underline{\underline{a}}_4 - \underline{\underline{S}}] - \tfrac{1}{2}|\tau|\underline{\underline{a}}_2^{-1}[(\hat{\tau}\underline{\underline{b}}_3 + \underline{\underline{b}}_4)\mathbf{x}_3 - (\hat{\tau}\underline{\underline{b}}_3 - \underline{\underline{b}}_4)\mathbf{x}_1] + \mathcal{O}\tau^2 \\ &= \tfrac{1}{2}[\mathbf{x}_3 + \mathbf{x}_1] - \tfrac{\tau}{2}\underline{\underline{a}}_2^{-1}\underline{\underline{b}}_3[\mathbf{x}_3 - \mathbf{x}_1] \\ &\equiv \tilde{\mathbf{x}}_2 + \tau\underline{\underline{\lambda}}[\mathbf{x}_3 - \mathbf{x}_1] \end{aligned} \tag{39}$$

This confirms that to leading order the optimum point is indeed the midpoint. When this is inserted into the second entropy for the sequential transition, the first-order correction cancels,

$$\begin{aligned} S^{(2)}(\mathbf{x}_3, \overline{\mathbf{x}}_2, \mathbf{x}_1) &= S^{(2)}(\mathbf{x}_3, \tilde{\mathbf{x}}_2, \mathbf{x}_1) + \mathbf{x}_1 \cdot \underline{\underline{B}}(\tau)[\overline{\mathbf{x}}_2 - \tilde{\mathbf{x}}_2] + [\overline{\mathbf{x}}_2 - \tilde{\mathbf{x}}_2] \cdot \underline{\underline{B}}(\tau)\mathbf{x}_3 \\ &\quad + \tfrac{1}{2}[\underline{\underline{A}}(\tau) + \underline{\underline{A}}'(\tau) - \underline{\underline{S}}] : \overline{\mathbf{x}}_2^2 - \tfrac{1}{2}[\underline{\underline{A}}(\tau) + \underline{\underline{A}}'(\tau) - \underline{\underline{S}}] : \tilde{\mathbf{x}}_2^2 \\ &\sim S^{(2)}(\mathbf{x}_3, \tilde{\mathbf{x}}_2, \mathbf{x}_1) + \frac{\hat{\tau}}{2}\mathbf{x}_1 \cdot \underline{\underline{b}}_3[\mathbf{x}_3 - \mathbf{x}_1] + \frac{\hat{\tau}}{2}[\mathbf{x}_3 - \mathbf{x}_1] \cdot \underline{\underline{b}}_3^{\mathrm{T}}\mathbf{x}_3 \\ &\quad - \frac{\hat{\tau}}{4}[\mathbf{x}_3 - \mathbf{x}_1] \cdot \underline{\underline{b}}_3^{\mathrm{T}}[\mathbf{x}_3 + \mathbf{x}_1] - \frac{\hat{\tau}}{4}[\mathbf{x}_3 + \mathbf{x}_1] \cdot \underline{\underline{b}}_3[\mathbf{x}_3 - \mathbf{x}_1] + \mathcal{O}\tau \\ &= S^{(2)}(\mathbf{x}_3, \tilde{\mathbf{x}}_2, \mathbf{x}_1) + \mathcal{O}\tau \end{aligned} \tag{40}$$

This means that all of the above expansions also hold for order $\mathcal{O}\tau^0$. Hence equating the coefficients of $\mathbf{x}_3^2|\tau|^0$ in Eqs. (28) and (29), it follows that

$$\tfrac{1}{8}[6\underline{\underline{a}}_4 - 4\underline{\underline{b}}_4 - \underline{\underline{S}}] = \tfrac{1}{2}\underline{\underline{a}}_4 \tag{41}$$

Since $\underline{\underline{a}}_4 + \underline{\underline{b}}_4 = \underline{\underline{S}}/2$, this has solution

$$\underline{\underline{a}}_4 = \underline{\underline{S}}/2 \quad \text{and} \quad \underline{\underline{b}}_4 = \underline{\underline{0}} \tag{42}$$

Equating the coefficient of $\mathbf{x}_1\mathbf{x}_3|\tau|^0$ in Eqs. (28) and (29) yields

$$\tfrac{1}{4}[2\underline{\underline{a}}_4 + 4\underline{\underline{b}}_4 + 4\hat{\tau}\underline{\underline{b}}_3 - \underline{\underline{S}}] = \hat{\tau}\underline{\underline{b}}_3 + \underline{\underline{b}}_4 \tag{43}$$

This is an identity and no information about $\underline{\underline{b}}_3$ can be extracted from it.

D. Regression Theorem

The most likely terminal position was given as Eq. (24), where it was mentioned that if the coefficient could be shown to scale linearly with time, then the Onsager regression hypothesis would emerge as a theorem. Hence the small-τ behavior of the coefficient is now sought. Postmultiply the most likely terminal position by $k_B^{-1}\mathbf{x}$ and take the average, which shows that the coefficient is related to the time correlation matrix defined in Eq. (17). Explicitly,

$$\underline{\underline{Q}}(\tau) = \underline{\underline{\epsilon}}\,\underline{\underline{A}}(\tau)^{-1}\underline{\underline{B}}(\tau)\underline{\underline{\epsilon}}\,\underline{\underline{S}}^{-1} \tag{44}$$

This invokes the result, $\langle \overline{\mathbf{x}}(\mathbf{x},\tau)\mathbf{x}\rangle_0 = \langle \mathbf{x}(t+\tau)\mathbf{x}(t)\rangle_0$, which is valid since the mode is equal to the mean for a Gaussian conditional probability. Inserting the expansion it follows that

$$\begin{aligned}\underline{\underline{Q}}(\tau)\underline{\underline{S}} &\sim \underline{\underline{\epsilon}}\left[\frac{1}{|\tau|}\underline{\underline{a}}_2 + \hat{\tau}\underline{\underline{a}}_3 + \underline{\underline{a}}_4\right]^{-1}\left[\frac{-1}{|\tau|}\underline{\underline{a}}_2 + \hat{\tau}\underline{\underline{b}}_3 + \underline{\underline{b}}_4\right]\underline{\underline{\epsilon}} \\ &\sim \underline{\underline{\epsilon}}\,[\underline{\underline{I}} - \tau\underline{\underline{a}}_2^{-1}\underline{\underline{a}}_3 - |\tau|\underline{\underline{a}}_2^{-1}\underline{\underline{a}}_4][-\underline{\underline{I}} + \tau\underline{\underline{a}}_2^{-1}\underline{\underline{b}}_3 + |\tau|\underline{\underline{a}}_2^{-1}\underline{\underline{b}}_4]\underline{\underline{\epsilon}} + \mathcal{O}\tau^2 \\ &\sim -\underline{\underline{I}} - \tau\underline{\underline{a}}_2^{-1}[\underline{\underline{a}}_3 + \underline{\underline{b}}_3] + |\tau|\underline{\underline{a}}_2^{-1}[\underline{\underline{a}}_4 + \underline{\underline{b}}_4] + \mathcal{O}\tau^2\end{aligned} \tag{45}$$

Here we have used the symmetry and commuting properties of the matrices to obtain the final line. This shows that the correlation matrix goes like

$$\underline{\underline{Q}}(\tau) \sim -\underline{\underline{S}}^{-1} + \tau\underline{\underline{Q}}^- + |\tau|\underline{\underline{Q}}^+ + \mathcal{O}\tau^2 \tag{46}$$

where $\underline{\underline{Q}}^-$ is block-adiagonal and $\underline{\underline{Q}}^+ = \underline{\underline{a}}_2^{-1}/2$ is block-diagonal. Since $\underline{\underline{Q}}(-\tau) = \underline{\underline{Q}}^T(\tau)$, the matrix $\underline{\underline{Q}}^+$ is symmetric, and the matrix $\underline{\underline{Q}}^-$ is asymmetric. This implies that

$$\underline{\underline{a}}_3 = \underline{\underline{0}} \quad \text{and} \quad \underline{\underline{Q}}^- = -\underline{\underline{a}}_2^{-1}\underline{\underline{b}}_3\underline{\underline{S}}^{-1} \tag{47}$$

since $\underline{\underline{a}}_3$ is block-adiagonal and symmetric, and $\underline{\underline{b}}_3$ is block-adiagonal and asymmetric. With these results, the expansions are

$$\underline{\underline{A}}(\tau) = \underline{\underline{A}}'(\tau) = \frac{1}{|\tau|}\underline{\underline{a}}_2 + \frac{1}{2}\underline{\underline{S}} + \mathcal{O}\tau \tag{48}$$

and

$$\underline{\underline{B}}(\tau) = \frac{-1}{|\tau|}\underline{\underline{a}}_2 + \hat{\tau}\underline{\underline{b}}_3 + \mathcal{O}\tau \tag{49}$$

The second entropy, Eq. (22), in the intermediate regime becomes

$$\begin{aligned}S^{(2)}(\mathbf{x}',\mathbf{x}|\tau) &= \tfrac{1}{2}\underline{\underline{A}}(\tau) : \mathbf{x}^2 + \tfrac{1}{2}\underline{\underline{A}}'(\tau) : \mathbf{x}'^2 + \mathbf{x}\cdot\underline{\underline{B}}(\tau)\mathbf{x}' \\ &= \frac{1}{2|\tau|}\underline{\underline{a}}_2 : [\mathbf{x}'-\mathbf{x}]^2 + \frac{1}{4}\underline{\underline{S}} : [\mathbf{x}'-\mathbf{x}]^2 + \frac{1}{2}\mathbf{x}\cdot\underline{\underline{S}}\mathbf{x}' + \hat{\tau}\mathbf{x}\cdot\underline{\underline{b}}_3\mathbf{x}' + \cdots \\ &= \frac{1}{2|\tau|}\underline{\underline{a}}_2 : [\mathbf{x}'-\mathbf{x}]^2 + \frac{1}{2}\mathbf{x}\cdot[\underline{\underline{S}} + 2\hat{\tau}\underline{\underline{b}}_3]\mathbf{x}' + \cdots\end{aligned} \tag{50}$$

Higher-order terms have been neglected in this small-τ expansion that is valid in the intermediate regime. This expression obeys exactly the symmetry relationships, and it obeys the reduction condition to leading order. (See Eq. (68) for a more complete expression that obeys the reduction condition fully.)

Assuming that the coarse velocity can be regarded as an intensive variable, this shows that the second entropy is extensive in the time interval. The time extensivity of the second entropy was originally obtained by certain Markov and integration arguments that are essentially equivalent to those used here [2]. The symmetric matrix $\underline{\underline{a}}_2$ controls the strength of the fluctuations of the coarse velocity about its most likely value. That the symmetric part of the transport matrix controls the fluctuations has been noted previously (see Section 2.6 of Ref. 35, and also Ref. 82).

The derivative with respect to $\mathbf{x}'$ is

$$\frac{\partial S^{(2)}(\mathbf{x}', \mathbf{x}|\tau)}{\partial \mathbf{x}'} = \frac{1}{|\tau|}\underline{\underline{a}}_2[\mathbf{x}' - \mathbf{x}] + \frac{1}{2}[\underline{\underline{S}} - 2\hat{\tau}\underline{\underline{b}}_3]\mathbf{x} \tag{51}$$

From this the most likely terminal position in the intermediate regime is

$$\begin{aligned}\bar{\mathbf{x}}(\mathbf{x}, \tau) &\sim \mathbf{x} - \frac{|\tau|}{2}\underline{\underline{a}}_2^{-1}\{\underline{\underline{S}} - 2\hat{\tau}\underline{\underline{b}}_3\}\mathbf{x} \\ &= \mathbf{x} - \frac{|\tau|}{2}\underline{\underline{a}}_2^{-1}\mathbf{X}(\mathbf{x}) + \tau\underline{\underline{a}}_2^{-1}\underline{\underline{b}}_3\underline{\underline{S}}^{-1}\mathbf{X}(\mathbf{x}) \\ &= \mathbf{x} - |\tau|[\underline{\underline{Q}}^{+} + \hat{\tau}\underline{\underline{Q}}^{-}]\mathbf{X}(\mathbf{x}) \\ &\equiv \mathbf{x} - |\tau|\underline{\underline{L}}(\hat{\tau})\mathbf{X}(\mathbf{x})\end{aligned} \tag{52}$$

It follows that the most likely coarse velocity is

$$\overset{\circ}{\bar{\mathbf{x}}}(\mathbf{x}, \tau) = -\hat{\tau}\underline{\underline{L}}(\hat{\tau})\mathbf{X}(\mathbf{x}) \tag{53}$$

and the most likely terminal velocity is

$$\dot{\bar{\mathbf{x}}}(\mathbf{x}, \tau) = -[\hat{\tau}\underline{\underline{Q}}^{+} + \underline{\underline{Q}}^{-}]\mathbf{X}(\mathbf{x}) = -\hat{\tau}\underline{\underline{L}}(\hat{\tau})\mathbf{X}(\mathbf{x}) \tag{54}$$

These indicate that the system returns to equilibrium at a rate proportional to the displacement, which is Onsager's famous regression hypothesis [10].

That the most likely coarse velocity is equal to the most likely terminal velocity can only be true in two circumstances: either the system began in the steady state and the most likely instantaneous velocity was constant throughout the interval, or else the system was initially in a dynamically disordered state, and τ was large enough that the initial inertial regime was relatively negligible. These equations are evidently untrue for $|\tau| \to 0$, since in this limit the most likely velocity is zero (if the system is initially dynamically disordered, as it is instantaneously following a fluctuation). In the limit that $|\tau| \to \infty$ these equations also break down, since then there can be no correlation between current position and future (or past) velocity. Also, since only the leading term or terms in the small-τ expansion have been retained above, the neglected terms must increasingly contribute as τ increases and the above explicit results must become increasingly inapplicable. Hence these results hold for τ in the intermediate regime, $\tau_{\text{short}} \lesssim \tau \lesssim \tau_{\text{long}}$.

The matrix $\underline{\underline{L}}(\hat{\tau})$ is called the transport matrix, and it satisfies

$$\underline{\underline{L}}(\hat{\tau}) = \underline{\underline{\epsilon}}\,\underline{\underline{L}}(\hat{\tau})^{\mathrm{T}}\underline{\underline{\epsilon}} = \underline{\underline{L}}(-\hat{\tau})^{\mathrm{T}} \tag{55}$$

This follows because, in the grouped representation, $\underline{\underline{Q}}^{+}$ contains nonzero blocks only on the diagonal and is symmetric, and $\underline{\underline{Q}}^{-}$ contains nonzero blocks only off the diagonal and is asymmetric. These symmetry rules are called the Onsager–Casimir reciprocal relations [10, 24]. They show that the magnitude of the coupling coefficient between a flux and a force is equal to that between the force and the flux.

1. *Asymmetry of the Transport Matrix*

A significant question is whether the asymmetric contribution to the transport matrix is zero or nonzero. That is, is there any coupling between the transport of variables of opposite parity? The question will recur in the discussion of the rate of entropy production later. The earlier analysis cannot decide the issue, since $\underline{\underline{b}}_3$ can be zero or nonzero in the earlier results. But some insight can be gained into the possible behavior of the system from the following analysis.

In the intermediate regime, $\tau_{\text{short}} \lesssim \tau \lesssim \tau_{\text{long}}$, the transport matrix is linear in τ and it follows that

$$\underline{\underline{L}}(\hat{\tau}) = \hat{\tau}\frac{\partial}{\partial\tau}\underline{\underline{Q}}(\tau) = \hat{\tau}k_{\mathrm{B}}\langle\dot{\mathbf{x}}(t+\tau)\mathbf{x}(t)\rangle_0 \tag{56}$$

or, equivalently,

$$\underline{\underline{L}}(\hat{\tau}) = \frac{1}{|\tau|}[\underline{\underline{Q}}(\tau) + \underline{\underline{S}}^{-1}] = \hat{\tau}k_{\mathrm{B}}\langle\overset{\circ}{\mathbf{x}}(t;\tau)\mathbf{x}(t)\rangle_0 \tag{57}$$

That the time correlation function is the same using the terminal velocity or the coarse velocity in the intermediate regime is consistent with Eqs (53) and (54).

Consider two variables, $\mathbf{x} = \{A, B\}$, where A has even parity and B has odd parity. Then using the terminal velocity it follows that

$$\underline{\underline{L}}(\hat{\tau}) = \hat{\tau}k_{\mathrm{B}}^{-1}\begin{pmatrix} \langle\dot{A}(t+\tau)A(t)\rangle_0 & \langle\dot{A}(t+\tau)B(t)\rangle_0 \\ \langle\dot{B}(t+\tau)A(t)\rangle_0 & \langle\dot{B}(t+\tau)B(t)\rangle_0 \end{pmatrix} \tag{58}$$

The matrix is readily shown to be antisymmetric, as it must be. In the intermediate regime, the transport matrix must be independent of $|\tau|$, which means that for nonzero τ,

$$\underline{\underline{0}} = \frac{\partial}{\partial \tau}\underline{\underline{L}}(\hat{\tau}) = \hat{\tau} k_{\mathrm{B}}^{-1} \begin{pmatrix} \langle \ddot{A}(t+\tau)A(t)\rangle_0 & \langle \ddot{A}(t+\tau)B(t)\rangle_0 \\ \langle \ddot{B}(t+\tau)A(t)\rangle_0 & \langle \ddot{B}(t+\tau)B(t)\rangle_0 \end{pmatrix} \tag{59}$$

That the terminal acceleration should most likely vanish is true almost by definition of the steady state; the system returns to equilibrium with a constant velocity that is proportional to the initial displacement, and hence the acceleration must be zero. It is stressed that this result only holds in the intermediate regime, for τ not too large. Hence and in particular, this constant velocity (linear decrease in displacement with time) is not inconsistent with the exponential return to equilibrium that is conventionally predicted by the Langevin equation, since the present analysis cannot be extrapolated directly beyond the small time regime where the exponential can be approximated by a linear function.

In the special case that $B = \dot{A}$, the transport matrix is

$$\underline{\underline{L}}'(\hat{\tau}) = \hat{\tau} k_{\mathrm{B}}^{-1} \begin{pmatrix} \langle \dot{A}(t+\tau)A(t)\rangle_0 & \langle \dot{A}(t+\tau)\dot{A}(t)\rangle_0 \\ \langle \ddot{A}(t+\tau)A(t)\rangle_0 & \langle \ddot{A}(t+\tau)\dot{A}(t)\rangle_0 \end{pmatrix} \tag{60}$$

Both entries on the second row of the transport matrix involve correlations with $\ddot{A}$, and hence they vanish. That is, the lower row of $\underline{\underline{L}}'$ equals the upper row of $\dot{\underline{\underline{L}}}$. By asymmetry, the upper right-hand entry of $\underline{\underline{L}}'$ must also vanish, and so the only nonzero transport coefficient is $L'_{AA} = \hat{\tau} k_{\mathrm{B}}^{-1} \langle \dot{A}(t+\tau)A(t)\rangle_0$. So this is one example when there is no coupling in the transport of variable of opposite parity. But there is no reason to suppose that this is true more generally.

E. Entropy Production

The constrained rate of first entropy production is

$$\dot{S}^{(1)}(\dot{\mathbf{x}}, \mathbf{x}) = \dot{\mathbf{x}} \cdot \mathbf{X}(\mathbf{x}) \tag{61}$$

This is a general result that holds for any structure $\mathbf{x}$, and any flux $\dot{\mathbf{x}}$.

In terms of the terminal velocity (the same result holds for the coarse velocity), the most likely rate of production of the first entropy is

$$\begin{aligned} \dot{\overline{S}}^{(1)}(\mathbf{x}) &= \dot{\overline{\mathbf{x}}}(\mathbf{x}, \hat{\tau}) \cdot \mathbf{X}(\mathbf{x}) \\ &= -[\hat{\tau}\underline{\underline{Q}}^{+} + \underline{\underline{Q}}^{-}] : \mathbf{X}(\mathbf{x})^2 \\ &= -\hat{\tau}\underline{\underline{Q}}^{+} : \mathbf{X}(\mathbf{x})^2 \end{aligned} \tag{62}$$

The asymmetric part of the transport matrix gives zero contribution to the scalar product and so does not contribute to the steady-state rate of first entropy production [7]. This was also observed by Casimir [24] and by Grabert et al. [25], Eq. (17).

As stressed at the end of the preceding section, there is no proof that the asymmetric part of the transport matrix vanishes. Casimir [24], no doubt motivated by his observation about the rate of entropy production, on p. 348 asserted that the antisymmetric component of the transport matrix had no observable physical consequence and could be set to zero. However, the present results show that the function makes an important and generally nonnegligible contribution to the dynamics of the steady state even if it does not contribute to the rate of first entropy production.

The optimum rate of first entropy production may also be written in terms of the fluxes,

$$\overline{\dot{S}}^{(1)}(\mathbf{x}) = -\hat{\tau}\underline{\underline{L}}(\hat{\tau})^{-1} : \overline{\dot{\mathbf{x}}}(\mathbf{x}, \hat{\tau})^2 \tag{63}$$

Only the symmetric part of the inverse of the transport matrix contributes to this (but of course this will involve products of the antisymmetric part of the transport matrix itself). Both these last two formulas are only valid in the optimum or steady state. They are not valid in a general constrained state, where Eq. (61) is the only formula that should be used. This distinction between formulas that are valid generally and those that are only valid in the optimum state is an example of the point of the second question posed in the introduction, Question (4). Unfortunately, most workers in the field regard the last two equations as general formulas for the rate of first entropy production and apply them to constrained, nonoptimum states where they have no physical meaning. Onsager, in his first two papers, combined the general expression, Eq. (61), with the restricted one, Eq. (63), to make a variational principle that is almost the same as the present second entropy and that is generally valid. However, in the later paper with Machlup [32], he proposes variational principles based solely on the two restricted expressions and applies these to the constrained states where they have no physical meaning. Prigogine [11, 83] uses both restricted expressions in his work and interprets them as the rate of entropy production, which is not correct for the constrained states to which he applies them.

The first question posed in the introduction, Question (3), makes the point that one cannot have a theory for the nonequilibrium state based on the first entropy or its rate of production. It ought to be clear that the steady state, which corresponds to the most likely flux, $\overline{\dot{\mathbf{x}}}(\mathbf{x}, \hat{\tau})$, gives neither the maximum nor the minimum of Eq. (61), the rate of first entropy production. From that equation, the extreme rates of first entropy production occur when $\dot{\mathbf{x}} = \pm\infty$. Theories that invoke the Principle of Minimum Dissipation, [10–12, 32] or the Principle of

Maximum Dissipation, [13–18] are fundamentally flawed from this point of view. These two principles are diametrically opposed, and it is a little surprising that they have both been advocated simultaneously.

Using the quadratic expression for the second entropy, Eq. (22), the reduction condition, Eq. (23), and the correlation function, Eq. (17), the second entropy may be written at all times as

$$S^{(2)}(\mathbf{x}', \mathbf{x}|\tau) = S^{(1)}(\mathbf{x}) + \tfrac{1}{2}[\underline{\underline{S}}^{-1} - \underline{\underline{Q}}(\tau)\underline{\underline{S}}\underline{\underline{Q}}(\tau)^{\mathrm{T}}]^{-1} : (\mathbf{x}' + \underline{\underline{Q}}(\tau)\underline{\underline{S}}\mathbf{x})^2 \tag{64}$$

It is evident from this that the most likely terminal position is $\overline{\mathbf{x}}(\mathbf{x}, \tau) = -\underline{\underline{Q}}(\tau)\underline{\underline{S}}\mathbf{x}$, as expected from the definition of the correlation function, and the fact that for a Gaussian probability means equal modes. This last point also ensures that the reduction condition is automatically satisfied, and that the maximum value of the second entropy is just the first entropy,

$$\overline{S^{(2)}}(\mathbf{x}, \tau) \equiv S^{(2)}(\overline{\mathbf{x}'}, \mathbf{x}|\tau) = S^{(1)}(\mathbf{x}) \tag{65}$$

This holds for all time intervals τ, and so in the optimum state the rate of production of second entropy vanishes. This is entirely analogous to the equilibrium situation, where at equilibrium the rate of change of first entropy vanishes.

The vanishing of the second term in the optimum state arises from a cancelation that lends itself to a physical interpretation. This expression for the second entropy may be rearranged as

$$\begin{aligned} S^{(2)}(\mathbf{x}', \mathbf{x}|\tau) = {} & S^{(1)}(\mathbf{x}) + \tfrac{1}{2}\left[\underline{\underline{S}}^{-1} - \underline{\underline{Q}}(\tau)\underline{\underline{S}}\underline{\underline{Q}}(\tau)^{\mathrm{T}}\right]^{-1} : (\mathbf{x}' - \mathbf{x})^2 \\ & + (\mathbf{x}' - \mathbf{x}) \cdot [\underline{\underline{S}}^{-1} - \underline{\underline{Q}}(\tau)\underline{\underline{S}}\underline{\underline{Q}}(\tau)^{\mathrm{T}}]^{-1}(\underline{\underline{I}} + \underline{\underline{Q}}(\tau)\underline{\underline{S}})\mathbf{x} \\ & + \tfrac{1}{2}[\underline{\underline{S}}^{-1} - \underline{\underline{Q}}(\tau)\underline{\underline{S}}\underline{\underline{Q}}(\tau)^{\mathrm{T}}]^{-1} : [(\underline{\underline{I}} + \underline{\underline{Q}}(\tau)\underline{\underline{S}})\mathbf{x}]^2 \end{aligned} \tag{66}$$

The first term on the right-hand side is the ordinary first entropy. It is negative and represents the cost of the order that is the constrained static state $\mathbf{x}$. The second term is also negative and is quadratic in the coarse velocity. It represents the cost of maintaining the dynamic order that is induced in the system for a nonzero flux $\overset{\circ}{\mathbf{x}}$. The third and fourth terms sum to a positive number, at least in the optimum state, where they cancel with the second term. As will become clearer shortly, they represent the production of first entropy as the system returns to equilibrium, and it is these terms that drive the flux.

Beyond the intermediate regime, in the long time limit the correlation function vanishes, $\underline{\underline{Q}}(\tau) \to \underline{\underline{0}}$. In this regime the second entropy is just the sum of the two first entropies, as is expected,

$$S^{(2)}(\mathbf{x}', \mathbf{x}|\tau) \to S^{(1)}(\mathbf{x}) + S^{(1)}(\mathbf{x}'), \quad |\tau| \to \infty \tag{67}$$

This follows directly by setting $\underline{\underline{Q}}$ to zero in either of the two previous expressions.

In the intermediate regime, $[\underline{\underline{S}}^{-1} - \underline{\underline{Q}}(\tau)\,\underline{\underline{S}}\underline{\underline{Q}}(\tau)^{\mathrm{T}}]^{-1} \sim (\underline{\underline{Q}}^{+})^{-1}/2|\tau| = \underline{\underline{a}}_2/|\tau|$, and $(\underline{\underline{I}} + \underline{\underline{Q}}(\tau)\underline{\underline{S}}) \sim \tau\underline{\underline{Q}}^{-}\underline{\underline{S}} + |\tau|\,\underline{\underline{Q}}^{+}\underline{\underline{S}}$. Hence the second entropy goes like

$$S^{(2)}(\mathbf{x}', \mathbf{x}|\tau) \sim S^{(1)}(\mathbf{x}) + \frac{|\tau|}{4}(\underline{\underline{Q}}^{+})^{-1} : \overset{\circ}{\mathbf{x}}{}^{2} + \frac{|\tau|}{2}\overset{\circ}{\mathbf{x}}\cdot(\underline{\underline{Q}}^{+})^{-1}[\underline{\underline{Q}}^{-} + \hat{\tau}\underline{\underline{Q}}^{+}]\underline{\underline{S}}\mathbf{x}$$
$$+ \frac{|\tau|}{4}(\underline{\underline{Q}}^{+})^{-1} : ([\underline{\underline{Q}}^{-} + \hat{\tau}\underline{\underline{Q}}^{+}]\underline{\underline{S}}\mathbf{x})^{2} \tag{68}$$

The terms that are linear in the time interval must add up to a negative number, and so as the flux spontaneously develops these terms approach zero from below. In the optimum or steady state, the third term on the right-hand side is minus twice the fourth term, and so the sum of these two terms is

$$\frac{|\tau|}{4}\overline{\overset{\circ}{\mathbf{x}}}\cdot(\underline{\underline{Q}}^{+})^{-1}[\underline{\underline{Q}}^{-} + \hat{\tau}\underline{\underline{Q}}^{+}]\underline{\underline{S}}\mathbf{x} \approx \frac{\tau}{4}\overline{\overset{\circ}{\mathbf{x}}}\cdot\mathbf{X}(\mathbf{x}) \tag{69}$$

where the asymmetric coupling has been neglected. For $\tau > 0$, this is one-quarter of the first entropy production and is positive, which justifies the above physical interpretation of these two terms.

In summary, following a fluctuation the system is initially dynamically disordered, and the flux is zero. If the flux were constrained to be zero for an intermediate time, the second entropy would be less than the first entropy of the fluctuation, and it would have decreased at a constant rate,

$$S^{(2)}(\overset{\circ}{\mathbf{x}} = \mathbf{0}, \mathbf{x}|\tau) = S^{(1)}(\mathbf{x}) + \frac{|\tau|}{4}\underline{\underline{Q}}^{+} : (\underline{\underline{S}}\mathbf{x})^{2} \tag{70}$$

ignoring the asymmetric term. If the constraint is relaxed, then the flux develops, the third positive term increases and produces first entropy at an increasing rate. It continues to increase until the cost of maintaining the dynamic order, the second negative term that is quadratic in the flux, begins to increase in magnitude at a greater rate than the third term that is linear in the flux. At this stage, the second, third, and fourth terms add to zero. The steady state occurs in the intermediate regime and is marked by the constancy of the flux, as is discussed in more detail in Section IIG.

F. Reservoir

If one now adds a reservoir with thermodynamic force $\mathbf{X}_{\mathrm{r}}$, then the subsystem macrostate $\mathbf{x}$ can change by internal processes $\Delta^{0}\mathbf{x}$, or by exchange with the reservoir, $\Delta^{\mathrm{r}}\mathbf{x} = -\Delta\mathbf{x}_{\mathrm{r}}$. Imagining that the transitions occur sequentially,

$\mathbf{x} \to \mathbf{x}' \to \mathbf{x}''$, with $\mathbf{x}' = \mathbf{x} + \Delta^0\mathbf{x}$ and $\mathbf{x}'' = \mathbf{x}' + \Delta^{\mathrm{r}}\mathbf{x}$, the second entropy for the stochastic transition is half the difference between the total first entropies of the initial and final states,

$$S_{\mathrm{r}}^{(2)}(\Delta^{\mathrm{r}}\mathbf{x}|\mathbf{x}, \mathbf{X}_{\mathrm{r}}) = \tfrac{1}{2}[S^{(1)}(\Delta^{\mathrm{r}}\mathbf{x} + \mathbf{x}) - (\Delta^{\mathrm{r}}\mathbf{x} + \mathbf{x}) \cdot \mathbf{X}_{\mathrm{r}} - S^{(1)}(\mathbf{x}) + \mathbf{x} \cdot \mathbf{X}_{\mathrm{r}}] \quad (71)$$

The factor of $\frac{1}{2}$ arises from the reversibility of such stochastic transitions. One can debate whether $\mathbf{x}$ or $\mathbf{x}'$ should appear here, but it does not affect the following results to leading order.

In the expression for the second entropy of the isolated system, Eq. (68), the isolated system first entropy appears, $S^{(1)}(\mathbf{x})$. In the present case this must be replaced by the total first entropy, $S^{(1)}(\mathbf{x}) - \mathbf{x} \cdot \mathbf{X}_{\mathrm{r}}$. With this replacement and adding the stochastic second entropy, the total second entropy is

$$\begin{aligned} S_{\mathrm{total}}^{(2)}&(\Delta^0\mathbf{x}, \Delta^{\mathrm{r}}\mathbf{x}, \mathbf{x}|\mathbf{X}_{\mathrm{r}}, \tau) \\ &= S^{(1)}(\mathbf{x}) - \mathbf{x} \cdot \mathbf{X}_{\mathrm{r}} + \frac{1}{2|\tau|}\underline{\underline{a}}_2 : (\Delta^0\mathbf{x})^2 + \frac{\hat{\tau}}{2}\Delta^0\mathbf{x} \cdot \underline{\underline{a}}_2[\hat{\tau}\underline{\underline{a}}_2^{-1} - 2\underline{\underline{a}}_2^{-1}\underline{\underline{b}}_3\underline{\underline{S}}^{-1}]\underline{\underline{S}}\mathbf{x} \\ &+ \frac{|\tau|}{8}\underline{\underline{a}}_2 : ([\hat{\tau}\underline{\underline{a}}_2^{-1} - 2\underline{\underline{a}}_2^{-1}\underline{\underline{b}}_3\underline{\underline{S}}^{-1}]\underline{\underline{S}}\mathbf{x})^2 + \frac{1}{2}[S^{(1)}(\Delta^{\mathrm{r}}\mathbf{x} + \mathbf{x}) - S^{(1)}(\mathbf{x}) - \Delta^{\mathrm{r}}\mathbf{x} \cdot \mathbf{X}_{\mathrm{r}}] \end{aligned} \quad (72)$$

Setting the derivative with respect to $\Delta^{\mathrm{r}}\mathbf{x}$ to zero, one finds

$$\overline{\mathbf{X}_{\mathrm{s}}''} = \mathbf{X}_{\mathrm{r}} \quad (73)$$

which is to say that in the steady state the subsystem force equals the reservoir force at the end of the transition. (Strictly speaking, the point at which the force is evaluated differs from $\mathbf{x}''$ by the adiabatic motion, but this is of higher order and can be neglected.)

The derivative with respect to $\mathbf{x}$ is

$$\begin{aligned} \frac{\partial S_{\mathrm{total}}^{(2)}}{\partial \mathbf{x}} &= \mathbf{X}_{\mathrm{s}} - \mathbf{X}_{\mathrm{r}} + \tfrac{1}{2}[\mathbf{X}_{\mathrm{s}}'' - \mathbf{X}_{\mathrm{s}}] + \mathcal{O}\Delta^0\mathbf{x} \\ &= \tfrac{1}{2}[\mathbf{X}_{\mathrm{s}} - \mathbf{X}_{\mathrm{r}}] \end{aligned} \quad (74)$$

To leading order this vanishes when the initial subsystem force equals that imposed by the reservoirs,

$$\overline{\mathbf{X}_{\mathrm{s}}} = \mathbf{X}_{\mathrm{r}} \quad (75)$$

Since the subsystem force at the end of the transition is also most likely equal to the reservoir force, this implies that the adiabatic change is canceled by the stochastic change, $\overline{\Delta^{\mathrm{r}}\mathbf{x}} = -\Delta^0\mathbf{x}$.

The derivative with respect to $\Delta^0\mathbf{x}$ is

$$\frac{\partial S^{(2)}_{\text{total}}}{\partial \Delta^0 \mathbf{x}} = \hat{\tau} \underline{\underline{a}}_2 \overset{\circ}{\mathbf{x}} + \frac{1}{2}(\underline{\underline{S}} - 2\hat{\tau}\underline{\underline{b}}_3)\mathbf{x} \tag{76}$$

Hence the most likely flux is

$$\overline{\overset{\circ}{\mathbf{x}}} = \frac{-\hat{\tau}}{2} \underline{\underline{a}}_2^{-1}[\underline{\underline{I}} - 2\hat{\tau}\underline{\underline{b}}_3\underline{\underline{S}}^{-1}]\mathbf{X}_{\text{r}} \tag{77}$$

This confirms Onsager's regression hypothesis, namely, that the flux following a fluctuation in an isolated system is the same as if that departure from equilibrium were induced by an externally applied force.

G. Intermediate Regime and Maximum Flux

Most of the previous analysis has concentrated on the intermediate regime, $\tau_{\text{short}} \lesssim \tau \lesssim \tau_{\text{long}}$. It is worth discussing the reasons for this in more detail, and to address the related question of how one chooses a unique transport coefficient since in general this is a function of τ.

At small time scales following a fluctuation, $\tau \lesssim \tau_{\text{short}}$, the system is dynamically disordered and the molecules behave essentially ballistically. This regime is the inertial regime as the dynamic order sets in and the flux becomes established. At long times, $\tau \gtrsim \tau_{\text{long}}$, the correlation function goes to zero and the flux dies out as the terminal position forgets the initial fluctuation. So it is clear that the focus for steady flow has to be on the intermediate regime.

To simplify the discussion, a scalar even variable x will be used. In this case the most likely terminal position is $\overline{x'}(x, \tau) = -Q(\tau)Sx$, where the correlation function is $Q(\tau) = k_{\text{B}}^{-1}\langle x(t+\tau)x(t)\rangle_0$. The most likely terminal velocity is

$$\overline{\dot{x}}(x, \tau) = \frac{\partial \overline{x'}(x, \tau)}{\partial \tau} = -\dot{Q}(\tau)Sx = -k_{\text{B}}^{-1}\langle \dot{x}(t+\tau)x(t)\rangle_0 Sx \tag{78}$$

The maximum terminal velocity is given by

$$0 = \left.\frac{\partial \overline{\dot{x}}(x, \tau)}{\partial \tau}\right|_{\tau=\tau^*} = -\ddot{Q}(\tau^*)Sx = -k_{\text{B}}^{-1}\langle \ddot{x}(t+\tau^*)x(t)\rangle_0 Sx \tag{79}$$

So the terminal velocity or flux is a maximum when the terminal acceleration is zero, which implies that the terminal velocity is a constant, which implies that $Q(\tau)$ is a linear function of τ. By definition, the steady state is the state of constant flux. This justifies the above focus on the terms that are linear in τ in the study of the steady state. The preceding equations show that the steady state corresponds to the state of maximum unconstrained flux. (By maximum is meant greatest magnitude; the sign of the flux is such that the first entropy increases.) The transport matrix is the one evaluated at this point of maximal flux. In the steady state during the adiabatic motion of the system, the only change is the steady decrease of the structure as it decays toward its equilibrium state.

The first role of a reservoir is to impose on the system a gradient that makes the subsystem structure nonzero. The adiabatic flux that consequently develops continually decreases this structure, but the second role of the reservoir is to cancel this decrement by exchange of variables conjugate to the gradient. This does not affect the adiabatic dynamics. Hence provided that the flux is maximal in the above sense, then this procedure ensures that both the structure and the dynamics of the subsystem are steady and unchanging in time. (See also the discussion of Fig. 9.) A corollary of this is that the first entropy of the reservoirs increases at the greatest possible rate for any unconstrained flux.

This last point suggests an alternative interpretation of the transport coefficient as the one corresponding to the correlation function evaluated at the point of maximum flux. The second entropy is maximized to find the optimum flux at each τ. Since the maximum value of the second entropy is the first entropy $S^{(1)}(\mathbf{x})$, which is independent of τ, one has no further variational principle to invoke based on the second entropy. However, one may assert that the optimal time interval is the one that maximizes the rate of production of the otherwise unconstrained first entropy, $\dot{S}(\overline{\dot{\mathbf{x}}'}(\mathbf{x},\tau),\mathbf{x}) = \overline{\dot{\mathbf{x}}'}(\mathbf{x},\tau)\cdot\mathbf{X}_s(\mathbf{x})$, since the latter is a function of the optimized fluxes that depend on τ.

Finally, a point can be made about using the present analysis to calculate a trajectory. Most of the above analysis invoked a small-τ expansion and kept terms up to linear order. This restricts the analysis to times not too long, $|\tau\overline{\dot{x}}(x,\tau)| \ll |x|$. If one has the entire correlation function, then one can predict the whole trajectory. The advantage of working in the intermediate regime is that one only need know a single quantity, namely, the transport coefficient corresponding to the maximal flux. Given this, one can of course still predict the behavior of the system over long time intervals by piecing together short segments in a Markov fashion. In this case the length of the segments should be τ^*, which is long enough to be beyond the inertial regime during which the system adjusts to the new structure, and short enough for the expansion to linear order in τ to be valid. In fact, it is of precisely the right length to be able to use constant, time-independent transport coefficients. And even if it weren't quite right, the first-order error in τ^* would create only a second-order error in the transport coefficient. Furthermore, choosing τ^* as the segment length will ensure that the system reaches equilibrium in the shortest possible time with unconstrained fluxes. Since $\overline{x'}(x,\tau) = -Q(\tau)Sx$ for τ intermediate, the Markov procedure predicts for a long time interval

$$\overline{x'}(x,t) = (-Q(\tau^*)S)^{t/\tau^*}x, \quad t \gtrsim \tau^* \tag{80}$$

This can be written as an exponential decay with relaxation time $\tau^*/\ln[-Q(\tau^*)S]$.

III. NONLINEAR THERMODYNAMICS

A. Quadratic Expansion

In the nonlinear regime, the thermodynamic force remains formally defined as the first derivative of the first entropy,

$$\mathbf{X}(\mathbf{x}) = \frac{\partial S^{(1)}(\mathbf{x})}{\partial \mathbf{x}} \tag{81}$$

However, it is no longer a linear function of the displacement. In other words, the second derivative of the first entropy, which is the first entropy matrix, is no longer constant:

$$\underline{\underline{S}}(\mathbf{x}) = \frac{\partial^2 S^{(1)}(\mathbf{x})}{\partial \mathbf{x}\, \partial \mathbf{x}} \tag{82}$$

For the second entropy, without assuming linearity, one can nevertheless take $\mathbf{x}'$ to be close to $\mathbf{x}$, which will be applicable for τ not too large. Define

$$E \equiv E(\mathbf{x};\tau) \equiv S^{(2)}(\mathbf{x},\mathbf{x}|\tau) \tag{83}$$

$$\mathbf{F} \equiv \mathbf{F}(\mathbf{x};\tau) \equiv \left.\frac{\partial S^{(2)}(\mathbf{x}',\mathbf{x}|\tau)}{\partial \mathbf{x}'}\right|_{\mathbf{x}'=\mathbf{x}} \tag{84}$$

and

$$\underline{\underline{G}} \equiv \underline{\underline{G}}(\mathbf{x};\tau) \equiv \left.\frac{\partial^2 S^{(2)}(\mathbf{x}',\mathbf{x}|\tau)}{\partial \mathbf{x}'\, \partial \mathbf{x}'}\right|_{\mathbf{x}'=\mathbf{x}} \tag{85}$$

With these the second entropy may be expanded about $\mathbf{x}$, and to second order it is

$$\begin{aligned} S^{(2)}(\mathbf{x}',\mathbf{x}|\tau) &= E + (\mathbf{x}'-\mathbf{x})\cdot\mathbf{F} + \tfrac{1}{2}\underline{\underline{G}} : (\mathbf{x}'-\mathbf{x})^2 \\ &= S^{(1)}(\mathbf{x}) + \tfrac{1}{2}\underline{\underline{G}} : [\mathbf{x}'-\mathbf{x}+\underline{\underline{G}}^{-1}\mathbf{F}]^2 \end{aligned} \tag{86}$$

The final equality comes about because this is in the form of a completed square, and hence the reduction condition, Eq. (11), immediately yields

$$S^{(1)}(\mathbf{x}) = E(\mathbf{x};\tau) - \tfrac{1}{2}\underline{\underline{G}}(\mathbf{x};\tau)^{-1} : \mathbf{F}(\mathbf{x};\tau)^2 \tag{87}$$

The right-hand side must be independent of τ. The asymmetry introduced by the expansion of $S^{(2)}(\mathbf{x}',\mathbf{x}|\tau)$ about $\mathbf{x}$ (it no longer obeys the symmetry rules, Eq. (12)) affects the neglected higher-order terms.

The second entropy is maximized by the most likely position, which from the completed square evidently is

$$\overline{\mathbf{x}'}(\mathbf{x}, \tau) = \mathbf{x} - \underline{\underline{G}}(\mathbf{x}, \tau)^{-1}\mathbf{F}(\mathbf{x}, \tau) \tag{88}$$

The parity and time scaling of these coefficients will be analyzed in the following subsection.

Before that, it is worth discussing the physical interpretation of the optimization. The first equality in expression (86) for the second entropy contains two terms involving $\mathbf{x}'$. The quadratic term is negative and represents the entropy cost of ordering the system dynamically; whether the departure from zero is positive or negative, any fluctuation in the flux represents order and is unfavorable. The linear term can be positive or negative; it is this term that encourages a nonzero flux that drives the system back toward equilibrium in the future, which obviously increases the first entropy. (The quantity $\mathbf{F}$ will be shown below to be related to the ordinary thermodynamic force $\mathbf{X}$.) Hence the optimization procedure corresponds to balancing these two terms, with the quadratic term that is unfavorable to dynamic order preventing large fluxes, where it dominates, and the linear term increasing the first entropy and dominating for small fluxes.

B. Parity

In addition to the coefficients for the nonlinear second entropy expansion defined earlier, Eqs. (83), (84), and (85) define

$$\mathbf{F}^{\dagger}(\mathbf{x}; \tau) \equiv \left. \frac{\partial S^{(2)}(\mathbf{x}, \mathbf{x}'|\tau)}{\partial \mathbf{x}'} \right|_{\mathbf{x}'=\mathbf{x}} \tag{89}$$

$$\underline{\underline{G}}^{\dagger}(\mathbf{x}; \tau) \equiv \left. \frac{\partial^2 S^{(2)}(\mathbf{x}, \mathbf{x}'|\tau)}{\partial \mathbf{x}'\, \partial \mathbf{x}'} \right|_{\mathbf{x}'=\mathbf{x}} \tag{90}$$

and

$$\underline{\underline{G}}^{\ddagger}(\mathbf{x}; \tau) \equiv \left. \frac{\partial^2 S^{(2)}(\mathbf{x}', \mathbf{x}|\tau)}{\partial \mathbf{x}'\, \partial \mathbf{x}} \right|_{\mathbf{x}'=\mathbf{x}} \tag{91}$$

Under the parity operator, these behave as

$$E(\mathbf{x}; \tau) = E(\underline{\underline{\epsilon}}\mathbf{x}; \tau) = E(\mathbf{x}; -\tau) \tag{92}$$

$$\mathbf{F}(\mathbf{x}; \tau) = \underline{\underline{\epsilon}}\mathbf{F}^{\dagger}(\underline{\underline{\epsilon}}\mathbf{x}; \tau) = \mathbf{F}^{\dagger}(\mathbf{x}; -\tau) \tag{93}$$

$$\underline{\underline{G}}(\mathbf{x}; \tau) = \underline{\underline{\epsilon}}\,\underline{\underline{G}}^{\dagger}(\underline{\underline{\epsilon}}\mathbf{x}; \tau)\underline{\underline{\epsilon}} = \underline{\underline{G}}^{\dagger}(\mathbf{x}; -\tau) \tag{94}$$

and

$$\underline{\underline{G}}^{\ddagger}(\mathbf{x};\tau) = \underline{\underline{\epsilon}}\,\underline{\underline{G}}^{\ddagger}(\underline{\underline{\epsilon}}\mathbf{x};\tau)^{\mathrm{T}}\underline{\underline{\epsilon}} = \underline{\underline{G}}^{\ddagger}(\mathbf{x};-\tau)^{\mathrm{T}} \tag{95}$$

The matrices $\underline{\underline{G}}$ and $\underline{\underline{G}}^{\dagger}$ are symmetric. If all the variables have the same parity, then $\underline{\underline{\epsilon}} = \pm\underline{\underline{I}}$, which simplifies these rules considerably.

C. Time Scaling

As in the linear regime, consider the sequential transition $\mathbf{x}_1 \xrightarrow{\tau} \mathbf{x}_2 \xrightarrow{\tau} \mathbf{x}_3$. Again Markovian behavior is assumed and the second entropy is added separately for the two transitions. In view of the previous results, in the nonlinear regime the second entropy for this may be written

$$\begin{aligned}
&S^{(2)}(\mathbf{x}_3,\mathbf{x}_2,\mathbf{x}_1|\tau,\tau)\\
&\quad= S^{(2)}(\mathbf{x}_3,\mathbf{x}_2|\tau) + S^{(2)}(\mathbf{x}_2,\mathbf{x}_1|\tau) - S^{(1)}(\mathbf{x}_2)\\
&\quad= \tfrac{1}{2}\underline{\underline{G}}^{\dagger}(\mathbf{x}_3;\tau) : [\mathbf{x}_2-\mathbf{x}_3]^2 + \mathbf{F}^{\dagger}(\mathbf{x}_3;\tau)\cdot[\mathbf{x}_2-\mathbf{x}_3] + E(\mathbf{x}_3;\tau)\\
&\qquad+\tfrac{1}{2}\underline{\underline{G}}(\mathbf{x}_1;\tau) : [\mathbf{x}_2-\mathbf{x}_1]^2 + \mathbf{F}(\mathbf{x}_1;\tau)\cdot[\mathbf{x}_2-\mathbf{x}_1] + E(\mathbf{x}_1;\tau) - S^{(1)}(\mathbf{x}_2)
\end{aligned} \tag{96}$$

The first three terms arise from the expansion of $S^{(2)}(\mathbf{x}_3,\mathbf{x}_2|\tau)$ about $\mathbf{x}_3$, which accounts for the appearance of the daggers, and the second three terms arise from the expansion of $S^{(2)}(\mathbf{x}_2,\mathbf{x}_1|\tau)$ about $\mathbf{x}_1$. This ansatz is only expected to be valid for large enough τ such that the two intervals may be regarded as independent. This restricts the following results to the intermediate time regime.

As in the linear case, the second entropy for the transition $\mathbf{x}_1 \xrightarrow{2\tau} \mathbf{x}_3$ is equal to the maximum value of that for the sequential transition,

$$S^{(2)}(\mathbf{x}_3,\mathbf{x}_1|2\tau) = S^{(2)}(\mathbf{x}_3,\overline{\mathbf{x}}_2,\mathbf{x}_1|\tau,\tau) \tag{97}$$

This result holds in so far as fluctuations about the most probable trajectory are relatively negligible. Writing twice the left-hand side as the expansion about the first argument plus the expansion about the second argument, it follows that

$$\begin{aligned}
2S^{(2)}(\mathbf{x}_3,\mathbf{x}_1|2\tau) &= \tfrac{1}{2}\underline{\underline{G}}^{\dagger}(\mathbf{x}_3;2\tau) : [\mathbf{x}_1-\mathbf{x}_3]^2 + \mathbf{F}^{\dagger}(\mathbf{x}_3;2\tau)\cdot[\mathbf{x}_1-\mathbf{x}_3] + E(\mathbf{x}_3;2\tau)\\
&\quad+\tfrac{1}{2}\underline{\underline{G}}(\mathbf{x}_1;2\tau) : [\mathbf{x}_3-\mathbf{x}_1]^2 + \mathbf{F}(\mathbf{x}_1;2\tau)\cdot[\mathbf{x}_3-\mathbf{x}_1] + E(\mathbf{x}_1;2\tau)
\end{aligned} \tag{98}$$

As in the linear case, the optimum point is approximated by the midpoint, and it is shown later that the shift is of second order. Hence the right-hand side of

Eq. (97) is given by Eq. (96) evaluated at the midpoint $\mathbf{x}_2 = \tilde{\mathbf{x}}_2$:

$$\begin{aligned} S^{(2)}(\mathbf{x}_3, \tilde{\mathbf{x}}_2, \mathbf{x}_1|\tau, \tau) = & \tfrac{1}{8}[\underline{\underline{G}}(\mathbf{x}_1;\tau) + \underline{\underline{G}}^\dagger(\mathbf{x}_3;\tau)] : [\mathbf{x}_3 - \mathbf{x}_1]^2 \\ & + \tfrac{1}{2}[\mathbf{F}(\mathbf{x}_1;\tau) - \mathbf{F}^\dagger(\mathbf{x}_3;\tau)] \cdot [\mathbf{x}_3 - \mathbf{x}_1] \\ & + E(\mathbf{x}_1;\tau) + E(\mathbf{x}_3;\tau) - \tfrac{1}{2}[S^{(1)}(\mathbf{x}_1) + S^{(1)}(\mathbf{x}_3)] \\ & + \tfrac{1}{16}[\mathbf{x}_3 - \mathbf{x}_1]^2 : [\underline{\underline{S}}(\mathbf{x}_1) + \underline{\underline{S}}(\mathbf{x}_3)] \end{aligned} \tag{99}$$

Here the first entropy $S^{(1)}(\tilde{\mathbf{x}}_2)$ has been expanded symmetrically about the terminal points to quadratic order.

Each term of this may be equated to half the corresponding one on the right-hand side of Eq. (98). From the quadratic term it follows that

$$\tfrac{1}{8}[\underline{\underline{G}}(\mathbf{x}_1;\tau) + \underline{\underline{G}}^\dagger(\mathbf{x}_3;\tau)] + \tfrac{1}{16}[\underline{\underline{S}}(\mathbf{x}_1) + \underline{\underline{S}}(\mathbf{x}_3)] = \tfrac{1}{4}[\underline{\underline{G}}(\mathbf{x}_1;2\tau) + \underline{\underline{G}}^\dagger(\mathbf{x}_3;2\tau)] \tag{100}$$

To satisfy this, $\underline{\underline{G}}$ must contain terms that scale inversely with the time interval, and terms that are independent of the time interval. As in the linear case, the expansion is nonanalytic, and it follows that

$$\underline{\underline{G}}(\mathbf{x};\tau) = \frac{1}{|\tau|}\underline{\underline{g}}_0(\mathbf{x}) + \frac{1}{\tau}\underline{\underline{g}}_1(\mathbf{x}) + \underline{\underline{g}}_2(\mathbf{x}) + \hat{\tau}\underline{\underline{g}}_3(\mathbf{x}) \tag{101}$$

and that, in view of the parity rule (94),

$$\underline{\underline{G}}^\dagger(\mathbf{x};\tau) = \frac{1}{|\tau|}\underline{\underline{g}}_0(\mathbf{x}) - \frac{1}{\tau}\underline{\underline{g}}_1(\mathbf{x}) + \underline{\underline{g}}_2(\mathbf{x}) - \hat{\tau}\underline{\underline{g}}_3(\mathbf{x}) \tag{102}$$

From Eq. (100),

$$\underline{\underline{g}}_2(\mathbf{x}) = \tfrac{1}{2}\underline{\underline{S}}(\mathbf{x}) \tag{103}$$

and $\underline{\underline{g}}_3(\mathbf{x}) = \underline{\underline{0}}$. It will be shown later that $\underline{\underline{g}}_1(\mathbf{x}) = \underline{\underline{0}}$. Note that neither $\underline{\underline{G}}(\mathbf{x};\tau)$ nor $\underline{\underline{G}}^\dagger(\mathbf{x};\tau)$ can contain any terms in the intermediate regime other than those explicitly indicated, unless the first-order contribution from the difference between the midpoint and the optimum point is included.

From the linear terms, the second entropy forces are

$$\mathbf{F}(\mathbf{x};2\tau) = \mathbf{F}(\mathbf{x};\tau) \quad \text{and} \quad \mathbf{F}^\dagger(\mathbf{x};2\tau) = \mathbf{F}^\dagger(\mathbf{x};\tau) \tag{104}$$

which imply that they are independent of the magnitude of the time interval. Hence

$$\mathbf{F}(\mathbf{x};\tau) \sim \mathbf{f}(\mathbf{x};\hat{\tau}), \quad \mathbf{f}(\mathbf{x};\hat{\tau}) \equiv \mathbf{f}_0(\mathbf{x}) + \hat{\tau}\mathbf{f}_1(\mathbf{x}) \tag{105}$$

and, from the parity rule (93),

$$\mathbf{F}^\dagger(\mathbf{x};\tau) \sim \mathbf{f}^\dagger(\mathbf{x};\hat{\tau}),\ \mathbf{f}^\dagger(\mathbf{x};\hat{\tau}) \equiv \mathbf{f}_0(\mathbf{x}) - \hat{\tau}\mathbf{f}_1(\mathbf{x}) \tag{106}$$

For the case of a system where the variables only have even parity, this implies $\mathbf{f}_1(\mathbf{x}) = \mathbf{0}$.

Finally,

$$\tfrac{1}{2}[E(\mathbf{x}_1;2\tau) + E(\mathbf{x}_3;2\tau)] = E(\mathbf{x}_1;\tau) + E(\mathbf{x}_3;\tau) - \tfrac{1}{2}[S^{(1)}(\mathbf{x}_1) + S^{(1)}(\mathbf{x}_3)] \tag{107}$$

This equation implies that

$$E(\mathbf{x};\tau) \sim S^{(1)}(\mathbf{x}) \tag{108}$$

to leading order.

These scaling relations indicate that in the intermediate regime the second entropy, Eq. (86), may be written

$$S^{(2)}(\mathbf{x}',\mathbf{x}|\tau) = \frac{1}{2|\tau|}\underline{\underline{g}}_0(\mathbf{x}) : [\mathbf{x}' - \mathbf{x}]^2 + \mathbf{f}(\mathbf{x};\hat{\tau})\cdot[\mathbf{x}' - \mathbf{x}] + S^{(1)}(\mathbf{x}) \tag{109}$$

(Here $\underline{\underline{g}}_1$ has been set to zero, as is justified later.) This shows that the fluctuations in the transition probability are determined by a symmetric matrix, $\underline{\underline{g}}_0$, in agreement with previous analyses [35, 82]. Written in this form, the second entropy satisfies the reduction condition upon integration over $\mathbf{x}'$ to leading order (c.f. the earlier discussion of the linear expression). One can make it satisfy the reduction condition identically by writing it in the form

$$S^{(2)}(\mathbf{x}',\mathbf{x}|\tau) = \frac{1}{2|\tau|}\underline{\underline{g}}_0(\mathbf{x}) : [\mathbf{x}' - \overline{\mathbf{x}'}]^2 + S^{(1)}(\mathbf{x}) \tag{110}$$

with the most likely terminal position given explicitly later.

The derivative of the left-hand side of Eq. (108) is

$$\frac{\partial E(\mathbf{x};\tau)}{\partial \mathbf{x}} \equiv \mathbf{F}(\mathbf{x};\tau) + \mathbf{F}^\dagger(\mathbf{x};\tau) \sim 2\mathbf{f}_0(\mathbf{x}) \tag{111}$$

Equating this to the derivative of the right-hand side shows that

$$\mathbf{f}_0(\mathbf{x}) = \tfrac{1}{2}\mathbf{X}(\mathbf{x}) \tag{112}$$

This relates the time-independent part of the natural nonlinear force to the thermodynamic force for a system of general parity in the intermediate time regime.

The scaling of $\underline{\underline{G}}(\mathbf{x};\tau)$ motivates writing

$$\underline{\underline{G}}^{\ddagger}(\mathbf{x};\tau) \sim \frac{1}{|\tau|}\underline{\underline{g}}_0^{\ddagger}(\mathbf{x}) + \frac{1}{\tau}\underline{\underline{g}}_1^{\ddagger}(\mathbf{x}) + \underline{\underline{g}}_2^{\ddagger}(\mathbf{x}) + \hat{\tau}\underline{\underline{g}}_3^{\ddagger}(\mathbf{x}) \tag{113}$$

In view of the parity rule (95), $\underline{\underline{g}}_0^{\ddagger}$ and $\underline{\underline{g}}_2^{\ddagger}$ are symmetric matrices and $\underline{\underline{g}}_1^{\ddagger}$ and $\underline{\underline{g}}_3^{\ddagger}$ are antisymmetric matrices.

The derivative of the second entropy force is

$$\frac{\partial \mathbf{F}(\mathbf{x};\tau)}{\partial \mathbf{x}} = \underline{\underline{G}}(\mathbf{x};\tau) + \underline{\underline{G}}^{\ddagger}(\mathbf{x};\tau) \tag{114}$$

which in the intermediate regime becomes

$$\frac{\partial \mathbf{f}_0(\mathbf{x})}{\partial \mathbf{x}} + \hat{\tau}\frac{\partial \mathbf{f}_1(\mathbf{x})}{\partial \mathbf{x}} = \frac{1}{|\tau|}[\underline{\underline{g}}_0(\mathbf{x}) + \underline{\underline{g}}_0^{\ddagger}(\mathbf{x})] + \frac{1}{\tau}[\underline{\underline{g}}_1(\mathbf{x}) + \underline{\underline{g}}_1^{\ddagger}(\mathbf{x})] + \underline{\underline{g}}_2(\mathbf{x}) + \underline{\underline{g}}_2^{\ddagger}(\mathbf{x}) + \hat{\tau}\underline{\underline{g}}_3^{\ddagger}(\mathbf{x}) \tag{115}$$

Clearly, the first two bracketed terms have to individually vanish. Since the first bracket contains two symmetric matrices, this implies that $\underline{\underline{g}}_0(\mathbf{x}) = -\underline{\underline{g}}_0^{\ddagger}(\mathbf{x})$, and since the second bracket contains a symmetric matrix and an antisymmetric matrix, this also implies that $\underline{\underline{g}}_1(\mathbf{x}) = \underline{\underline{g}}_1^{\ddagger}(\mathbf{x}) = \underline{\underline{0}}$. Furthermore, since $\underline{\underline{g}}_2(\mathbf{x}) = \underline{\underline{S}}(\mathbf{x})/2 = \partial \mathbf{f}_0(\mathbf{x})/\partial \mathbf{x}^{\mathrm{T}}$, it is also concluded that $\underline{\underline{g}}_2^{\ddagger}(\mathbf{x}) = \underline{\underline{0}}$. Explicitly then, in the intermediate regime it follows that

$$\underline{\underline{G}}(\mathbf{x};\tau) = \frac{1}{|\tau|}\underline{\underline{g}}_0(\mathbf{x}) + \frac{1}{2}\underline{\underline{S}}(\mathbf{x}) \tag{116}$$

$$\underline{\underline{G}}^{\dagger}(\mathbf{x};\tau) = \frac{1}{|\tau|}\underline{\underline{g}}_0(\mathbf{x}) + \frac{1}{2}\underline{\underline{S}}(\mathbf{x}) \tag{117}$$

and

$$\underline{\underline{G}}^{\ddagger}(\mathbf{x};\tau) = \frac{-1}{|\tau|}\underline{\underline{g}}_0(\mathbf{x}) + \hat{\tau}\underline{\underline{g}}_3^{\ddagger}(\mathbf{x}) \tag{118}$$

with

$$\underline{\underline{g}}_3^{\ddagger}(\mathbf{x}) = \partial \mathbf{f}_1(\mathbf{x})/\partial \mathbf{x} \tag{119}$$

which will be used later. These further expansions justify the second entropy given earlier, Eq. (109).

In view of the reduction condition, Eq. (87), and the earlier scaling of $\underline{\underline{G}}$ and $\mathbf{F}$, the reduction condition (108) can be written to higher order:

$$E(\mathbf{x};\tau) \sim S^{(1)}(\mathbf{x}) + \frac{|\tau|}{2}\underline{\underline{g}}_0^{-1}(\mathbf{x}) : \mathbf{f}(\mathbf{x};\hat{\tau})^2 + \mathcal{O}\tau^2 \tag{120}$$

Since $E(\mathbf{x};\tau)$ is an even function of τ, this shows that

$$\mathbf{f}_1(\mathbf{x}) \cdot \underline{\underline{g}}_0^{-1}(\mathbf{x})\mathbf{X}(\mathbf{x}) = 0 \tag{121}$$

This says that $\mathbf{f}_1$ is orthogonal to the usual thermodynamic force $\mathbf{X}$ (using the inner product with metric $\underline{\underline{g}}_0$).

It is always possible to write

$$\mathbf{f}_1(\mathbf{x}) = \underline{\underline{g}}_0(\mathbf{x})\underline{\underline{\Phi}}(\mathbf{x})\mathbf{X}(\mathbf{x}) \tag{122}$$

The matrix $\underline{\underline{\Phi}}$ is underdetermined by this equation. If the matrix is taken to be antisymmetric,

$$\underline{\underline{\Phi}}(\mathbf{x}) = -\underline{\underline{\Phi}}(\mathbf{x})^{\mathrm{T}} \tag{123}$$

then the quasi-orthogonality condition (121) is automatically satisfied. That same condition shows that

$$\underline{\underline{\Phi}}(\mathbf{x}) = \underline{\underline{\epsilon}}\,\underline{\underline{\Phi}}(\underline{\underline{\epsilon}}\mathbf{x})^{\mathrm{T}}\underline{\underline{\epsilon}} \tag{124}$$

since $\mathbf{f}_1(\underline{\underline{\epsilon}}\mathbf{x}) = -\underline{\underline{\epsilon}}\mathbf{f}_1(\mathbf{x})$.

The derivative of the second entropy, Eq. (109), is

$$\frac{\partial S^{(2)}(\mathbf{x}',\mathbf{x}|\tau)}{\partial \mathbf{x}'} = \hat{\tau}\underline{\underline{g}}_0(\mathbf{x})\,\overset{\circ}{\mathbf{x}} + \tfrac{1}{2}[\mathbf{X}(\mathbf{x}) + 2\hat{\tau}\mathbf{f}_1(\mathbf{x})] \tag{125}$$

Hence the most likely terminal position is

$$\begin{aligned}\overline{\mathbf{x}}(\mathbf{x},\tau) &= \mathbf{x} - |\tau|\underline{\underline{g}}_0(\mathbf{x})^{-1}[\tfrac{1}{2}\mathbf{X}(\mathbf{x}) + \hat{\tau}\mathbf{f}_1(\mathbf{x})] \\ &= \mathbf{x} - \frac{|\tau|}{2}\underline{\underline{g}}_0(\mathbf{x})^{-1}\mathbf{X}(\mathbf{x}) - \tau\underline{\underline{\Phi}}(\mathbf{x})\mathbf{X}(\mathbf{x}) \\ &\equiv \mathbf{x} - |\tau|\underline{\underline{L}}(\mathbf{x};\hat{\tau})\mathbf{X}(\mathbf{x})\end{aligned} \tag{126}$$

The antisymmetric part of nonlinear transport matrix is not uniquely defined (due to the nonuniqueness of $\underline{\underline{\Phi}}$). However, the most likely terminal position is given uniquely by any $\underline{\underline{\Phi}}$ that satisfies Eq. (122).

The nonlinear transport matrix satisfies the reciprocal relation

$$\underline{\underline{L}}(\mathbf{x};\hat{\tau}) = \underline{\underline{\epsilon}}\,\underline{\underline{L}}(\underline{\underline{\epsilon}}\mathbf{x};\hat{\tau})^{\mathrm{T}}\underline{\underline{\epsilon}} = \underline{\underline{L}}(\mathbf{x};-\hat{\tau})^{\mathrm{T}} \tag{127}$$

These relations are the same as the parity rules obeyed by the second derivative of the second entropy, Eqs. (94) and (95). This effectively is the nonlinear version of Casimir's [24] generalization to the case of mixed parity of Onsager's reciprocal relation [10] for the linear transport coefficients, Eq. (55). The nonlinear result was also asserted by Grabert et al., (Eq. (2.5) of Ref. 25), following the assertion of Onsager's regression hypothesis with a state-dependent transport matrix.

The symmetric and asymmetric most likely positions can be defined:

$$\begin{aligned}\mathbf{x}_{\pm}(\mathbf{x};\tau) &\equiv \tfrac{1}{2}[\overline{\mathbf{x}}(\mathbf{x};\tau) \pm \overline{\mathbf{x}}(\mathbf{x};-\tau)] \\ &= \begin{cases} \mathbf{x} - \frac{|\tau|}{2}\underline{\underline{g}}_0(\mathbf{x})^{-1}\mathbf{X}(\mathbf{x}) \\ -\tau \underline{\underline{g}}_0(\mathbf{x})^{-1}\mathbf{f}_1(\mathbf{x}) \end{cases}\end{aligned} \tag{128}$$

This shows that the even temporal development of the system is governed directly by the thermodynamic force, and that the odd temporal development is governed by $\mathbf{f}_1$.

Analogous to the linear case, the most likely velocity is

$$\dot{\mathbf{x}}(\mathbf{x},\tau) = \overset{\circ}{\bar{x}}(\mathbf{x},\tau) = -\underline{\underline{L}}(\mathbf{x};\hat{\tau})\mathbf{X}(\mathbf{x}) \tag{129}$$

In terms of this, the most likely rate of production of the first entropy is

$$\begin{aligned}\overline{\overset{\circ}{S}}{}^{(1)}(\mathbf{x}) &= \overline{\overset{\circ}{\mathbf{x}}}(\mathbf{x},\hat{\tau})\cdot\mathbf{X}(\mathbf{x}) \\ &= \frac{-\hat{\tau}}{2}[\mathbf{X}(\mathbf{x}) + 2\hat{\tau}\mathbf{f}_1(\mathbf{x})]\cdot\underline{\underline{g}}_0(\mathbf{x})^{-1}\mathbf{X}(\mathbf{x}) \\ &= \frac{-\hat{\tau}}{2}\mathbf{X}(\mathbf{x})\cdot\underline{\underline{g}}_0(\mathbf{x})^{-1}\mathbf{X}(\mathbf{x})\end{aligned} \tag{130}$$

where the orthogonality condition (121) has been used. This shows that, as in the linear case, only the even part of the regression contributes to the most likely rate of first entropy production, which is equivalent to retaining only the symmetric part of the transport matrix. This was also observed by Grabert et al. (Eq. (2.13) of Ref. 25).

As in the linear case, it should be stressed that the asymmetric part of the transport matrix (equivalently $\mathbf{f}_1$) cannot be neglected just because it does not contribute to the steady rate of first entropy production. The present results show that the function makes an important and generally nonnegligible contribution to the dynamics of the steady state.

As in the linear case, the most likely value of the second entropy is $S^{(1)}(\mathbf{x})$, provided that the reduction condition is satisfied. However, Eq. (109) only satisfies the reduction condition to leading order, and instead its maximum is

$$\begin{aligned}\overline{S^{(2)}}(\mathbf{x}|\tau) &= S^{(2)}(\overline{\mathbf{x}'},\mathbf{x}|\tau)\\ &= \frac{|\tau|}{8}\underline{\underline{g}}_0(\mathbf{x})^{-1}:[\mathbf{X}(\mathbf{x})+2\hat{\tau}\mathbf{f}_1(\mathbf{x})]^2-\frac{|\tau|}{4}\underline{\underline{g}}_0(\mathbf{x})^{-1}:[\mathbf{X}(\mathbf{x})+2\hat{\tau}\mathbf{f}_1(\mathbf{x})]^2+S^{(1)}(\mathbf{x})\\ &= \frac{-|\tau|}{8}\underline{\underline{g}}_0(\mathbf{x})^{-1}:\mathbf{X}(\mathbf{x})^2-\frac{|\tau|}{2}\underline{\underline{g}}_0(\mathbf{x})^{-1}:\mathbf{f}_1(\mathbf{x})^2+S^{(1)}(\mathbf{x})\end{aligned} \tag{131}$$

As in the linear case, Eq. (66), one can identify three terms on the right-hand side of the second equality: the first term, which is negative and scales with τ, represents the ongoing cost of maintaining the dynamic order; the second term, which is positive, scales with τ, and is larger in magnitude than the first, represents the ongoing first entropy produced by the flux; and the third term, which is negative and independent of τ, represents the initial cost of erecting the static structure. In the final equality, one can see that, due to the orthogonality condition, the even and odd parts of the regression contribute separately to the maximum value of the second entropy. Note that in the expression for the second entropy a term independent of $\mathbf{x}'$ has been neglected in this section. Hence the second entropy does not here reduce to the first entropy as it does in Section 2.

1. *Optimum Intermediate Point*

The optimum intermediate point of the sequential transition may be obtained by maximizing the corresponding second entropy. Using the expansion (96), the derivative is

$$\begin{aligned}\frac{\partial S^{(2)}(\mathbf{x}_3,\mathbf{x}_2,\mathbf{x}_1|\tau,\tau)}{\partial \mathbf{x}_2} &= \mathbf{F}^{\dagger}(\mathbf{x}_3;\tau)+\underline{\underline{G}}^{\dagger}(\mathbf{x}_3;\tau)[\mathbf{x}_2-\mathbf{x}_3]\\ &\quad +\mathbf{F}(\mathbf{x}_1;\tau)+\underline{\underline{G}}(\mathbf{x}_1;\tau)[\mathbf{x}_2-\mathbf{x}_1]-\mathbf{X}(\mathbf{x}_2)\end{aligned} \tag{132}$$

Setting this to zero at the optimum point $\overline{\mathbf{x}}_2$, and writing it as the departure from the midpoint $\tilde{\mathbf{x}}_2$, it follows that

$$\begin{aligned}\mathbf{0} &= \mathbf{F}(\mathbf{x}_1;\tau)+\mathbf{F}^{\dagger}(\mathbf{x}_3;\tau)-\mathbf{X}(\tilde{\mathbf{x}}_2)-\underline{\underline{S}}(\tilde{\mathbf{x}}_2)[\overline{\mathbf{x}}_2-\tilde{\mathbf{x}}_2]\\ &\quad +[\underline{\underline{G}}(\mathbf{x}_1;\tau)+\underline{\underline{G}}^{\dagger}(\mathbf{x}_3;\tau)][\overline{\mathbf{x}}_2-\tilde{\mathbf{x}}_2]+\tfrac{1}{2}[\underline{\underline{G}}(\mathbf{x}_1;\tau)-\underline{\underline{G}}^{\dagger}(\mathbf{x}_3;\tau)][\mathbf{x}_3-\mathbf{x}_1]\end{aligned} \tag{133}$$

Hence to linear order in the differences, in the intermediate regime this is

$$\mathbf{0} = \hat{\tau}\mathbf{f}_1(\mathbf{x}_1) - \hat{\tau}\mathbf{f}_1(\mathbf{x}_3) - \underline{\underline{S}}(\tilde{\mathbf{x}}_2)[\overline{\mathbf{x}}_2 - \tilde{\mathbf{x}}_2] + \frac{2}{|\tau|}\underline{\underline{g}}_0(\tilde{\mathbf{x}}_2)[\overline{\mathbf{x}}_2 - \tilde{\mathbf{x}}_2] \tag{134}$$

which has solution

$$\begin{aligned}\overline{\mathbf{x}}_2 - \tilde{\mathbf{x}}_2 &= \hat{\tau}\left[\underline{\underline{S}}(\tilde{\mathbf{x}}_2) - \frac{2}{|\tau|}\underline{\underline{g}}_0(\tilde{\mathbf{x}}_2)\right]^{-1}\left[\frac{\partial \mathbf{f}_1(\tilde{\mathbf{x}}_2)}{\partial \tilde{\mathbf{x}}_2}[\mathbf{x}_1 - \mathbf{x}_3]\right] \\ &= \frac{\tau}{2}\underline{\underline{g}}_0(\tilde{\mathbf{x}}_2)^{-1}\underline{\underline{g}}_3^{\ddagger}(\tilde{\mathbf{x}}_2)[\mathbf{x}_3 - \mathbf{x}_1]\end{aligned} \tag{135}$$

using Eq. (119). This shows that the departure of the optimum point from the midpoint is of second order (linear in τ and in $\mathbf{x}_3 - \mathbf{x}_1$), and that $\underline{\underline{\lambda}}$ is of linear order in τ. Neglecting it in the previous analysis yields the leading contributions.

D. Linear Limit of Nonlinear Coefficients

In the linear limit, differentiation of the second entropy, Eq. (19), gives the relationship between the two sets of coefficients. One obtains

$$E(\mathbf{x}, \tau) = \tfrac{1}{2}[\underline{\underline{A}} + \underline{\underline{A}}' + \underline{\underline{B}} + \underline{\underline{B}}^{\mathrm{T}}] : \mathbf{x}^2 \tag{136}$$

$$\mathbf{F}(\mathbf{x}, \tau) = [\underline{\underline{A}}' + \underline{\underline{B}}^{\mathrm{T}}] : \mathbf{x} \tag{137}$$

$$\mathbf{F}^{\dagger}(\mathbf{x}, \tau) = [\underline{\underline{A}} + \underline{\underline{B}}] : \mathbf{x} \tag{138}$$

$$\underline{\underline{G}}(\mathbf{x}, \tau) = \underline{\underline{A}}' \tag{139}$$

$$\underline{\underline{G}}^{\dagger}(\mathbf{x}, \tau) = \underline{\underline{A}} \tag{140}$$

and

$$\underline{\underline{G}}^{\ddagger}(\mathbf{x}, \tau) = \underline{\underline{B}}^{\mathrm{T}} \tag{141}$$

These expressions can be used to confirm the consistency between the linear and the nonlinear results given earlier.

E. Exchange with a Reservoir

Now the previously isolated subsystem is allowed to exchange $\mathbf{x}$ with an external reservoir that applies a thermodynamic force $\mathbf{X}_{\mathrm{r}}$. Following the linear analysis (Section II F), denote the subsystem thermodynamic force by $\mathbf{X}_{\mathrm{s}}(\mathbf{x})$ (this was denoted simply $\mathbf{X}(\mathbf{x})$ for the earlier isolated system). At equilibrium and in the steady state, it is expected that $\mathbf{X}_s(\overline{\mathbf{x}}) = \mathbf{X}_r$, as will be shown explicitly. The reservoir force and the subsystem force must always be kept conceptually distinct even though they are numerically equal in the optimum state.

In the event that the external reservoir actually comprises two spatially separated reservoirs with a thermodynamic difference between them, then an external thermodynamic gradient is being applied across the subsystem, and in the optimum state there is a steady flux of $\mathbf{x}$ from one external reservoir to the other through the subsystem, which remains in the state $\overline{\mathbf{x}}$. For example, in the case of two temperature reservoirs, the thermodynamic gradient is related to the difference between their inverse temperatures divided by their separation, and the conjugate thermodynamic variable turns out to be the first energy moment of the subsystem. In this case the flux of energy from one reservoir to the other turns out to be related to the adiabatic rate of change of the first energy moment of the subsystem.

Let $\mathbf{x}' = \mathbf{x} + \Delta^0\mathbf{x}$ be the state of the subsystem after the internal change (i.e., as if it were isolated), and let $\mathbf{x}'' = \mathbf{x}' + \Delta^{\mathrm{r}}\mathbf{x}$ be the state after the externally induced changes. By conservation the change in the reservoir is $\Delta\mathbf{x}_r = -\Delta^{\mathrm{r}}\mathbf{x}$, and the change in the reservoir first entropy is $\Delta\mathbf{x}_r \cdot \mathbf{X}_r = -\Delta^{\mathrm{r}}\mathbf{x} \cdot \mathbf{X}_r$. Even when the subsystem force is nonlinear, the reservoir force remains independent of $\mathbf{x}$. The analysis will be carried out using the nonlinear results for the subsystem; in the linear regime the correspondence given in the preceding subsection can be applied.

The second entropy then is a function of three constrained variables, $S^{(2)}(\Delta^0\mathbf{x}, \Delta^{\mathrm{r}}\mathbf{x}, \mathbf{x}|\tau, \mathbf{X}_r)$. The internal part, which accounts for the resistance to the flux, is as given earlier; it characterizes the transition $\mathbf{x} \to \mathbf{x}'$. The external part is entirely influenced by the reservoir, and it consists of two parts: the equilibration with the reservoir for the initial state $\mathbf{x}$, and the transition $\mathbf{x}' \to \mathbf{x}''$.

The singlet probability, representing the equilibration with the reservoir, is

$$\wp(\mathbf{x}|\mathbf{X}_r) \propto \exp[S^{(1)}(\mathbf{x}) - \mathbf{x} \cdot \mathbf{X}_r]/k_B \tag{142}$$

with the most likely state evidently satisfying $\mathbf{X}_s(\overline{\mathbf{x}}) = \mathbf{X}_r$. The exponent is the total first entropy and replaces the isolated system first entropy that appears in the expression for the second entropy of the isolated system, Eq. (86).

The transition $\mathbf{x}' \to \mathbf{x}''$ is determined by one-half of the external change in the total first entropy. The factor of $\frac{1}{2}$ occurs for the conditional transition probability with no specific correlation between the terminal states, as this preserves the singlet probability during the reservoir induced transition [4, 8, 80]. The implicit assumption underlying this is that the conductivity of the reservoirs is much greater than that of the subsystem. The second entropy for the stochastic transition is the same as in the linear case, Eq. (71). In the expression for the second entropy of the isolated system, Eq. (109), the isolated system first entropy, $S^{(1)}(\mathbf{x})$, must be replaced by the total first entropy, $S^{(1)}(\mathbf{x}) - \mathbf{x}\cdot\mathbf{X}_r$. With this replacement and adding the stochastic second entropy, the total second entropy is

$$S^{(2)}_{\text{total}}(\Delta^0\mathbf{x}, \Delta^r\mathbf{x}, \mathbf{x}|\mathbf{X}_r, \tau) = \frac{1}{2|\tau|}\underline{\underline{g}}_0(\mathbf{x}) : [\Delta^0\mathbf{x}]^2 + \mathbf{f}(\mathbf{x};\hat{\tau})\cdot[\Delta^0\mathbf{x}] + S^{(1)}(\mathbf{x}) - \mathbf{x}\cdot\mathbf{X}_r + \tfrac{1}{2}[S^{(1)}(\mathbf{x}'') - S^{(1)}(\mathbf{x}') - \Delta^r\mathbf{x}\cdot\mathbf{X}_r] \tag{143}$$

Setting the derivative with respect to $\Delta^r\mathbf{x}$ to zero, one finds

$$\overline{\mathbf{X}''_s} = \mathbf{X}_r \tag{144}$$

As in the linear case, in the steady state the subsystem force equals the reservoir force at the end of the transition.

The derivative with respect to $\mathbf{x}$ is

$$\frac{\partial S^{(2)}_{\text{total}}}{\partial \mathbf{x}} = \mathbf{X}_s - \mathbf{X}_r + \tfrac{1}{2}[\mathbf{X}''_s - \mathbf{X}'_s] + \mathcal{O}\Delta^0\mathbf{x} \tag{145}$$

To leading order this vanishes when the initial subsystem force equals that imposed by the reservoirs,

$$\overline{\mathbf{X}_s} = \mathbf{X}_r \tag{146}$$

Since the subsystem force at the end of the transition is also most likely equal to the reservoir force, this implies that the adiabatic change is canceled by the stochastic change, $\overline{\Delta^r\mathbf{x}} = -\overline{\Delta^0\mathbf{x}}$.

The derivative with respect to $\Delta^0\mathbf{x}$ is

$$\frac{\partial S^{(2)}_{\text{total}}}{\partial \Delta^0\mathbf{x}} = \frac{1}{|\tau|}\underline{\underline{g}}_0(\mathbf{x})\Delta^0\mathbf{x} + \mathbf{f}(\mathbf{x};\hat{\tau}) + \frac{1}{2}[\mathbf{X}''_s - \mathbf{X}'_s] \tag{147}$$

The difference between the two forces is on the order of the perturbation and can again be neglected. Hence the most likely flux is

$$\overline{\overset{\circ}{\mathbf{x}}} = \frac{-\hat{\tau}}{2}\underline{\underline{g}}_0(\mathbf{x})^{-1}[\mathbf{X}_r + 2\hat{\tau}\overline{\mathbf{f}_1}] \tag{148}$$

Even in the nonlinear regime Onsager's regression hypothesis holds: the flux following a (rare) fluctuation in an isolated system is the same as if that departure from equilibrium were induced by an externally applied force.

F. Rate of Entropy Production

The constrained rate of first entropy production for the total system is

$$\begin{aligned}\dot{S}^{(1)}_{\text{total}}(\dot{\mathbf{x}}, \mathbf{x}) &= \dot{S}^{(1)}_{\text{s}}(\dot{\mathbf{x}}, \mathbf{x}) + \dot{\mathbf{x}}_{\text{r}} \cdot \mathbf{X}_{\text{r}} \\ &= [\dot{\mathbf{x}}^0 - \dot{\mathbf{x}}_{\text{r}}] \cdot \mathbf{X}_{\text{s}}(\mathbf{x}) + \dot{\mathbf{x}}_{\text{r}} \cdot \mathbf{X}_{\text{r}}\end{aligned} \tag{149}$$

This is a general result that holds for any structure $\mathbf{x}$, and any internal flux $\dot{\mathbf{x}} = [\Delta^0\mathbf{x} - \Delta\mathbf{x}_{\text{r}}]/\tau$, and any external flux $\dot{\mathbf{x}}_{\text{r}} = \Delta\mathbf{x}_{\text{r}}/\tau$. In the most likely state, the internal force equals the reservoir force, $\mathbf{X}_{\text{s}}(\overline{\mathbf{x}}) = \mathbf{X}_{\text{r}}$, and the internal flux vanishes because the change in the reservoir exactly compensates the change in the subsystem. Hence the most likely rate of production of the total first entropy is

$$\begin{aligned}\dot{\overline{S}}^{(1)}_{\text{total}}(\mathbf{x}) &= \dot{\overline{\mathbf{x}}}_{\text{r}}(\mathbf{x}, \hat{\tau}) \cdot \mathbf{X}_{\text{r}} \\ &= -\hat{\tau}\underline{\underline{L}}(\mathbf{x}; \hat{\tau})^{\text{T}} : \mathbf{X}_{\text{r}}^2 \\ &= \frac{-\hat{\tau}}{2}\underline{\underline{g}}_0(\overline{\mathbf{x}})^{-1} : \mathbf{X}_{\text{r}}^2\end{aligned} \tag{150}$$

As in the case of the isolated system, the asymmetric part of the transport matrix does not contribute to the scalar product or to the steady-state rate of first entropy production. All of the first entropy produced comes from the reservoirs, as it must since in the steady state the structure of the subsystem and hence its first entropy doesn't change. The rate of first entropy production is of course positive.

IV. NONEQUILIBRIUM STATISTICAL MECHANICS

A. Steady-State Probability Distribution

The aim of this section is to give the steady-state probability distribution in phase space. This then provides a basis for nonequilibrium statistical mechanics, just as the Boltzmann distribution is the basis for equilibrium statistical mechanics. The connection with the preceding theory for nonequilibrium thermodynamics will also be given.

The generic case is a subsystem with phase function $\hat{\mathbf{x}}(\boldsymbol{\Gamma})$ that can be exchanged with a reservoir that imposes a thermodynamic force $\mathbf{X}_{\text{r}}$. (The circumflex denoting a function of phase space will usually be dropped, since the argument $\boldsymbol{\Gamma}$ distinguishes the function from the macrostate label $\mathbf{x}$.) This case includes the standard equilibrium systems as well as nonequilibrium systems in steady flux. The probability of a state $\boldsymbol{\Gamma}$ is the exponential of the associated entropy, which is the total entropy. However, as usual it is assumed (it can be shown) [9] that the points in the phase space of the subsystem have equal weight and therefore that the associated subsystem entropy may be set to zero. Hence it follows that

$$S^{(1)}_{\text{total}}(\mathbf{\Gamma}|\mathbf{X}_r) = S^{(1)}_s(\mathbf{\Gamma}) + S^{(1)}_r(\mathbf{x}(\mathbf{\Gamma})|\mathbf{X}_r) = \text{const.} - \mathbf{x}(\mathbf{\Gamma}) \cdot \mathbf{X}_r \tag{151}$$

The final term is the subsystem-dependent part of the reservoir entropy, which arises from exchanging $\mathbf{x}$ between the two.

Using this the so-called static probability distribution is

$$\wp_{st}(\mathbf{\Gamma}|\mathbf{X}_r) = \frac{1}{Z_{st}(\mathbf{X}_r)} e^{-\mathbf{x}(\mathbf{\Gamma})\cdot\mathbf{X}_r/k_B} \tag{152}$$

where k_B is Boltzmann's constant. This is the analogue of Boltzmann's distribution and hence will yield the usual equilibrium results. However, it is dynamically disordered; under time reversal, $\mathbf{x}(\mathbf{\Gamma}) \Rightarrow \mathbf{x}(\mathbf{\Gamma}^\dagger) = \underline{\underline{\epsilon}}\mathbf{x}(\mathbf{\Gamma})$, and $\mathbf{X}_r \Rightarrow \mathbf{X}^\dagger_r \equiv \underline{\underline{\epsilon}}\mathbf{X}_r$, it remains unchanged:

$$\wp_{st}(\mathbf{\Gamma}|\mathbf{X}_r) = \wp_{st}(\mathbf{\Gamma}^\dagger|\mathbf{X}^\dagger_r) \tag{153}$$

It is clear that the true nonequilibrium probability distribution requires an additional factor of odd parity. Figure 1 sketches the origin of the extra term.

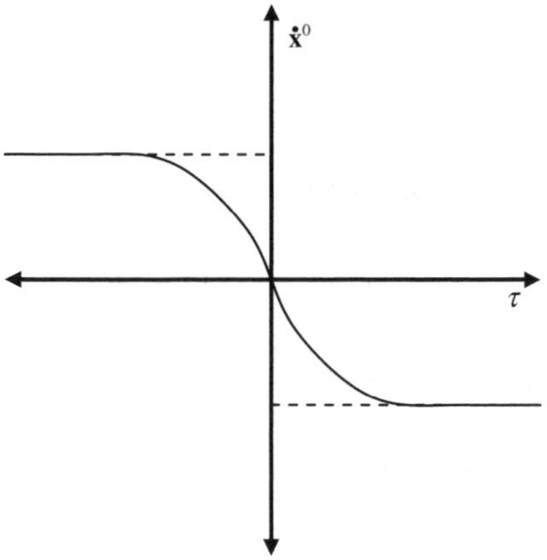

Figure 1. Sketch describing the origin of the odd work. The solid curve passing through the origin is the adiabatic velocity that results if the current point in phase space is dynamically disordered. The dashed lines are the most likely steady-state velocity corresponding to the current macrostate. The area between the curve and the line is the expected excess change in the macrostate of the isolated subsystem in the future and in the past, and this is equal to the putative change in the reservoir macrostate. The entropy of the latter gives the additional weight to be factored into the weight of the disordered macrostate.

The static probability places the subsystem in a dynamically disordered state, $\mathbf{\Gamma}_1$ so that at $\tau = 0$ the flux most likely vanishes, $\dot{\mathbf{x}}(\mathbf{\Gamma}_1) = \mathbf{0}$. If the system is constrained to follow the adiabatic trajectory, then as time increases the flux will become nonzero and approach its optimum or steady-state value, $\dot{\mathbf{x}}(\tau) \rightarrow \underline{\underline{L}}(\mathbf{x}_1, +1)\mathbf{X}_1$, where $\mathbf{x}_1 \equiv \mathbf{x}(\mathbf{\Gamma}_1)$ and $\mathbf{X}_1 \equiv \mathbf{X}_s(\mathbf{\Gamma}_1)$. Conversely, if the adiabatic trajectory is followed back into the past, then the flux would asymptote to its optimum value, $\dot{\mathbf{x}}(-\tau) \rightarrow -\underline{\underline{L}}(\mathbf{x}_1, -1)\mathbf{X}_1$.

Fix $\tau > 0$ at some intermediate value. (Soon it will be shown that the results are insensitive to the magnitude of τ.) The quantity

$$\mathbf{x}(\mathbf{\Gamma}_1; 0) - \mathbf{x}(\mathbf{\Gamma}_1; -\tau) \equiv \int_{-\tau}^{0} \mathrm{d}t\, \dot{\mathbf{x}}(\mathbf{\Gamma}_0(t|\mathbf{\Gamma}_1)) \tag{154}$$

is the past adiabatic change in the subsystem macrostate, and

$$\begin{aligned} \mathbf{x}(\mathbf{\Gamma}_1; \tau) - \mathbf{x}(\mathbf{\Gamma}_1; 0) &\equiv \int_{0}^{\tau} \mathrm{d}t\, \dot{\mathbf{x}}(\mathbf{\Gamma}_0(t|\mathbf{\Gamma}_1)) \\ &= -\int_{-\tau}^{0} \mathrm{d}t\, \underline{\underline{\varepsilon}}\dot{\mathbf{x}}(\mathbf{\Gamma}_0(t|\mathbf{\Gamma}_1^{\dagger})) \end{aligned} \tag{155}$$

is the future adiabatic change in the subsystem macrostate. In the second equality the minus sign arises from the fact that $\dot{\mathbf{x}}$ has opposite parity to $\mathbf{x}$, $\dot{\mathbf{x}}(\mathbf{\Gamma}) = -\underline{\underline{\varepsilon}}\dot{\mathbf{x}}(\mathbf{\Gamma}^{\dagger})$. The quantity $[\mathbf{x}(\mathbf{\Gamma}_1; \tau) - \mathbf{x}(\mathbf{\Gamma}_1; -\tau)]/2$ is half of the total adiabatic change in the subsystem macrostate associated with the current phase space point $\mathbf{\Gamma}_1$, and it may be written

$$\begin{aligned} \mathbf{x}_{\Delta}(\mathbf{\Gamma}_1) \equiv \frac{1}{2}[\mathbf{x}(\mathbf{\Gamma}_1; \tau) - \mathbf{x}(\mathbf{\Gamma}_1; -\tau)] &\equiv \frac{1}{2}\int_{-\tau}^{\tau} \mathrm{d}t\, \dot{\mathbf{x}}(\mathbf{\Gamma}_0(t|\mathbf{\Gamma}_1)) \\ &= \frac{1}{2}\int_{-\tau}^{0} \mathrm{d}t\, [\dot{\mathbf{x}}(\mathbf{\Gamma}_0(t|\mathbf{\Gamma}_1)) - \underline{\underline{\varepsilon}}\dot{\mathbf{x}}(\mathbf{\Gamma}_0(t|\mathbf{\Gamma}_1^{\dagger}))] \end{aligned} \tag{156}$$

From the second expression, it is easy to see that

$$\mathbf{x}_{\Delta}(\mathbf{\Gamma}^{\dagger}) = -\underline{\underline{\varepsilon}}\mathbf{x}_{\Delta}(\mathbf{\Gamma}) \tag{157}$$

This says that the phase function $\mathbf{x}_{\Delta}$ has opposite parity to the original function $\mathbf{x}$.

As mentioned, $\mathbf{x}_{\Delta}(\mathbf{\Gamma})$ is half of the total adiabatic change in the subsystem macrostate associated with the current phase space point $\mathbf{\Gamma}$. The factor of $\frac{1}{2}$ is used to compensate for double counting of the past and future changes. In the steady state, the subsystem most likely does not change macrostate, and hence this change has to be compensated by the change in the reservoir, $\Delta\mathbf{x}_r = \mathbf{x}_{\Delta}(\mathbf{\Gamma})$.

Accordingly, the change in the entropy of the reservoir associated with the current point in phase space is

$$\begin{aligned}\Delta S_{\rm r}(\mathbf{\Gamma}_1;\tau) &= \mathbf{X}_{\rm r}\cdot\mathbf{x}_\Delta(\mathbf{\Gamma}_1)\\ &= \frac{1}{2}\int_{-\tau}^{\tau}\mathrm{d}t\,[\mathbf{X}_{\rm r}\cdot\dot{\mathbf{x}}(\mathbf{\Gamma}_0(t|\mathbf{\Gamma}_1))+\hat{t}\,\mathbf{X}_{\rm r}\cdot\underline{\underline{L}}^{\rm s}(\mathbf{x}_1)\mathbf{X}_{\rm r}]\\ &= \frac{1}{2}\int_{-\tau}^{\tau}\mathrm{d}t\,[\mathbf{X}_{\rm r}\cdot\dot{\mathbf{x}}(\mathbf{\Gamma}_0(t|\mathbf{\Gamma}_1))+\hat{t}\,\mathbf{X}_{\rm r}\cdot\underline{\underline{L}}(\mathbf{x}_1,\hat{t})\mathbf{X}_{\rm r}]\end{aligned} \tag{158}$$

where $\hat{t} = \text{sign}(t)$. In the second equality, the symmetric part of the transport matrix appears, $\underline{\underline{L}}^{\rm s}(\mathbf{x}) = [\underline{\underline{L}}(\mathbf{x},+1)+\underline{\underline{L}}(\mathbf{x},-1)]/2$, which is independent of $\hat{t}$. This means that the second term in the integrand is odd in time and it gives zero contribution to the integral. In the third equality, the full transport matrix has been invoked, because the asymmetric part gives zero contribution to the scalar product (c.f. Eq. (130)).

The reason for adding this second term is to show that the integrand is short-ranged and hence that the integral is independent of τ. Since the most likely subsystem force equals that imposed by the reservoir, $\overline{\mathbf{X}}_{\rm s} \approx \mathbf{X}_{\rm r}$, with relatively negligible error, the asymptote can be written

$$\mathbf{X}_{\rm r}\cdot\dot{\mathbf{x}}(\mathbf{\Gamma}_0(t|\mathbf{\Gamma}_1)) \rightarrow \mp\mathbf{X}_{\rm r}\cdot\underline{\underline{L}}(\mathbf{x}_1,\pm1)\mathbf{X}_1 \approx \mp\underline{\underline{L}}(\mathbf{x}_1,\pm1):\mathbf{X}_{\rm r}^2, \quad t\rightarrow\pm\tau \tag{159}$$

This is equal and opposite to the term added to the integrand and so cancels with it leaving a short-ranged integrand. Since this asymptote gives zero contribution to the integral, because of this cancelation or, equivalently, because it is odd, the integral is insensitive to the value of τ provided that it is in the intermediate regime.

The odd contribution to the nonequilibrium steady-state probability distribution is just the exponential of this entropy change. Hence the full nonequilibrium steady-state probability distribution is

$$\wp_{\rm ss}(\mathbf{\Gamma}|\mathbf{X}_{\rm r}) = \frac{1}{Z_{\rm ss}(\mathbf{X}_{\rm r})}e^{-\mathbf{x}(\mathbf{\Gamma})\cdot\mathbf{X}_{\rm r}/k_{\rm B}}e^{\mathbf{x}_\Delta(\mathbf{\Gamma})\cdot\mathbf{X}_{\rm r}/k_{\rm B}} \tag{160}$$

Whereas the static probability distribution was invariant under time reversal, Eq. (153), the actual steady-state probability satisfies

$$\frac{\wp_{\rm ss}(\mathbf{\Gamma}|\mathbf{X}_{\rm r})}{\wp_{\rm ss}(\mathbf{\Gamma}^\dagger|\mathbf{X}_{\rm r}^\dagger)} = e^{2\mathbf{x}_\Delta(\mathbf{\Gamma})\cdot\mathbf{X}_{\rm r}/k_{\rm B}} \tag{161}$$

This tells the odds of the current phase space point compared to its conjugate. The quantity $\mathbf{x}_\Delta(\mathbf{\Gamma}) \cdot \mathbf{X}_r$ may be called the odd work.

B. Relationship with Green–Kubo Theory

It is now shown that the steady-state probability density, Eq. (160), gives the Green–Kubo expression for the linear transport coefficient. Linearizing the exponents for small applied forces, $\mathbf{X}_r \cdot \mathbf{x}_s \ll 1$, and taking the transport coefficient to be a constant, gives

$$\begin{aligned}
\langle \dot{\mathbf{x}}(\mathbf{\Gamma}) \rangle_{ss} &= \frac{\int d\mathbf{\Gamma}\, e^{-\mathbf{x}(\mathbf{\Gamma})\cdot\mathbf{X}_r/k_B} e^{\mathbf{x}_\Delta(\mathbf{\Gamma})\cdot\mathbf{X}_r/k_B} \dot{\mathbf{x}}(\mathbf{\Gamma})}{\int d\mathbf{\Gamma}\, e^{-\mathbf{x}(\mathbf{\Gamma})\cdot\mathbf{X}_r/k_B} e^{\mathbf{x}_\Delta(\mathbf{\Gamma})\cdot\mathbf{X}_r/k_B}} \\
&= \frac{\int d\mathbf{\Gamma}\, e^{-\mathcal{H}(\mathbf{\Gamma})/k_B T}[1 - \mathbf{x}'(\mathbf{\Gamma}) \cdot \mathbf{X}'_r/k_B][1 + \mathbf{x}_\Delta(\mathbf{\Gamma}) \cdot \mathbf{X}_r/k_B]\dot{\mathbf{x}}(\mathbf{\Gamma})}{\int d\mathbf{\Gamma}\, e^{-\mathcal{H}(\mathbf{\Gamma})/k_B T}[1 - \mathbf{x}'(\mathbf{\Gamma}) \cdot \mathbf{X}'_r/k_B][1 + \mathbf{x}_\Delta(\mathbf{\Gamma}) \cdot \mathbf{X}_r/k_B]} + \mathcal{O}\mathbf{X}_r^2 \\
&= \frac{\int d\mathbf{\Gamma}\, e^{-\mathcal{H}(\mathbf{\Gamma})/k_B T} \dot{\mathbf{x}}(\mathbf{\Gamma}) \mathbf{x}_\Delta(\mathbf{\Gamma}) \cdot \mathbf{X}_r/k_B}{\int d\mathbf{\Gamma}\, e^{-\mathcal{H}(\mathbf{\Gamma})/k_B T}} + \mathcal{O}\mathbf{X}_r^2 \\
&= \frac{1}{k_B} \langle \dot{\mathbf{x}}(\mathbf{\Gamma}) \mathbf{x}_\Delta(\mathbf{\Gamma}) \rangle_0 \cdot \mathbf{X}_r \\
&\equiv \underline{\underline{L}}\mathbf{X}_r
\end{aligned} \tag{162}$$

All the other linear terms vanish because they have opposite parity to the flux, $\langle \dot{\mathbf{x}}(\mathbf{\Gamma})\mathbf{x}(\mathbf{\Gamma}) \rangle_0 = \underline{\underline{0}}$. (This last statement is only true if the vector has pure even or pure odd parity, $\mathbf{x}(\mathbf{\Gamma}) = \pm\mathbf{x}(\mathbf{\Gamma}^\dagger)$. The following results are restricted to this case.) The static average is the same as an equilibrium average to leading order. That is, it is supposed that the exponential may be linearized with respect to all the reservoir forces except the zeroth one, which is the temperature, $X_{0,r} = 1/T$, and hence $x_0(\mathbf{\Gamma}) = \mathcal{H}(\mathbf{\Gamma})$, the Hamiltonian. From the definition of the adiabatic change, the linear transport coefficient may be written

$$\begin{aligned}
\underline{\underline{L}} &\equiv \frac{1}{k_B} \langle \dot{\mathbf{x}}(\mathbf{\Gamma}) \mathbf{x}_\Delta(\mathbf{\Gamma}) \rangle_0 \\
&= \frac{1}{2k_B} \int_{-\tau}^{\tau} dt' \, \langle \dot{\mathbf{x}}(t) \dot{\mathbf{x}}(t+t') \rangle_0
\end{aligned} \tag{163}$$

In the intermediate regime, this may be recognized as the Green–Kubo expression for the thermal conductivity [84], which in turn is equivalent to the Onsager expression for the transport coefficients [2].

This result is a very stringent test of the present expression for the steady-state probability distribution, Eq. (160). There is one, and only one, exponent that is odd, linear in $\mathbf{X}_r$, and that satisfies the Green–Kubo relation.

Likewise the even part of the probability can be tested by taking the steady-state average force. Using similar linearization arguments to the earlier ones, it may be shown that

$$\langle \mathbf{X}_s(\mathbf{\Gamma}) \rangle_{ss} = \mathbf{X}_r \tag{164}$$

which is the expected result. Again this result is true for vector components of pure parity. This confirms that the even part of the steady-state probability distribution, Eq. (160), is correct since there is one, and only one, even exponent that is linear in $\mathbf{X}_r$ that will yield this result.

In the case of mixed parity, it is expected that the steady-state probability distribution, Eq. (160), will remain valid, but the results in this section require clarification. Take, for example, the case of a subsystem with mobile charges on which is imposed crossed electric and magnetic fields. A steady current flows in the direction of the electric field, and an internal voltage is induced transverse to the electric field such that there is no net transverse force or flux (Hall effect). The induced transverse force is not equal to the imposed magnetic force (but it is equal and opposite to the induced Lorentz force), and hence the earlier result for the average internal force equalling the imposed force no longer holds. Equally, in the linear regime, the current induced parallel to the applied electric field is independent of the applied magnetic field, which implies that the cross component of the linear transport matrix vanishes.

C. Microstate Transitions

1. *Adiabatic Evolution*

Let $\mathbf{\Gamma}' = \mathbf{\Gamma} + \Delta_t \dot{\mathbf{\Gamma}}$ be the adiabatic evolution of $\mathbf{\Gamma}$ after an infinitesimal time step $\Delta_t > 0$. The adiabatic evolution of $\mathbf{x}_\Delta(\mathbf{\Gamma})$ can be obtained from

$$\begin{aligned} \mathbf{x}_\Delta(\mathbf{\Gamma}) &= \frac{1}{2}\int_{-\tau}^{\tau} dt' \dot{\mathbf{x}}(\mathbf{\Gamma}_0(t'|\mathbf{\Gamma})) \\ &= \frac{1}{2}\int_{-\tau-\Delta_t}^{\tau-\Delta_t} dt'' \dot{\mathbf{x}}(\mathbf{\Gamma}_0(t''+\Delta_t|\mathbf{\Gamma})) \\ &= \frac{1}{2}\int_{-\tau-\Delta_t}^{\tau-\Delta_t} dt'' \dot{\mathbf{x}}(\mathbf{\Gamma}_0(t''|\mathbf{\Gamma}')) \\ &= \mathbf{x}_\Delta(\mathbf{\Gamma}') - \frac{\Delta_t}{2}[\dot{\mathbf{x}}(\mathbf{\Gamma}_0(\tau|\mathbf{\Gamma}')) - \dot{\mathbf{x}}(\mathbf{\Gamma}_0(-\tau|\mathbf{\Gamma}'))] \\ &= \mathbf{x}_\Delta(\mathbf{\Gamma}') + \Delta_t \underline{\underline{L}}^s(\mathbf{x}(\mathbf{\Gamma}'))\mathbf{X}_s(\mathbf{\Gamma}') \end{aligned} \tag{165}$$

In the final equality the asymptote has been used. Although this result has been derived for infinitesimal Δ_t, it is also valid in the intermediate regime, $\Delta_t \sim \mathcal{O}\tau$.

This follows because the flux is constant and equal to its asymptotic value at the extremity of the intervals (recall $\tau > 0$ and $\Delta_t > 0$),

$$\begin{aligned} \frac{1}{2}\int_{\tau}^{\tau-\Delta_t} dt''\dot{\mathbf{x}}(\boldsymbol{\Gamma}_0(t''|\boldsymbol{\Gamma}')) &= \frac{1}{2}\int_{\tau}^{\tau-\Delta_t} dt''[-\underline{\underline{L}}(\mathbf{x}(\boldsymbol{\Gamma}'),+1)\mathbf{X}_s(\boldsymbol{\Gamma}')] \\ &= \frac{\Delta_t}{2}\underline{\underline{L}}(\mathbf{x}(\boldsymbol{\Gamma}'),+1)\mathbf{X}_s(\boldsymbol{\Gamma}') \end{aligned} \tag{166}$$

and similarly at the other extremity,

$$\begin{aligned} \frac{1}{2}\int_{-\tau-\Delta_t}^{-\tau} dt''\dot{\mathbf{x}}(\boldsymbol{\Gamma}_0(t''|\boldsymbol{\Gamma}')) &= \frac{1}{2}\int_{-\tau-\Delta_t}^{-\tau} dt''[\underline{\underline{L}}(\mathbf{x}(\boldsymbol{\Gamma}'),-1)\mathbf{X}_s(\boldsymbol{\Gamma}')] \\ &= \frac{\Delta_t}{2}\underline{\underline{L}}(\mathbf{x}(\boldsymbol{\Gamma}'),-1)\mathbf{X}_s(\boldsymbol{\Gamma}') \end{aligned} \tag{167}$$

Adding these together gives formally the same result as for the infinitesimal time step. Hence whether the time step is infinitesimal or intermediate, the adiabatic change in the odd work is

$$\begin{aligned} [\mathbf{x}_\Delta(\boldsymbol{\Gamma}') - \mathbf{x}_\Delta(\boldsymbol{\Gamma})]\cdot\mathbf{X}_r &= -\Delta_t\mathbf{X}_r\cdot\underline{\underline{L}}^s(\mathbf{x}(\boldsymbol{\Gamma}'))\mathbf{X}_s(\boldsymbol{\Gamma}') \\ &\approx -\Delta_t\underline{\underline{L}}^s(\mathbf{x}(\boldsymbol{\Gamma})):\mathbf{X}_r^2 \end{aligned} \tag{168}$$

The final approximation is valid if the adiabatic change in the macrostate is relatively negligible, $|\mathbf{x}' - \mathbf{x}| \ll |\mathbf{x}|$, and if the departure of the internal force from the reservoir force is relatively negligible, $|\mathbf{X}_s' - \mathbf{X}_r| \ll |\mathbf{X}_r|$.

Again denoting the adiabatic evolution over the intermediate time Δ_t by a prime, $\boldsymbol{\Gamma}' = \boldsymbol{\Gamma}(\Delta_t|\boldsymbol{\Gamma})$, the adiabatic change in the even exponent that appears in the steady-state probability distribution is

$$\begin{aligned} -[\mathbf{x}(\boldsymbol{\Gamma}') - \mathbf{x}(\boldsymbol{\Gamma})]\cdot\mathbf{X}_r &\approx \Delta_t\mathbf{X}_r\cdot\underline{\underline{L}}(\mathbf{x}(\boldsymbol{\Gamma}),+1)\mathbf{X}_s(\boldsymbol{\Gamma}) \\ &\approx \Delta_t\underline{\underline{L}}^s(\mathbf{x}(\boldsymbol{\Gamma})):\mathbf{X}_r^2 \end{aligned} \tag{169}$$

This is equal and opposite to the adiabatic change in the odd exponent. (More detailed analysis shows that the two differ at order Δ_t^2, provided that the asymmetric part of the transport matrix may be neglected.) It follows that the steady-state probability distribution is unchanged during adiabatic evolution over intermediate time scales:

$$\wp_{ss}(\boldsymbol{\Gamma}_0(\Delta_t|\boldsymbol{\Gamma})|\mathbf{X}_r) \approx \wp_{ss}(\boldsymbol{\Gamma}|\mathbf{X}_r) + \mathcal{O}\Delta_t^2, \quad \tau_{\text{short}} \lesssim \Delta_t \lesssim \tau_{\text{long}} \tag{170}$$

This confirms the earlier interpretation that the exponent reflects the entropy of the reservoirs only, and that the contribution from internal changes of the subsystem has been correctly removed. During the adiabatic transition the reservoirs do not change, and so the probability density must be constant. Obviously there is an upper limit on the time interval over which this result holds since the assumption that $\mathbf{X}'_{\rm s} \approx \mathbf{X}_{\rm s}$ implies that $\Delta_t|\underline{\underline{S}}_{\rm s}(\mathbf{x})\dot{\mathbf{x}}| \ll |\mathbf{X}_{\rm s}|$.

2. *Stochastic Transition*

In addition to the adiabatic transitions that would occur if the subsystem were isolated, stochastic perturbations from the reservoirs are also present. Hence the transition between the microstates $\mathbf{\Gamma} \rightarrow \mathbf{\Gamma}''$ in the intermediate time step Δ_t comprises the deterministic adiabatic transition $\mathbf{\Gamma} \rightarrow \mathbf{\Gamma}'$ due to the internal forces of the subsystem, followed by a stochastic transition $\mathbf{\Gamma}' \rightarrow \mathbf{\Gamma}''$ due to the perturbations by the reservoir.

The stochastic transition probability due to the reservoir can be taken to be

$$\Lambda_{\rm r}(\mathbf{\Gamma}''|\mathbf{\Gamma}';\mathbf{X}_{\rm r}) = \Theta_\Delta(|\mathbf{\Gamma}'' - \mathbf{\Gamma}'|)e^{-[\mathbf{x}''-\mathbf{x}']\cdot\mathbf{X}_{\rm r}/2k_{\rm B}}e^{[\mathbf{x}''_\Delta-\mathbf{x}'_\Delta]\cdot\mathbf{X}_{\rm r}/2k_{\rm B}} \tag{171}$$

The first factor is an appropriately normalized short-ranged function, such as a Gaussian or Heaviside step function, that represents the strength of the reservoir perturbations over the time interval. To leading order the normalization factor does not depend on the reservoir forces or the points in phase space. The exponent represents half the change in reservoir entropy during the transition. Such stochastic transition probabilities were originally used in equilibrium contexts [80, 81]. Half the difference between the final and initial reservoir entropies that appears in the first exponent is the same term as appears in Glauber or Kawasaki dynamics [75–78], where it guarantees detailed balance in the equilibrium context.

The unconditional stochastic transition probability is reversible,

$$\begin{aligned}\Lambda_{\rm r}(\mathbf{\Gamma}''|\mathbf{\Gamma}';\mathbf{X}_{\rm r})\wp_{\rm ss}(\mathbf{\Gamma}'|\mathbf{X}_{\rm r}) &= \Theta_\Delta(|\mathbf{\Gamma}''-\mathbf{\Gamma}'|)e^{-[\mathbf{x}''-\mathbf{x}']\cdot\mathbf{X}_{\rm r}/2k_{\rm B}}e^{[\mathbf{x}''_\Delta-\mathbf{x}'_\Delta]\cdot\mathbf{X}_{\rm r}/2k_{\rm B}}\frac{e^{-\mathbf{x}'\cdot\mathbf{X}_{\rm r}/k_{\rm B}}e^{\mathbf{x}'_\Delta\cdot\mathbf{X}_{\rm r}/k_{\rm B}}}{Z_{\rm ss}(\mathbf{X}_{\rm r})}\\ &= \Theta_\Delta(|\mathbf{\Gamma}''-\mathbf{\Gamma}'|)e^{-[\mathbf{x}''+\mathbf{x}']\cdot\mathbf{X}_{\rm r}/2k_{\rm B}}e^{[\mathbf{x}''_\Delta+\mathbf{x}'_\Delta]\cdot\mathbf{X}_{\rm r}/2k_{\rm B}}\frac{1}{Z_{\rm ss}(\mathbf{X}_{\rm r})}\\ &= \Lambda_{\rm r}(\mathbf{\Gamma}'|\mathbf{\Gamma}'';\mathbf{X}_{\rm r})\wp_{\rm ss}(\mathbf{\Gamma}''|\mathbf{X}_{\rm r})\end{aligned} \tag{172}$$

This is reasonable, since the arrow of time is provided by the adiabatic evolution of the subsystem in the intermediate regime. The perturbative influence of the reservoir ought to be insensitive to the direction of time or the direction of motion.

3. *Stationary Steady-State Probability*

For steady heat flow the probability density should not depend explicitly on time, and so it must be stationary under the combined transition probability given earlier. This can be verified directly:

$$\begin{aligned}
\wp(\mathbf{\Gamma}''|\mathbf{X}_{\rm r};\Delta_t) &= \int d\mathbf{\Gamma}'\, d\mathbf{\Gamma}\, \Lambda_{\rm r}(\mathbf{\Gamma}''|\mathbf{\Gamma}';\mathbf{X}_{\rm r})\delta(\mathbf{\Gamma}' - \mathbf{\Gamma}_0(\Delta_t|\mathbf{\Gamma}))\wp_{\rm ss}(\mathbf{\Gamma}|\mathbf{X}_{\rm r}) \\
&= \int d\mathbf{\Gamma}'\, \Lambda_{\rm r}(\mathbf{\Gamma}''|\mathbf{\Gamma}';\mathbf{X}_{\rm r})\wp_{\rm ss}(\mathbf{\Gamma}'|\mathbf{X}_{\rm r}) \\
&= \int d\mathbf{\Gamma}'\, \Lambda_{\rm r}(\mathbf{\Gamma}'|\mathbf{\Gamma}'';\mathbf{X}_{\rm r})\wp_{\rm ss}(\mathbf{\Gamma}''|\mathbf{X}_{\rm r}) \\
&= \wp_{\rm ss}(\mathbf{\Gamma}''|\mathbf{X}_{\rm r}), \quad \tau_{\rm short} \lesssim \Delta_t \lesssim \tau_{\rm long}
\end{aligned} \tag{173}$$

The final equality follows from the normalization of the conditional stochastic transition probability. This is the required result, which shows the stationarity of the steady-state probability under the present transition probability. This result invokes the preservation of the steady-state probability during adiabatic evolution over intermediate time scales.

4. *Forward and Reverse Transitions*

The previous results for the transition probability held over intermediate time scales. On infinitesimal time scales the adiabatic evolution of the steady-state probability has to be accounted for. The unconditional transition probability over an infinitesimal time step is given by

$$\begin{aligned}
\wp(\mathbf{\Gamma}'' \leftarrow \mathbf{\Gamma}|\Delta_t, \mathbf{X}_{\rm r}) &= \Lambda_{\rm r}(\mathbf{\Gamma}''|\mathbf{\Gamma}';\mathbf{X}_{\rm r})\wp_{\rm ss}(\mathbf{\Gamma}|\mathbf{X}_{\rm r}) \\
&= \Lambda_{\rm r}(\mathbf{\Gamma}''|\mathbf{\Gamma}';\mathbf{X}_{\rm r})\wp_{\rm ss}(\mathbf{\Gamma}'|\mathbf{X}_{\rm r})e^{\Delta_t\dot{\mathbf{x}}\cdot\mathbf{X}_{\rm r}/k_{\rm B}}e^{-\Delta_t\dot{\mathbf{x}}_\Delta\cdot\mathbf{X}_{\rm r}/k_{\rm B}} \\
&= \frac{\Theta_\Delta(|\mathbf{\Gamma}'' - \mathbf{\Gamma}'|)}{Z_{\rm ss}(\mathbf{X}_{\rm r})} e^{-[\mathbf{x}''+\mathbf{x}']\cdot\mathbf{X}_{\rm r}/2k_{\rm B}} e^{[\mathbf{x}''_\Delta+\mathbf{x}'_\Delta]\cdot\mathbf{X}_{\rm r}/2k_{\rm B}} e^{\Delta_t[\dot{\mathbf{x}}-\dot{\mathbf{x}}_\Delta]\cdot\mathbf{X}_{\rm r}/k_{\rm B}}
\end{aligned} \tag{174}$$

Now consider the forward transition, $\mathbf{\Gamma} \rightarrow \mathbf{\Gamma}' \rightarrow \mathbf{\Gamma}''$, and its reverse, $\mathbf{\Gamma}''^{\dagger} \rightarrow \mathbf{\Gamma}''' \rightarrow \mathbf{\Gamma}^{\dagger}$ (see Fig. 2). Note that $\mathbf{\Gamma}''' \neq \mathbf{\Gamma}'^{\dagger}$, but that $|\mathbf{\Gamma}'' - \mathbf{\Gamma}'| = |\mathbf{\Gamma}^{\dagger} - \mathbf{\Gamma}'''|$. The

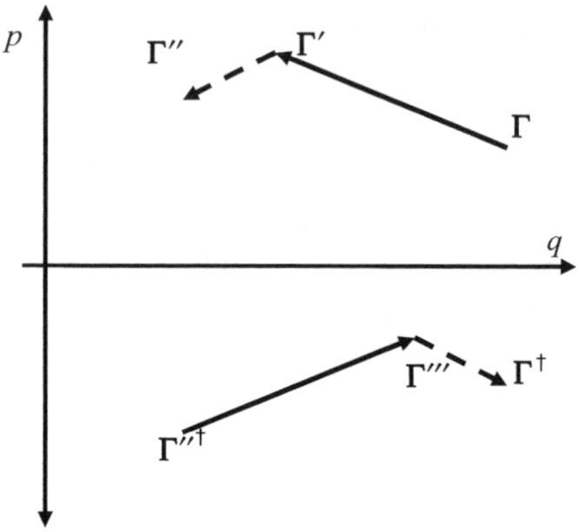

Figure 2. Forward and reverse transitions. The solid lines are the adiabatic trajectories over an infinitesimal time step, and the dashed lines are the stochastic transitions.

ratio of the forward to the reverse transition probabilities is

$$\begin{aligned}
&\frac{\wp(\mathbf{\Gamma}'' \leftarrow \mathbf{\Gamma}|\Delta_t, \mathbf{X}_{\rm r})}{\wp(\mathbf{\Gamma}^{\dagger} \leftarrow \mathbf{\Gamma}''^{\dagger}|\Delta_t, \mathbf{X}_{\rm r}^{\dagger})} \\
&\quad = \frac{\Theta_\Delta(|\mathbf{\Gamma}'' - \mathbf{\Gamma}'|) e^{-[\mathbf{x}''+\mathbf{x}']\cdot\mathbf{X}_{\rm r}/2k_{\rm B}} e^{[\mathbf{x}''_\Delta+\mathbf{x}'_\Delta]\cdot\mathbf{X}_{\rm r}/2k_{\rm B}} e^{\Delta_t[\dot{\mathbf{x}}-\dot{\mathbf{x}}_\Delta]\cdot\mathbf{X}_{\rm r}/k_{\rm B}} / Z_{\rm ss}(\mathbf{X}_{\rm r})}{\Theta_\Delta(|\mathbf{\Gamma}^{\dagger} - \mathbf{\Gamma}'''|) e^{-[\mathbf{x}^{\dagger}+\mathbf{x}''']\cdot\mathbf{X}_{\rm r}^{\dagger}/2k_{\rm B}} e^{[\mathbf{x}_\Delta^{\dagger}+\mathbf{x}'''_\Delta]\cdot\mathbf{X}_{\rm r}^{\dagger}/2k_{\rm B}} e^{\Delta_t[\dot{\mathbf{x}}''^{\dagger}-\dot{\mathbf{x}}''^{\dagger}_\Delta]\cdot\mathbf{X}_{\rm r}^{\dagger}/k_{\rm B}} / Z_{\rm ss}(\mathbf{X}_{\rm r}^{\dagger})} \\
&\quad = \frac{e^{-[\mathbf{x}''+\mathbf{x}]\cdot\mathbf{X}_{\rm r}/2k_{\rm B}} e^{[\mathbf{x}''_\Delta+\mathbf{x}_\Delta]\cdot\mathbf{X}_{\rm r}/2k_{\rm B}} e^{\Delta_t[\dot{\mathbf{x}}-\dot{\mathbf{x}}_\Delta]\cdot\mathbf{X}_{\rm r}/k_{\rm B}}}{e^{-[\mathbf{x}^{\dagger}+\mathbf{x}''^{\dagger}]\cdot\mathbf{X}_{\rm r}^{\dagger}/2k_{\rm B}} e^{[\mathbf{x}_\Delta^{\dagger}+\mathbf{x}''^{\dagger}_\Delta]\cdot\mathbf{X}_{\rm r}^{\dagger}/2k_{\rm B}} e^{\Delta_t[\dot{\mathbf{x}}''^{\dagger}-\dot{\mathbf{x}}''^{\dagger}_\Delta]\cdot\mathbf{X}_{\rm r}^{\dagger}/k_{\rm B}}} \\
&\quad = \frac{e^{-[\mathbf{x}''+\mathbf{x}]\cdot\mathbf{X}_{\rm r}/2k_{\rm B}} e^{[\mathbf{x}''_\Delta+\mathbf{x}_\Delta]\cdot\mathbf{X}_{\rm r}/2k_{\rm B}} e^{\Delta_t[\dot{\mathbf{x}}-\dot{\mathbf{x}}_\Delta]\cdot\mathbf{X}_{\rm r}/k_{\rm B}}}{e^{-[\mathbf{x}+\mathbf{x}'']\cdot\mathbf{X}_{\rm r}/2k_{\rm B}} e^{-[\mathbf{x}_\Delta+\mathbf{x}''_\Delta]\cdot\mathbf{X}_{\rm r}/2k_{\rm B}} e^{-\Delta_t[\dot{\mathbf{x}}''+\dot{\mathbf{x}}''_\Delta]\cdot\mathbf{X}_{\rm r}/k_{\rm B}}} \\
&\quad = e^{[\mathbf{x}''_\Delta+\mathbf{x}_\Delta]\cdot\mathbf{X}_{\rm r}/k_{\rm B}} e^{\Delta_t[\dot{\mathbf{x}}+\dot{\mathbf{x}}'']\cdot\mathbf{X}_{\rm r}/k_{\rm B}}
\end{aligned} \tag{175}$$

the even terms canceling. Not that this uses the fact that $\mathbf{x}$ and $\dot{\mathbf{x}}_\Delta$ have the same parity and that this is opposite to that of $\dot{\mathbf{x}}$ and of $\mathbf{x}_\Delta$, and also the fact that $\mathbf{x}^{\dagger} \cdot \mathbf{X}_{\rm r}^{\dagger} = \mathbf{x} \cdot \mathbf{X}_{\rm r}$, and similar facts. This result shows that the ratio of the forward to the reverse transition probability over an infinitesimal time step is given by the exponential of the odd work, plus the adiabatic evolution of the reversible work.

Now consider a trajectory $[\mathbf{\Gamma}] = \{\mathbf{\Gamma}_0, \mathbf{\Gamma}_1, \ldots, \mathbf{\Gamma}_f\}$, in the presence of the reservoir force $\mathbf{X}_r$, over a time interval $t_f = f\Delta_t$. Denote the reverse trajectory with force $\mathbf{X}_r^\dagger \equiv \underline{\underline{\epsilon}}\mathbf{X}_r$ by $[\mathbf{\Gamma}^\ddagger] = \{\mathbf{\Gamma}_f^\dagger, \mathbf{\Gamma}_{f-1}^\dagger, \ldots, \mathbf{\Gamma}_0^\dagger\}$. Let $\mathbf{\Gamma}_i' \equiv \mathbf{\Gamma}_0(\Delta_t|\mathbf{\Gamma}_i)$ denote the adiabatic development, and let $\mathbf{x}_i \equiv \mathbf{x}(\mathbf{\Gamma}_i)$ and $\mathbf{x}_i' \equiv \mathbf{x}(\mathbf{\Gamma}_i')$. The total adiabatic change in state over the trajectory is

$$\Delta^0\mathbf{x}[\mathbf{\Gamma}] = \sum_{i=0}^{f}[\mathbf{x}_i' - \mathbf{x}_i] - \tfrac{1}{2}[\mathbf{x}_f' - \mathbf{x}_f + \mathbf{x}_0' - \mathbf{x}_0] \tag{176}$$

and the total adiabatic change in the odd work parameter is

$$\Delta^0\mathbf{x}_\Delta[\mathbf{\Gamma}] = \sum_{i=0}^{f}[\mathbf{x}_{\Delta,i}' - \mathbf{x}_{\Delta,i}] - \tfrac{1}{2}[\mathbf{x}_{\Delta,f}' - \mathbf{x}_{\Delta,f} + \mathbf{x}_{\Delta,0}' - \mathbf{x}_{\Delta,0}] \tag{177}$$

Clearly, the adiabatic work done on the forward trajectory is equal and opposite to that done on the reverse trajectory, $\Delta^0\mathbf{x}[\mathbf{\Gamma}^\ddagger] \cdot \mathbf{X}_r^\dagger = -\Delta^0\mathbf{x}[\mathbf{\Gamma}] \cdot \mathbf{X}_r$, and the adiabatic odd work is the same on both trajectories, $\Delta^0\mathbf{x}_\Delta[\mathbf{\Gamma}^\ddagger] \cdot \mathbf{X}_r^\dagger = \Delta^0\mathbf{x}_\Delta[\mathbf{\Gamma}] \cdot \mathbf{X}_r$. Note that for large f the end corrections may be neglected.

The change in entropy of the reservoir over the trajectory is

$$\begin{aligned}\Delta S_r[\mathbf{\Gamma}] &= \mathbf{X}_r \cdot \Delta\mathbf{x}_r = -\mathbf{X}_r \cdot \Delta^r\mathbf{x} = -\mathbf{X}_r \cdot (\Delta\mathbf{x} - \Delta^0\mathbf{x}) \\ &= -\mathbf{X}_r \cdot (\mathbf{x}_f - \mathbf{x}_0 - \Delta^0\mathbf{x}[\mathbf{\Gamma}])\end{aligned} \tag{178}$$

This is equal and opposite to the change on the reverse trajectory, $\Delta S_r[\mathbf{\Gamma}] = -\Delta S_r[\mathbf{\Gamma}^\ddagger]$.

The unconditional probability of the trajectory is

$$\begin{aligned}&\wp([\mathbf{\Gamma}]|\mathbf{X}_r) \\ &= \prod_{i=1}^{f}[\Lambda_r(\mathbf{x}_i|\mathbf{x}_{i-1}', \mathbf{X}_r)]\wp_{ss}(\mathbf{\Gamma}_0|\mathbf{X}_r) \\ &= \prod_{i=1}^{f}[\Theta_\Delta(|\mathbf{\Gamma}_i - \mathbf{\Gamma}_{i-1}'|)e^{-[\mathbf{x}_i - \mathbf{x}_{i-1}']\cdot\mathbf{X}_r/2k_B}e^{[\mathbf{x}_{\Delta,i} - \mathbf{x}_{\Delta,i-1}']\cdot\mathbf{X}_r/2k_B}]\wp_{ss}(\mathbf{\Gamma}_0|\mathbf{X}_r) \\ &= \prod_{i=1}^{f}[\Theta_\Delta(|\mathbf{\Gamma}_i - \mathbf{\Gamma}_{i-1}'|)e^{-[\mathbf{x}_i - \mathbf{x}_{i-1}]\cdot\mathbf{X}_r/2k_B}e^{[\mathbf{x}_{\Delta,i} - \mathbf{x}_{\Delta,i-1}]\cdot\mathbf{X}_r/2k_B}e^{\Delta_t[\dot{\mathbf{x}}_{i-1} - \dot{\mathbf{x}}_{\Delta,i-1}]\cdot\mathbf{X}_r/2k_B}]\wp_{ss}(\mathbf{\Gamma}_0|\mathbf{X}_r) \\ &= \prod_{i=1}^{f}[\Theta_\Delta(|\mathbf{\Gamma}_i - \mathbf{\Gamma}_{i-1}'|)]\sqrt{\wp_{ss}(\mathbf{\Gamma}_f|\mathbf{X}_r)\wp_{ss}(\mathbf{\Gamma}_0|\mathbf{X}_r)}e^{\Delta^0\mathbf{x}[\mathbf{\Gamma}]\cdot\mathbf{X}_r/2k_B}e^{-\Delta^0\mathbf{x}_\Delta[\mathbf{\Gamma}]\cdot\mathbf{X}_r/2k_B} \\ &\quad \times e^{-[\mathbf{x}_f' - \mathbf{x}_f - \mathbf{x}_0' + \mathbf{x}_0]\cdot\mathbf{X}_r/4k_B}e^{[\mathbf{x}_{\Delta,f}' - \mathbf{x}_{\Delta,f} - \mathbf{x}_{\Delta,0}' + \mathbf{x}_{\Delta,0}]\cdot\mathbf{X}_r/4k_B}\end{aligned} \tag{179}$$

This only depends on the probability of the termini, the total adiabatic works, and the total weight of stochastic transitions. The first of these is for uncorrelated motion and is the one that occurs in Glauber or Kawasaki dynamics [75–78]. The last term is, of course, very sensitive to the specified trajectory and the degree to which it departs from the adiabatic motion. However, the stochastic transitions are the same on the forward and on the reverse trajectory, and the ratio of the probabilities of these is

$$\begin{aligned}
\frac{\wp([\boldsymbol{\Gamma}]|\mathbf{X}_{\rm r})}{\wp([\boldsymbol{\Gamma}^{\ddagger}]|\mathbf{X}_{\rm r}^{\dagger})} &= \frac{\sqrt{\wp_{\rm ss}(\boldsymbol{\Gamma}_f|\mathbf{X}_{\rm r})\wp_{\rm ss}(\boldsymbol{\Gamma}_0|\mathbf{X}_{\rm r})}}{\sqrt{\wp_{\rm ss}(\boldsymbol{\Gamma}_0^{\dagger}|\mathbf{X}_{\rm r}^{\dagger})\wp_{\rm ss}(\boldsymbol{\Gamma}_f^{\dagger}|\mathbf{X}_{\rm r}^{\dagger})}} \frac{e^{\Delta^0\mathbf{x}[\boldsymbol{\Gamma}]\cdot\mathbf{X}_{\rm r}/2k_{\rm B}} e^{-\Delta^0\mathbf{x}_{\Delta}[\boldsymbol{\Gamma}]\cdot\mathbf{X}_{\rm r}/2k_{\rm B}}}{e^{\Delta^0\mathbf{x}[\boldsymbol{\Gamma}^{\ddagger}]\cdot\mathbf{X}_{\rm r}^{\dagger}/2k_{\rm B}} e^{-\Delta^0\mathbf{x}_{\Delta}[\boldsymbol{\Gamma}^{\ddagger}]\cdot\mathbf{X}_{\rm r}^{\dagger}/2k_{\rm B}}} \\
&\quad \times \frac{e^{-[\mathbf{x}_f'-\mathbf{x}_f-\mathbf{x}_0'+\mathbf{x}_0]\cdot\mathbf{X}_{\rm r}/4k_{\rm B}} e^{[\mathbf{x}_{\Delta,f}'-\mathbf{x}_{\Delta,f}-\mathbf{x}_{\Delta,0}'+\mathbf{x}_{\Delta,0}]\cdot\mathbf{X}_{\rm r}/4k_{\rm B}}}{e^{[\mathbf{x}_f'^{\dagger}-\mathbf{x}_f^{\dagger}-\mathbf{x}_0'^{\dagger}+\mathbf{x}_0^{\dagger}]\cdot\mathbf{X}_{\rm r}^{\dagger}/4k_{\rm B}} e^{-[\mathbf{x}_{\Delta,f}'^{\dagger}-\mathbf{x}_{\Delta,f}^{\dagger}-\mathbf{x}_{\Delta,0}'^{\dagger}+\mathbf{x}_{\Delta,0}^{\dagger}]\cdot\mathbf{X}_{\rm r}^{\dagger}/4k_{\rm B}}} \\
&= e^{\Delta^0\mathbf{x}[\boldsymbol{\Gamma}]\cdot\mathbf{X}_{\rm r}/k_{\rm B}} e^{[\mathbf{x}_{\Delta}(\boldsymbol{\Gamma}_0)+\mathbf{x}_{\Delta}(\boldsymbol{\Gamma}_f)]\cdot\mathbf{X}_{\rm r}/k_{\rm B}} e^{\Delta_t[\dot{\mathbf{x}}_{\Delta,f}-\dot{\mathbf{x}}_{\Delta,0}]\cdot\mathbf{X}_{\rm r}/2k_{\rm B}} \\
&\approx e^{\Delta^0\mathbf{x}[\boldsymbol{\Gamma}]\cdot\mathbf{X}_{\rm r}/k_{\rm B}} e^{[\mathbf{x}_{\Delta}(\boldsymbol{\Gamma}_0)+\mathbf{x}_{\Delta}(\boldsymbol{\Gamma}_f)]\cdot\mathbf{X}_{\rm r}/k_{\rm B}} \\
&\approx e^{\Delta^0\mathbf{x}[\boldsymbol{\Gamma}]\cdot\mathbf{X}_{\rm r}/k_{\rm B}}
\end{aligned} \tag{180}$$

the even terms canceling. In the limit of large f, this ratio depends only on the adiabatic work done along the path.

The probability of observing the entropy of the reservoirs change by ΔS over a period t_f is related to the probability of observing the opposite change by

$$\begin{aligned}
\wp(\Delta S_{\rm r}|\mathbf{X}_{\rm r}, t_f) &= \int {\rm d}[\boldsymbol{\Gamma}]\, \delta(\Delta S_{\rm r} - \Delta S_{\rm r}[\boldsymbol{\Gamma}])\, \wp([\boldsymbol{\Gamma}]|\mathbf{X}_{\rm r}) \\
&= \int {\rm d}[\boldsymbol{\Gamma}^{\ddagger}]\, \delta(\Delta S_{\rm r} - \mathbf{X}_{\rm r}\cdot\Delta^0\mathbf{x}[\boldsymbol{\Gamma}])\wp([\boldsymbol{\Gamma}^{\ddagger}]|\mathbf{X}_{\rm r}) e^{\Delta^0\mathbf{x}[\boldsymbol{\Gamma}]\cdot\mathbf{X}_{\rm r}/k_{\rm B}} \\
&= e^{\Delta S_{\rm r}/k_B}\wp(-\Delta S_{\rm r}|\mathbf{X}_{\rm r}, t_f)
\end{aligned} \tag{181}$$

The end effects have been neglected here, including in the expression for change in reservoir entropy, Eq. (178). This result says in essence that the probability of a positive increase in entropy is exponentially greater than the probability of a decrease in entropy during heat flow. In essence this is the thermodynamic gradient version of the fluctuation theorem that was first derived by Bochkov and Kuzovlev [60] and subsequently by Evans et al. [56, 57]. It should be stressed that these versions relied on an adiabatic trajectory, macrovariables, and mechanical work. The present derivation explicitly accounts for interactions with the reservoir during the thermodynamic (here) or mechanical (later) work, and begins from the more fundamental result for the microscopic transition probability at the molecular level [4].

Closely related to the fluctuation theorem is the work theorem. Consider the average of the exponential of the negative of the entropy change,

$$
\begin{aligned}
\langle e^{-\mathbf{X}_r \cdot \Delta^0 \mathbf{x}[\boldsymbol{\Gamma}]} \rangle_{\mathrm{ss},t_f} &= \int \mathrm{d}[\boldsymbol{\Gamma}]\, \wp([\boldsymbol{\Gamma}]|\mathbf{X}_r) e^{-\mathbf{X}_r \cdot \Delta^0 \mathbf{x}[\boldsymbol{\Gamma}]} \\
&= \int \mathrm{d}[\boldsymbol{\Gamma}^{\ddagger}]\, \wp([\boldsymbol{\Gamma}^{\ddagger}]|\mathbf{X}_r^{\dagger}) \\
&= 1
\end{aligned}
\tag{182}
$$

Note that the exponent on the left-hand side is the negative of the change in reservoir entropy over the trajectory neglecting end effects, Eq. (178). Similarly, in going from the first to the second equality, it has been assumed that the trajectory is long enough that the end effects may be neglected. Although the averand on the left-hand side scales exponentially with time, this result shows that this particular average is not extensive in time (i.e., it does not depend on t_f).

The work theorem in the case of mechanical work was first given by Bochkov and Kuzovlev (for the case of a long cyclic trajectory) [60] and later by Jarzynski [55]. Both derivations invoked an adiabatic trajectory, in contrast to the present result that is valid for a system that can exchange with a reservoir during the performance of the work.

V. TIME-DEPENDENT MECHANICAL WORK

A. Phase Space Probability

An important class of nonequilibrium systems are those in which mechanical work, either steady or varying, is performed on the subsystem while it is in contact with a heat reservoir. Such work is represented by a time-dependent Hamiltonian, $\mathcal{H}_\mu(\boldsymbol{\Gamma}, t)$, where $\mu(t)$ is the work parameter. (For example, this could represent the position of a piston or the strength of an electric field.)

Yamada and Kawasaki [68, 69] proposed a nonequilibrium probability distribution that is applicable to an adiabatic system. *If* the system were isolated from the thermal reservoir during its evolution, and *if* the system were Boltzmann distributed at $t - \tau$, *then* the probability distribution at time t would be

$$
\begin{aligned}
\tilde{\wp}_\mu(\boldsymbol{\Gamma}|\beta, t) &= \tilde{Z}^{-1} e^{-\beta \mathcal{H}_\mu(\boldsymbol{\Gamma}_-, t-\tau)} \\
&= \tilde{Z}^{-1} e^{-\beta \mathcal{H}_\mu(\boldsymbol{\Gamma}, t)} e^{\beta W_\mu(\boldsymbol{\Gamma}, t)}
\end{aligned}
\tag{183}
$$

where $\beta = 1/k_B T$ is the inverse temperature of the reservoir. Here $\boldsymbol{\Gamma}_- = \boldsymbol{\Gamma}_\mu(t-\tau|\boldsymbol{\Gamma}, t)$ is the starting point of the adiabatic trajectory, and the work done is the change in energy of the system, $W_\mu(\boldsymbol{\Gamma}, t) = \mathcal{H}_\mu(\boldsymbol{\Gamma}, t) - \mathcal{H}_\mu(\boldsymbol{\Gamma}_-, t-\tau)$. This result invokes Liouville's theorem, namely, that the probability density is conserved on an adiabatic trajectory [9]. This expression, which is intended to represent the probability of the phase point $\boldsymbol{\Gamma}$ at t, suffers from several deficiencies when it is applied to a realistic system that is in contact with a heat reservoir during the mechanical work. First, it depends rather sensitively on the magnitude of τ, whereas no such artificial time parameter is expected in a realistic system. Second, it does not take into account the influence of the heat reservoir while the work is being performed, which means that the structure of the subsystem evolves adiabatically away from the true nonequilibrium structure. A modified thermostatted form of the Yamada–Kawasaki distribution has been given, but it is said to be computationally intractable [53, 70, 71]. It is unclear whether artifacts are introduced by the artificial thermostat. Based on calculations performed with the Yamada–Kawasaki distribution, some have even come to doubt the very existence of a nonequilibrium probability distribution [85–87].

The problems with the adiabatic Yamada–Kawasaki distribution and its thermostatted versions can be avoided by developing a nonequilibrium phase space probability distribution for the present case of mechanical work that is analogous to the one developed in Section IVA for thermodynamic fluxes due to imposed thermodynamic gradients. The odd work is required. To obtain this, one extends the work path into the future by making it even about t:

$$\tilde{\mu}_t(t') \equiv \begin{cases} \mu(t'), & t' \leq t \\ \mu(2t - t'), & t' > t \end{cases} \tag{184}$$

The corresponding adiabatic trajectory that is at $\boldsymbol{\Gamma}$ at time t is denoted by

$$\boldsymbol{\Gamma}_{\tilde{\mu}_t}(t'|\boldsymbol{\Gamma}, t) = \begin{cases} \boldsymbol{\Gamma}_\mu(t'|\boldsymbol{\Gamma}, t), & t' \leq t \\ \boldsymbol{\Gamma}_\mu(2t - t'|\boldsymbol{\Gamma}^\dagger, t)^\dagger, & t' > t \end{cases} \tag{185}$$

One defines $\boldsymbol{\Gamma}_\pm(\boldsymbol{\Gamma}) \equiv \boldsymbol{\Gamma}_{\tilde{\mu}_t}(t \pm \tau|\boldsymbol{\Gamma}, t)$, which have the property $\boldsymbol{\Gamma}_\pm(\boldsymbol{\Gamma}^\dagger) = (\boldsymbol{\Gamma}_\mp(\boldsymbol{\Gamma}))^\dagger$.

With these the odd work is [5]

$$\begin{aligned} \tilde{W}_\mu(\boldsymbol{\Gamma}, t) &= \frac{1}{2}[\mathcal{H}_{\tilde{\mu}_t}(\boldsymbol{\Gamma}_+, t+\tau) - \mathcal{H}_{\tilde{\mu}_t}(\boldsymbol{\Gamma}_-, t-\tau)] \\ &= \frac{1}{2}\int_{t-\tau}^{t+\tau} \mathrm{d}t'\, \dot{\mathcal{H}}_{\tilde{\mu}_t}(\boldsymbol{\Gamma}_{\tilde{\mu}_t}(t'|\boldsymbol{\Gamma}, t), t') \\ &= \frac{1}{2}\int_{t-\tau}^{t} \mathrm{d}t'\, [\dot{\mathcal{H}}_\mu(\boldsymbol{\Gamma}_\mu(t'|\boldsymbol{\Gamma}, t), t') - \dot{\mathcal{H}}_\mu(\boldsymbol{\Gamma}_\mu(t'|\boldsymbol{\Gamma}^\dagger, t), t')] \end{aligned} \tag{186}$$

(In this section, $\dot{\mathcal{H}}_\mu$ means the adiabatic rate of change of the Hamiltonian.) By construction, the odd work has odd phase space parity, $\tilde{W}_\mu(\mathbf{\Gamma}, t) = -\tilde{W}_\mu(\mathbf{\Gamma}^\dagger, t)$. With it the nonequilibrium probability distribution for mechanical work is [5]

$$\wp_\mu(\mathbf{\Gamma}|\beta, t) = \frac{e^{-\beta\mathcal{H}_\mu(\mathbf{\Gamma},t)} e^{\beta\tilde{W}_\mu(\mathbf{\Gamma},t)}}{h^{3N}N!Z_\mu(\beta, t)} \tag{187}$$

The probability distribution is normalized by $Z_\mu(\beta, t)$, which is a time-dependent partition function whose logarithm gives the nonequilibrium total entropy, which may be used as a generating function.

Asymptotically the rate of change of energy is

$$\dot{\mathcal{H}}_{\tilde{\mu}_t}(\mathbf{\Gamma}_{\tilde{\mu}_t}(t'|\mathbf{\Gamma}, t), t') \sim \text{sign}(t')\overline{\dot{\mathcal{H}}}_\mu(t), \quad |t'| \gtrsim \tau_{\text{short}} \tag{188}$$

where the most likely rate of doing work at time t appears. This assumes that the change in energy is negligible on the relevant time scales, $\tau|\overline{\dot{\mathcal{H}}}_\mu| \ll |\mathcal{H}_\mu|$. Since this asymptote is odd in time, one concludes that the odd work is independent of τ for τ in the intermediate regime, and that $\tilde{W}_\mu$ is dominated by the region $t' \approx t$.

From this and the fact that $\mathrm{d}\mathbf{\Gamma}_\mu(t'|\mathbf{\Gamma}, t)/\mathrm{d}t = 0$, it follows that the rate of change of the odd work along a Hamiltonian trajectory is

$$\dot{\tilde{W}}_{\tilde{\mu}}(\mathbf{\Gamma}, t) = \overline{\dot{\mathcal{H}}}_\mu(t) \tag{189}$$

This shows that the nonequilibrium probability distribution is stationary during adiabatic evolution on the most likely points of phase space.

B. Transition Probability

The evolution of the probability distribution over time consists of adiabatic development and stochastic transitions due to perturbations from the reservoir. As above, use a single prime to denote the adiabatic development in time Δ_t, $\mathbf{\Gamma} \to \mathbf{\Gamma}'$, and a double prime to denote the final stochastic position due to the influence of the reservoir, $\mathbf{\Gamma}' \to \mathbf{\Gamma}''$. The conditional stochastic transition probability may be taken to be

$$\Lambda^{\text{r}}_\mu(\mathbf{\Gamma}''|\mathbf{\Gamma}', t, \beta) = \Theta_\Delta(|\mathbf{\Gamma}'' - \mathbf{\Gamma}'|)e^{-\beta(\mathcal{H}''_\mu - \mathcal{H}'_\mu)/2} e^{\beta(\tilde{W}''_\mu - \tilde{W}'_\mu)/2} \tag{190}$$

This makes the unconditional stochastic transition probability reversible,

$$\Lambda^{\text{r}}_\mu(\mathbf{\Gamma}''|\mathbf{\Gamma}', t, \beta)\wp_\mu(\mathbf{\Gamma}'|\beta, t) = \Lambda^{\text{r}}_\mu(\mathbf{\Gamma}'|\mathbf{\Gamma}'', t, \beta)\wp_\mu(\mathbf{\Gamma}''|\beta, t) \tag{191}$$

The unconditional microscopic transition probability is

$$\wp_\mu(\mathbf{\Gamma}'' \leftarrow \mathbf{\Gamma}|t, \Delta_t, \beta) \tag{192}$$

$$\begin{aligned}
&= \Lambda^{\mathrm{r}}_\mu(\mathbf{\Gamma}''|\mathbf{\Gamma}', t, \beta)\wp_\mu(\mathbf{\Gamma}|\beta, t)\\
&= \Theta_\Delta(|\mathbf{\Gamma}'' - \mathbf{\Gamma}'|)e^{-\beta(\mathcal{H}''_\mu - \mathcal{H}'_\mu)/2}e^{\beta(\tilde{W}''_\mu - \tilde{W}'_\mu)/2}\frac{e^{-\beta\mathcal{H}_\mu}e^{\beta\tilde{W}_\mu}}{h^{3N}N!Z_\mu(\beta, t)}\\
&= \Theta_\Delta(|\mathbf{\Gamma}'' - \mathbf{\Gamma}'|)e^{-\beta(\mathcal{H}''_\mu + \mathcal{H}'_\mu)/2}e^{\beta(\tilde{W}''_\mu + \tilde{W}'_\mu)/2}e^{\beta\Delta_t\dot{\mathcal{H}}_\mu}e^{-\beta\Delta_t\dot{\tilde{W}}_\mu}/h^{3N}N!Z_\mu(\beta, t)\\
&= \Theta_\Delta(|\mathbf{\Gamma}'' - \mathbf{\Gamma}'|)e^{-\beta(\mathcal{H}''_\mu + \mathcal{H}_\mu)/2}e^{\beta(\tilde{W}''_\mu + \tilde{W}_\mu)/2}e^{\beta\Delta_t\dot{\mathcal{H}}_\mu/2}e^{-\beta\Delta_t\dot{\tilde{W}}_\mu/2}/h^{3N}N!Z_\mu(\beta, t)
\end{aligned} \tag{193}$$

for an infinitesimal Δ_t. The $\dot{\mathcal{H}}_\mu$ can be replaced by $(\dot{\mathcal{H}}''_\mu + \dot{\mathcal{H}}_\mu)/2$ to this order, and similarly for $\dot{\tilde{W}}_\mu$.

Consider the forward transition $\mathbf{\Gamma} \rightarrow \mathbf{\Gamma}' \rightarrow \mathbf{\Gamma}''$ and its reverse $\mathbf{\Gamma}''^{\dagger} \rightarrow \mathbf{\Gamma}''' \rightarrow \mathbf{\Gamma}^{\dagger}$ on the mirror path. (It is necessary to use the mirror path centered at $t + \Delta_t$ because time goes from $t + \Delta_t \rightarrow t + 2\Delta_t$, and $\mu(t + \Delta_t) \rightarrow \mu(t)$.) The ratio of the forward to the reverse transition probabilities is

$$\begin{aligned}
&\frac{\wp_\mu(\mathbf{\Gamma}'' \leftarrow \mathbf{\Gamma}|t, \Delta_t, \beta)}{\wp_{\tilde{\mu}_{t+\Delta_t}}(\mathbf{\Gamma}^{\dagger} \leftarrow \mathbf{\Gamma}''^{\dagger}|t + \Delta_t, \Delta_t, \beta)}\\
&= e^{\beta[\tilde{W}_\mu(\mathbf{\Gamma}'', t+\Delta_t) + \tilde{W}_\mu(\mathbf{\Gamma}, t)]}e^{\Delta_t\beta[\dot{\mathcal{H}}_\mu(\mathbf{\Gamma}'', t+\Delta_t) + \dot{\mathcal{H}}_\mu(\mathbf{\Gamma}, t)]/2}\frac{Z_\mu(\beta, t + \Delta_t)}{Z_\mu(\beta, t)}
\end{aligned} \tag{194}$$

the even terms canceling. The ratio of partition functions can be written as the exponential of the change in free energy.

Now consider a trajectory $[\mathbf{\Gamma}] = \{\mathbf{\Gamma}_0, \mathbf{\Gamma}_1, \ldots, \mathbf{\Gamma}_f\}$, at times $t_n = n\Delta_t$, $n = 0, 1, \ldots, f$, with parameter path $\mu_n = \mu(t_n)$. The reverse trajectory and path is $\mathbf{\Gamma}^{\ddagger}_n = \mathbf{\Gamma}^{\dagger}_{f-n}$, $\mu^{\ddagger}_n = \mu_{f-n}$, $n = 0, 1, \ldots, f$. The path is running backwards in time, and hence the time derivatives of the Hamiltonian and of the odd work change sign. The work done on the subsystem on the trajectory is just the total adiabatic change in the Hamiltonian,

$$\Delta^0\mathcal{H}_\mu[\mathbf{\Gamma}] = \frac{\Delta_t}{2}[\dot{\mathcal{H}}_\mu(\mathbf{\Gamma}_0) + \dot{\mathcal{H}}_\mu(\mathbf{\Gamma}_f)] + \Delta_t\sum_{i=1}^{f-1}\dot{\mathcal{H}}_\mu(\mathbf{\Gamma}_i) \tag{195}$$

Clearly, $\Delta^0\mathcal{H}_{\mu^{\ddagger}}[\mathbf{\Gamma}^{\ddagger}] = -\Delta^0\mathcal{H}_\mu[\mathbf{\Gamma}]$. For large $f\Delta_t$ this gives the change in the reservoirs' entropy over the trajectory, $\Delta S_{\mathrm{r}}/k_{\mathrm{B}} = -\beta\Delta\mathcal{H}_\mu + \beta\Delta^0\mathcal{H}_\mu \approx \beta\Delta^0\mathcal{H}_\mu$. The odd work done adiabatically is

$$\Delta^0\tilde{W}_\mu[\mathbf{\Gamma}] = \frac{\Delta_t}{2}[\dot{\tilde{W}}_\mu(\mathbf{\Gamma}_0) + \dot{\tilde{W}}_\mu(\mathbf{\Gamma}_f)] + \Delta_t\sum_{i=1}^{f-1}\dot{\tilde{W}}_\mu(\mathbf{\Gamma}_i) \tag{196}$$

In this case, $\Delta^0 \tilde{W}_{\mu^\ddagger}[\mathbf{\Gamma}^\ddagger] = \Delta^0 \tilde{W}_\mu[\mathbf{\Gamma}]$, because $\tilde{W}$ changes sign both with parity and with the time derivative.

The unconditional probability of the trajectory is

$$\begin{aligned}
\wp_\mu([\mathbf{\Gamma}],\beta) &= \prod_{i=1}^{f} \{\Theta_\Delta(|\mathbf{\Gamma}_i - \mathbf{\Gamma}'_{i-1}|) e^{-\beta(\mathcal{H}_{\mu,i}-\mathcal{H}'_{\mu,i-1})/2} e^{\beta(\tilde{W}_{\mu,i}-\tilde{W}'_{\mu,i-1})/2}\} \wp_\mu(\mathbf{\Gamma}_0|\beta,0) \\
&= \prod_{i=1}^{f} \{\Theta_\Delta(|\mathbf{\Gamma}_{i+1} - \mathbf{\Gamma}'_i|)\} e^{-\beta\mathcal{H}_{\mu,f}/2} e^{\beta\tilde{W}_{\mu,f}/2} e^{\beta(\mathcal{H}_{\mu,0}+\Delta_t\dot{\mathcal{H}}_{\mu,0})/2} e^{-\beta(\tilde{W}_{\mu,0}+\Delta_t\dot{\tilde{W}}_{\mu,0})/2} \\
&\quad \times e^{\beta\Delta^0\mathcal{H}_\mu[\mathbf{\Gamma}]/2} e^{-\beta\Delta_t(\dot{\mathcal{H}}_{\mu,f}+\dot{\mathcal{H}}_{\mu,0})/4} e^{-\beta\Delta^0\tilde{W}_\mu[\mathbf{\Gamma}]/2} e^{\beta\Delta_t(\dot{\tilde{W}}_{\mu,f}+\dot{\tilde{W}}_{\mu,0})/4} \frac{e^{-\beta\mathcal{H}_{\mu,0}} e^{\beta\tilde{W}_{\mu,0}}}{h^{3N} N! Z_{\mu,0}} \\
&= \prod_{i=1}^{f} \{\Theta\Delta(|\mathbf{\Gamma}_{i+1} - \mathbf{\Gamma}'_i|)\} e^{\beta\Delta^0\mathcal{H}_\mu[\mathbf{\Gamma}]/2} e^{-\beta\Delta^0\tilde{W}_\mu[\mathbf{\Gamma}]/2} e^{-\beta\Delta_t(\dot{\mathcal{H}}_{\mu,f}-\dot{\mathcal{H}}_{\mu,0})/4} e^{\beta\Delta_t(\dot{\tilde{W}}_{\mu,f}-\dot{\tilde{W}}_{\mu,0})/4} \\
&\quad \times \sqrt{\wp_\mu(\mathbf{\Gamma}_f|\beta,t_f)\wp_\mu(\mathbf{\Gamma}_0|\beta,0)} \sqrt{Z_\mu(\beta,t_f)/Z_\mu(\beta,0)}
\end{aligned} \tag{197}$$

since $\mathcal{H}'_{\mu,i-1} = \mathcal{H}_{\mu,i-1} + \Delta_t \dot{\mathcal{H}}_{\mu,i-1}$. The terms involving Δ_t may be neglected. In order to calculate the probabilities that appear in the final line, the behavior of $\mu(t)$ for $t < 0$ must be known. An explicit simplifying form is given shortly.

The ratio of the probabilities of the forward and reverse trajectories is

$$\frac{\wp_\mu([\mathbf{\Gamma}],\beta)}{\wp_{\mu^\ddagger}([\mathbf{\Gamma}^\ddagger],\beta)} = e^{\beta\tilde{W}_\mu(\mathbf{\Gamma}_f,t_f)} e^{\beta\tilde{W}_\mu(\mathbf{\Gamma}_0,t_0)} \frac{Z_\mu(\beta,t_f)}{Z_\mu(\beta,0)} e^{\beta\Delta^0\mathcal{H}_\mu[\mathbf{\Gamma}]} \tag{198}$$

This is the mechanical version of the *Reverse Transition Theorem* [4]. The first three factors are boundary terms, whereas the final exponent scales with the length of the time interval.

It is possible to simplify these results for the case when it is valid to draw the start and end points of the trajectories from a probability density with even parity such as the Boltzmann distribution,

$$\wp_B(\mathbf{\Gamma}|\beta,0) = \frac{e^{-\beta\mathcal{H}_\mu(\mathbf{\Gamma},0)}}{Z_B(\beta,\mu(0))} \quad \text{and} \quad \wp_B(\mathbf{\Gamma}|\beta,t_f) = \frac{e^{-\beta\mathcal{H}_\mu(\mathbf{\Gamma},t_f)}}{Z_B(\beta,\mu(t_f))} \tag{199}$$

This would be valid if the parameter $\mu(t)$ were held constant on some initial part and on some final part of the path (equilibration phase), during which periods the odd work vanishes. In this case the ratio of the forward and reverse trajectories is

$$\frac{\wp_B([\mathbf{\Gamma}],\beta)}{\wp_B([\mathbf{\Gamma}^\ddagger],\beta)} = \frac{Z_B(\beta,\mu(t_f))}{Z_B(\beta,\mu(0))} e^{\beta\Delta^0\mathcal{H}_\mu[\mathbf{\Gamma}]} \tag{200}$$

This modification of the termini of the trajectory is identical to that assumed in the fluctuation [56, 57, 60] and work [55, 60] theorems discussed next. The type of work paths for which this modification is valid is discussed at the end of this section.

The probability of observing the entropy of the reservoir change by ΔS over a period t_f is related to the probability of observing the opposite change by

$$\begin{aligned}\wp_\mu(\Delta S|\beta, t_f) &= \int \mathrm{d}[\mathbf{\Gamma}]\delta(\Delta S - \Delta S[\mathbf{\Gamma}])\wp_\mu([\mathbf{\Gamma}], \beta) \\ &\approx k_\mathrm{B}^{-1} \int \mathrm{d}[\mathbf{\Gamma}^\ddagger]\delta(\Delta S/k_\mathrm{B} - \beta\Delta^0\mathcal{H}_\mu[\mathbf{\Gamma}])\wp_{\mu^\ddagger}([\mathbf{\Gamma}]^\ddagger, \beta)e^{\beta\tilde{W}_\mu(\mathbf{\Gamma}_f, t_f)}e^{\beta\tilde{W}_\mu(\mathbf{\Gamma}_0, t_0)} \\ &\quad \times \frac{Z_\mu(\beta, t_f)}{Z_\mu(\beta, 0)} e^{\beta\Delta^0\mathcal{H}_\mu[\mathbf{\Gamma}]} \\ &\approx \frac{Z_\mu(\beta, t_f)}{Z_\mu(\beta, 0)} e^{\Delta S/k_B}\wp_{\mu^\ddagger}(-\Delta S|\beta, t_f)\end{aligned} \tag{201}$$

the odd work being neglected in the final equality. Using instead trajectories that begin and end with the Boltzmann distribution, the exact result is

$$\wp_\mathrm{B}(\Delta^0\mathcal{H}_\mu|\beta, t_f) = e^{\beta\Delta^0\mathcal{H}_\mu}\wp_\mathrm{B}(-\Delta^0\mathcal{H}_\mu|\beta, t_f)Z_\mathrm{B}(\beta, t_f)/Z_\mathrm{B}(\beta, 0) \tag{202}$$

This result says in essence that the probability of a positive increase in entropy is exponentially greater than the probability of a decrease in entropy during mechanical work. This is in essence the fluctuation theorem that was first derived by Bochkov and Kuzovlev [58–60] and later by Evans et al. [56, 57]. A derivation has also been given by Crooks [61, 62], and the theorem has been verified experimentally [63]. The present derivation is based on the author's microscopic transition probability [4].

For the case of the work theorem [55], consider the average of the exponential of the negative of the heat flow,

$$\begin{aligned}\langle e^{-\beta\Delta^0\mathcal{H}_\mu}\rangle_{[\mu], t_f} &= \int \mathrm{d}[\mathbf{\Gamma}]\, e^{-\beta\Delta^0\mathcal{H}_\mu[\mathbf{\Gamma}]}\wp_\mu([\mathbf{\Gamma}]|\beta, t_f) \\ &= \int \mathrm{d}[\mathbf{\Gamma}^\ddagger]\, \wp_{\mu^\ddagger}([\mathbf{\Gamma}^\ddagger]|\beta, t_f)e^{\beta(\tilde{W}_\mu(\mathbf{\Gamma}_f, t_f) + \tilde{W}_\mu(\mathbf{\Gamma}_0, t_0))}Z_\mu(\beta, t_f)/Z_\mu(\beta, 0) \\ &\approx \int \mathrm{d}\mathbf{\Gamma}_f^\dagger\, \wp_\mu(\mathbf{\Gamma}_f^\dagger|\beta, t_f)e^{-\beta\tilde{W}_\mu(\mathbf{\Gamma}_f^\dagger, t_f)}Z_\mu(\beta, t_f) \\ &\quad \times \int \mathrm{d}\mathbf{\Gamma}_0^\dagger\, \wp_\mu(\mathbf{\Gamma}_0^\dagger|\beta, t_0)e^{-\beta\tilde{W}_\mu(\mathbf{\Gamma}_0^\dagger, 0)}/Z_\mu(\beta, 0) \\ &= \frac{Z_\mathrm{B}(\beta, \mu(t_f))Z_\mathrm{B}(\beta, \mu(0))}{Z_\mu(\beta, 0)^2}\end{aligned} \tag{203}$$

Here it has been assumed that the trajectory is long enough that the ends are uncorrelated. It has also been assumed that the path beyond the interval is such that at both ends the nonequilibrium probability distribution has the form given in Eq. (187). This result shows that this particular average is not extensive in time (i.e., it does not scale with t_f). If the termini of the trajectories are drawn from a Boltzmann distribution, the result becomes

$$\langle e^{-\beta\Delta^0\mathcal{H}_\mu}\rangle_{B,[\mu],t_f} = \frac{Z_B(\beta,\mu(t_f))}{Z_B(\beta,\mu(0))} = e^{-\beta\Delta F(\beta)} \tag{204}$$

where the exponent is the difference between the Helmholtz free energies of the system at the final and initial values of the work parameter. This is the work theorem in the form given by Jarzynski [55], which has been rederived in different fashions [57, 64, 65] and verified experimentally [66]. The original version of the work theorem due to Bochkov and Kuzovlev [58–60] is valid for work in a closed cycle, $\mu(t_f) = \mu(0)$, in which case the right-hand side is unity.

When are the simplified results valid? If the work path has buffer regions at its beginning and end during which the work parameter is fixed for a time $\gtrsim \tau_{\text{short}}$, then the subsystem will have equilibrated at the initial and final values of μ in each case. Hence the odd work vanishes because $\dot{\mathcal{H}} = 0$, and the probability distribution reduces to Boltzmann's.

VI. NONEQUILIBRIUM QUANTUM STATISTICAL MECHANICS

A theory for nonequilibrium quantum statistical mechanics can be developed using a time-dependent, Hermitian, Hamiltonian operator $\hat{\mathcal{H}}(t)$. In the quantum case it is the wave functions ψ that are the microstates analogous to a point in phase space. The complex conjugate ψ^* plays the role of the conjugate point in phase space, since, according to Schrödinger, it has equal and opposite time derivative to ψ.

Define the odd work operator,

$$\tilde{W}(t) = [\hat{E}_+(t) - \hat{E}_-(t)]/2 \tag{205}$$

where the past and future energy operators are

$$\hat{E}_\pm(t) = \hat{\Theta}(\mp\tau;t)\tilde{\mathcal{H}}_t(t\pm\tau)\hat{\Theta}(\pm\tau;t) \tag{206}$$

and where the time-shift operator is

$$\hat{\Theta}(\tau;t) = \exp\left[\frac{-i}{\hbar}\int_t^{t+\tau} dt'\tilde{\mathcal{H}}_t(t')\right] \tag{207}$$

The odd Hamiltonian operator has been continued into the future,

$$\tilde{\mathcal{H}}_t(t') \equiv \begin{cases} \hat{\mathcal{H}}(t'), & t' \leq t \\ \hat{\mathcal{H}}(2t - t'), & t' > t \end{cases} \tag{208}$$

and the manipulation of the operators derived from it is facilitated by the symmetry about t, $\tilde{\mathcal{H}}_t(t') = \tilde{\mathcal{H}}_t(2t - t')$.

With these definitions, the nonequilibrium density operator for a subsystem of a thermal reservoir of inverse temperature β is [5]

$$\hat{\rho}(t) = \frac{1}{Z(t)} \exp -\beta[\hat{\mathcal{H}}(t) - \tilde{W}(t)] \tag{209}$$

where the normalization factor is $Z(t) = \text{TR}\{e^{-\beta[\hat{\mathcal{H}}(t) - \tilde{W}(t)]}\}$. Accordingly, the average of an observable at time t is

$$\langle \hat{O} \rangle_t = \text{TR}\{\hat{\rho}(t)\hat{O}(t)\} \tag{210}$$

and the present density operator can be said to provide a basis for nonequilibrium quantum statistical mechanics.

VII. HEAT FLOW

A. Thermodynamic Variables

The canonical nonequilibrium system consists of a subsystem sandwiched between two thermal reservoirs of different temperatures, with heat flowing steadily through the subsystem from the hot reservoir to the cold reservoir. Application of the general theory to this canonical problem illustrates the theory and serves to make the analysis more concrete. The first task is to identify explicitly the thermodynamic variables appropriate for this problem.

Consider a subsystem connected to thermal reservoirs of temperatures $T_\pm$ and located at $z = \pm L/2$. (To simplify the notation, only this scalar one-dimensional case is treated.) It is expected that the imposed temperature gradient, $(T_+ - T_-)/L$, will induce a corresponding temperature gradient in the subsystem, and also a gradient in the energy density.

To see this quantitatively the first entropy is required. Let the energies of the respective reservoirs be $E_{\text{r}\pm}$. Imagine a fixed region of the subsystem adjacent to each boundary and denote the energy of these regions by $E_{\text{s}\pm}$. Now impose the energy conservation laws

$$\Delta E_{\text{s}+} = -\Delta E_{\text{r}+}, \text{ and } \Delta E_{\text{s}-} = -\Delta E_{\text{r}-} \tag{211}$$

This constraint is not exact because the boundary regions can also exchange energy with the interior of the subsystem. However, it may be assumed that the rate of this exchange is much slower than that with the reservoir. This point is addressed further below.

Using the definition of temperature, $T^{-1} = \partial S/\partial E$, the conservation laws, and a Taylor expansion, the reservoirs' entropy may be written

$$S_{r\pm}(E_{r\pm}) = S_{r\pm}(E_{r0\pm}) - E_{s\pm}/T_{\pm} \tag{212}$$

The first term on the right-hand side is independent of the subsystem and may be neglected. With this, the entropy of the total system constrained to be in the macrostate $E_{s\pm}$ is the sum of that of the isolated subsystem in that macrostate and that of the reservoirs,

$$S_{\text{total}}(E_{s+}, E_{s-}|T_+, T_-) = S(E_{s+}, E_{s-}) - E_{s+}/T_+ - E_{s-}/T_- \tag{213}$$

Here $S(E_{s+}, E_{s-})$ is the entropy of the isolated subsystem. The derivative with respect to the constraints yields

$$\frac{\partial S_{\text{total}}}{\partial E_{s\pm}} = \frac{1}{T_{s\pm}} - \frac{1}{T_{\pm}} \tag{214}$$

The *most likely* state, which is denoted throughout by an overbar, is the one that maximizes the entropy. In this case the entropy derivative vanishes when the boundary temperatures of the subsystem equal that of the respective reservoirs,

$$\overline{T}_{s\pm} = T_{\pm} \tag{215}$$

This result is intuitively appealing and not unexpected.

The same problem can be treated from a slightly different perspective. Motivated by the fact that it is the inverse temperature that is thermodynamically conjugate to the energy, define the zeroth temperature [1, 3] as

$$\frac{1}{T_0} \equiv \frac{1}{2}\left[\frac{1}{T_+} + \frac{1}{T_-}\right] \tag{216}$$

and the first temperature as

$$\frac{1}{T_1} \equiv \frac{1}{L}\left[\frac{1}{T_+} - \frac{1}{T_-}\right] \tag{217}$$

The zeroth temperature is essentially the average temperature of the two reservoirs, and the first temperature is essentially the applied temperature gradient, ${T_1}^{-1} \equiv \nabla T^{-1} = -\nabla T/T_0^2$. The subsystem temperatures T_{s0} and T_{s1} can be defined identically in terms of T_{s+} and T_{s-}.

The subsystem energies used above can be rearranged in the form of energy moments,

$$E_0 = [E_{s+} + E_{s-}] \tag{218}$$

and

$$E_1 = (L/2)[E_{s+} - E_{s-}] \tag{219}$$

It may be shown that with these definitions the zeroth and first temperatures of the subsystem are thermodynamically conjugate to the zeroth and first energy moments,

$$\frac{1}{T_{s0}} = \frac{\partial S(E_0, E_1)}{\partial E_0}, \quad \text{and} \quad \frac{1}{T_{s1}} = \frac{\partial S(E_0, E_1)}{\partial E_1} \tag{220}$$

This is in fact the more fundamental definition of the temperatures, but it is entirely consistent with the preceding expressions.

The above definitions can be used to rearrange the constrained total entropy as

$$S_{\text{total}}(E_0, E_1|T_0, T_1) = S(E_0, E_1) - E_0/T_0 - E_1/T_1 \tag{221}$$

Maximizing the total entropy with respect to its arguments shows that the most likely state is one in which these subsystem temperatures equal their reservoir counterparts, $\overline{T}_{s0} = T_0$ and $\overline{T}_{s1} = T_1$.

This second formulation is more general than the first. The energy moments are always well defined, for example,

$$E_\alpha = \int_V d\mathbf{r}\; z^\alpha\; \epsilon(\mathbf{r}) \tag{222}$$

where $\epsilon(\mathbf{r})$ is the energy density. In this formulation there is no need to invoke an arbitrary boundary region. It will turn out that the rate of change of the first energy moment is related to the heat flux through the subsystem. The zeroth and first temperatures are *defined* to be conjugate to the respective energy moments via the derivatives of the entropy, and again one can avoid having to invoke a boundary region. A formal derivation of the conjugate relations may be found in Ref. 1. The idea that moments and gradients are conjugate is due to Onsager [10].

B. Gaussian Approximation

The first energy moment of the isolated system is not conserved and it fluctuates about zero. According to the general analysis of Section IIB, the entropy of the isolated system may be written as a quadratic form,

$$S(E_0, E_1) = S_0(E_0) + \tfrac{1}{2} E_1 S E_1 \tag{223}$$

with the correlation coefficient satisfying

$$S^{-1} = \frac{-1}{k_B} \langle E_1 E_1 \rangle \tag{224}$$

In this approximation the first temperature is given by

$$T_{s1}^{-1} = S E_1 \tag{225}$$

In view of this, an applied temperature gradient induces an energy moment in the subsystem that is given by

$$\overline{E}_1 = S^{-1} T_1^{-1} \tag{226}$$

C. Second Entropy

1. Isolated System

Now consider the transition $E_1 \rightarrow E_1'$ in time $\tau > 0$. Most interest lies in the linear regime, and so for an isolated system the results of Section IID apply. From Eq. (50), the second entropy in the intermediate regime is

$$\begin{aligned} S^{(2)}(E_1', E_1|\tau) &= \frac{|\tau|}{2} a_2 \left[\overset{\circ}{E}{}_1^0 + \frac{\hat{\tau}}{2} a_2^{-1} S E_1 \right]^2 + \frac{1}{2} S E_1^2 - \frac{|\tau|}{8} a_2^{-1} S^2 E_1^2 \\ &= \frac{1}{2|\tau|} a_2 [E_1' - E_1]^2 + \frac{1}{2} E_1' S E_1 \end{aligned} \tag{227}$$

where the coarse velocity is $\overset{\circ}{E}{}_1^0 \equiv (E_1' - E_1)/\tau$. The zero superscript is appended to make it clear that this is the internal or adiabatic rate of change of moment (i.e.,the system is isolated from any reservoirs). In the case of a reservoir it will be necessary to distinguish this internal change from the total change. Note that because E_1 has even parity and is the only variable to appear, $b_3 = 0$.

If the final term in the first equality were to be neglected, then the reduction condition would be satisfied exactly with respect to E_1', but not with respect the E_1, and the symmetry between the two variables would be broken. In the present form the reduction condition is satisfied by both variables to first order, and this expansion of the reduction condition is applied consistently.

The derivative with respect to E_1' is

$$\frac{\partial S^{(2)}(E_1', E_1|\tau)}{\partial E_1'} = \hat{\tau} a_2 [\overset{\circ}{E}{}_1^0 + \hat{\tau} a_2^{-1} S E_1/2] \tag{228}$$

Hence the most likely rate of change of moment due to internal processes is

$$\overline{\overset{\circ}{E}{}_1^0} = \frac{-\hat{\tau}}{2} a_2^{-1} S E_1 = \frac{-\hat{\tau}}{2} a_2^{-1} T_{s1}^{-1} \tag{229}$$

The most likely rate of change of the first entropy of the isolated system is

$$\overline{\dot{S}}^{(1)}(E_1) \equiv \overline{\dot{E}}_1 T_{s1}^{-1} = \frac{-\hat{\tau}}{2} a_2^{-1} S^2 E_1^2 \tag{230}$$

Hence the first entropy most likely increases linearly with time. The most likely value of the second entropy of the isolated system is

$$\begin{aligned} \overline{S}^{(2)}(E_1|\tau) &= \frac{|\tau|}{2} a_2 \overline{\overset{\circ}{E}{}_1^0}^2 + \frac{\tau}{2} \overline{\overset{\circ}{E}{}_1^0} S E_1 + \frac{1}{2} S E_1^2 \\ &= \frac{|\tau|}{8} a_2^{-1} S^2 E_1^2 - \frac{|\tau|}{4} a_2^{-1} S^2 E_1^2 + \frac{1}{2} S E_1^2 \end{aligned} \tag{231}$$

The final term is negative and independent of the time interval. It represents the entropy cost of establishing the structural order in the isolated system. The first term is negative and scales with the length of the time interval. It represents the ongoing cost of maintaining the dynamic order of the isolated system. The middle term is positive and scales with the length of the time interval. This term represents the ongoing first entropy gained as the system relaxes toward equilibrium. The dynamic order evidently costs less than the disorder it produces.

2. *Reservoirs*

Now place the subsystem in thermal contact with the two reservoirs discussed at the start of this section. In this case the energy moment can change by the internal processes just discussed, or by exchange with the reservoirs,

$$E_1' = E_1 + \Delta^0 E_1 + \Delta^{\mathrm{r}} E_1 \tag{232}$$

By energy conservation, $\Delta^{\mathrm{r}} E_1 = -\Delta E_{\mathrm{r}1}$. The second entropy now depends on $\Delta^0 E_1$, $\Delta^{\mathrm{r}} E_1$, and E_1. The second entropy due to the reservoirs is

$$S_{\mathrm{r}}^{(2)}(\Delta^{\mathrm{r}} E_1, E_1|T_1) = -\frac{1}{2}\left[\frac{E_1}{T_1} + \frac{E_1 + \Delta^{\mathrm{r}} E_1}{T_1}\right] \tag{233}$$

This is half the sum of the first entropy of the initial and final states, as is usual for a stochastic transition. Adding this to the isolated system second entropy given earlier, one has

$$\begin{aligned} & S^{(2)}_{\text{total}}(\Delta^0 E_1, \Delta^{\text{r}} E_1, E_1 | T_1, \tau) \\ & = \frac{1}{2|\tau|} a_2 [\Delta^0 E_1]^2 + \frac{1}{2} [E_1 + \Delta^0 E_1 + \Delta^{\text{r}} E_1]\ SE_1 - \frac{1}{2} \frac{2E_1 + \Delta^{\text{r}} E_1}{T_1} \end{aligned} \tag{234}$$

Setting the derivative with respect to the external change $\Delta^{\text{r}} E_1$ to zero shows that

$$S\overline{E}_1 = 1/T_1 \tag{235}$$

which is to say that in the steady state the induced first temperature is equal to the applied first temperature, $\overline{T}_{\text{s1}} = T_1$. The derivative with respect to E_1 yields

$$\frac{\partial S^{(2)}_{\text{total}}}{\partial E_1} = \frac{1}{2} SE_1 + \frac{1}{2} SE_1' + \frac{1}{T_1} \tag{236}$$

which vanishes when $E_1 = \overline{E}_1$ and $E_1' = \overline{E}_1$. This says that in the steady state the change in the subsystem moment due to the internal energy flows within the subsystem is exactly canceled by that due to energy flow from the reservoir, and so the subsystem structure remains unchanged, $\overline{\Delta^0 E_1} = -\overline{\Delta^{\text{r}} E_1}$. Optimizing with respect to $\Delta^0 E_1$ yields the isolated system result,

$$\overline{\overset{\circ}{E}{}^0_1} = \frac{-\hat{\tau}}{2} a_2^{-1} S\overline{E}_1 = \frac{-\hat{\tau}}{2} a_2^{-1} T_1^{-1} \tag{237}$$

This result shows that the most likely rate of change of the moment due to internal processes is linearly proportional to the imposed temperature gradient. This is a particular form of the linear transport law, Eq. (54), with the imposed temperature gradient providing the thermodynamic driving force for the flux. Note that for driven transport τ is taken to be positive because it is assumed that the system has been in a steady state for some time already (i.e., the system is *not* time reversible).

This result confirms Onsager's regression hypothesis. The most likely velocity in an isolated system following a fluctuation from equilibrium, Eq. (229), is equal to the most likely velocity due to an externally imposed force, Eq. (237), when the internal force is equal to the external force, $\overline{T}_{\text{s1}} = T_1$.

This result for the most likely change in moment is equivalent to Fourier's law of heat conduction. To see this take note of the fact that in the steady state the total rate of change of moment is zero, $\overline{\dot{E}_1} = 0$, so that the internal change is canceled by that due to the reservoirs, $\overline{\dot{E}_1^0} = \overline{\dot{E}_{r1}}$. (In the steady state $\overline{\dot{E}_1^0} = \overline{\overset{\circ}{E}{}_1^0}$.) The rate of change of the reservoir moment is just the energy flux through the system,

$$J = \frac{\pm \dot{E}_{r\pm}}{A} = \frac{\dot{E}_{r1}}{V} = \frac{\overline{\dot{E}_1^0}}{V} \tag{238}$$

where A is the cross-sectional area of the subsystem and $V = AL$ is the subsystem volume. Now Fourier's law says

$$J = -\lambda \nabla T = \lambda T_0^2 / T_1 \tag{239}$$

where λ is the thermal conductivity. Accordingly,

$$\overline{\dot{E}_1^0} = \lambda V T_0^2 / T_1 \tag{240}$$

This means that thermal conductivity and second entropy transport coefficient are related by $\lambda = -1/2VT_0^2 a_2$.

3. *Rate of Entropy Production*

The optimum values of the first and second entropies, and the steady rates of production of these by the reservoirs may readily be obtained. The maximum value of the total first entropy is

$$\overline{S_{\text{total}}^{(1)}}(T_1) = \frac{1}{2} S \overline{E_1}^2 - \frac{\overline{E_1}}{T_1} = \frac{-1}{2ST_1^2} \tag{241}$$

The rate of change of the total first entropy, which, since the subsystem's structure does not change with time, is the same as that of the reservoirs, is given by

$$\overline{\dot{S}_r^{(1)}}(T_1) = \frac{\overline{\Delta E_{r1}}}{T_1 \tau} = \frac{\overline{\dot{E}_1^0}}{T_1} = \frac{-a_2^{-1}}{2T_1^2} \tag{242}$$

This is evidently positive (since $a_2 < 0$, the second entropy being concave down).

It should be clear that the most likely or physical rate of first entropy production is neither minimal nor maximal; these would correspond to values of the heat flux of $\pm\infty$. The conventional first entropy does not provide any variational principle for heat flow, or for nonequilibrium dynamics more generally. This is consistent with the introductory remarks about the second law of equilibrium thermodynamics, Eq. (1), namely, that this law and the first entropy that in invokes are independent of time. In the literature one finds claims for both extreme theorems: some claim that the rate of entropy production is minimal and others claim that it is maximal, whereas the present results indicate that it is neither. One has to distinguish the first or structural entropy, which conventionally is what is meant by the word "entropy" in the literature, from the second or dynamic entropy, which was introduced in Ref. 2 and which provides the basis for the present nonequilibrium theory. It is the second entropy that obeys a variational principle that allows the physical nonequilibrium state to be obtained by maximizing it with respect to constraints.

The most likely value of the total second entropy, which is of course its maximum value, is

$$\begin{aligned}\overline{S_{\text{total}}^{(2)}}(T_1,\tau) &= S_{\text{total}}^{(2)}(\overline{\Delta^0 E_1},\overline{\Delta^{\text{r}} E_1},\overline{E_1}|T_1,\tau)\\ &= \frac{1}{2|\tau|}a_2\overline{\Delta^0 E_1}^2+\frac{1}{2}\overline{E_1}S\overline{E_1}-\frac{1}{2}\frac{2\overline{E_1}+\overline{\Delta^{\text{r}} E_1}}{T_1}\\ &= \frac{|\tau|}{8}a_2^{-1}\frac{1}{T_1^2}+\frac{1}{2}S^{-1}\frac{1}{T_1^2}-S^{-1}\frac{1}{T_1^2}-\frac{|\tau|}{4}a_2^{-1}\frac{1}{T_1^2}\end{aligned} \tag{243}$$

The first term, which is negative and scales linearly with time, is the ongoing cost of maintaining dynamic order in the subsystem. The second term is the cost of ordering the static structure of the subsystem, and this is more than compensated by the third term, which is the entropy gain that comes from exchanging energy with the reservoirs to establish that static order. The fourth term, which is positive and scales linearly with time, is the ongoing first entropy produced by the reservoirs as they exchange energy by a constant flux through the subsystem.

Physically the variational procedure based on the second entropy may be interpreted like this. If the flux $\dot{E}_1^0$ were increased beyond its optimum value, then the rate of entropy consumption by the subsystem would be increased due to its increased dynamic order by a greater amount than the entropy production of the reservoirs would be increased due to the faster transfer of heat. The converse holds for a less than optimum flux. In both cases the total rate of second entropy production would fall from its maximum value.

D. Phase Space Probability Distribution

The steady-state probability distribution for a system with an imposed temperature gradient, $\wp_{\text{ss}}(\mathbf{\Gamma}|\beta_0,\beta_1)$, is now given. This is the microstate probability density for the phase space of the subsystem. Here the reservoirs enter by the zeroth, $\beta_0 \equiv 1/k_{\text{B}}T_0$, and the first, $\beta_1 \equiv 1/k_{\text{B}}T_1$, temperatures. The zeroth energy moment is the ordinary Hamiltonian,

$$E_0(\mathbf{\Gamma}) = \mathcal{H}(\mathbf{\Gamma}) = \sum_{i=1}^{N}\epsilon_i \tag{244}$$

and the first energy moment in the z-direction is just

$$E_1(\mathbf{\Gamma}) = \sum_{i=1}^{N} \epsilon_i z_i \tag{245}$$

where ϵ_i is the total energy of particle i. The adiabatic rate of change of the energy moment, which is the natural or Hamiltonian motion where no heat flows to the isolated system, is denoted as $\dot{E}_1^0(\mathbf{\Gamma}) = \dot{\mathbf{\Gamma}}^0 \cdot \nabla E_1(\mathbf{\Gamma})$. Both moments have even phase space parity, $E_0(\mathbf{\Gamma}) = E_0(\mathbf{\Gamma}^\dagger)$ and $E_1(\mathbf{\Gamma}) = E_1(\mathbf{\Gamma}^\dagger)$, since it is assumed that there are no velocity-dependent forces in the Hamiltonian. The rate of change of the energy moment necessarily has odd parity, $\dot{E}_1^0(\mathbf{\Gamma}^\dagger) = -\dot{E}_1^0(\mathbf{\Gamma})$.

From Eq. (152), the static or time-reversible phase space probability density is

$$\wp_{\text{st}}(\mathbf{\Gamma}|\beta_0, \beta_1) = \frac{1}{Z_{\text{ss}}(\beta_0, \beta_1)} e^{-\beta_0 E_0(\mathbf{\Gamma})} e^{-\beta_1 E_1(\mathbf{\Gamma})} \tag{246}$$

and from Eq. (160), the steady-state probability is

$$\wp_{\text{ss}}(\mathbf{\Gamma}|\beta_0, \beta_1) = \frac{1}{Z_{\text{ss}}(\beta_0, \beta_1)} e^{-\beta_0 E_0(\mathbf{\Gamma})} e^{-\beta_1 E_1(\mathbf{\Gamma})} e^{\beta_1 E_{1\Delta}(\mathbf{\Gamma})} \tag{247}$$

where the odd work is

$$\beta_1 E_{1\Delta}(\mathbf{\Gamma}) = \frac{\beta_1}{2} \int_{\tau}^{\tau} \mathrm{d}t\, \dot{E}_1^0(\mathbf{\Gamma}_0(t|\mathbf{\Gamma})), \quad \tau_{\text{short}} < \tau < \tau_{\text{long}} \tag{248}$$

The odd work should be insensitive to the value of τ, provided that it is in the intermediate regime, $\tau_{\text{short}} \lesssim \tau \lesssim \tau_{\text{long}}$. An estimate of the inertial time is [2]

$$\begin{aligned} \tau_{\text{short}} &= \frac{-2\left\langle \dot{E}_1^0(t) E_1(0) \right\rangle_0}{\langle (\dot{E}_1^0)^2 \rangle_0}, \quad t \gg \tau_{\text{short}} \\ &= \frac{2\lambda V k_{\text{B}} T_0^2}{\left\langle (\dot{E}_1^0)^2 \right\rangle_0} \end{aligned} \tag{249}$$

The long time limit may be estimated from [2]

$$\tau_{\text{long}} = \frac{\langle E_1^2 \rangle_0}{2\lambda V k_{\text{B}} T_0^2} \tag{250}$$

From Eq. (174) the unconditional transition probability is

$$\wp(\mathbf{\Gamma}'' \leftarrow \mathbf{\Gamma}|\Delta_t, \beta_0, \beta_1) = \frac{\Theta_\Delta(|\mathbf{\Gamma}'' - \mathbf{\Gamma}'|)}{Z_{ss}(\beta_0, \beta_1)} e^{-\beta_0[E_0''+E_0']/2} e^{-\beta_1[E_1''+E_1']/2} e^{\beta_1[E_{1\Delta}''+E_{1\Delta}']/2} e^{\Delta_t\beta_1[\dot{E}_1^0-\dot{E}_{1\Delta}^0]} \tag{251}$$

where the adiabatic development is denoted by a prime, $\mathbf{\Gamma}' = \mathbf{\Gamma}_0(\Delta_t|\mathbf{\Gamma})$ and so on. The zeroth energy moment is of course a constant of the adiabatic motion.

From Eq. (175), the ratio of the forward and reverse transitions is

$$\frac{\wp(\mathbf{\Gamma}'' \leftarrow \mathbf{\Gamma}|\Delta_t, \beta_0, \beta_1)}{\wp(\mathbf{\Gamma} \leftarrow \mathbf{\Gamma}''|\Delta_t, \beta_0, \beta_1)} = e^{\beta_1[E_{1\Delta}''+E_{1\Delta}]} e^{\Delta_t\beta_1[\dot{E}_1+\dot{E}_1'']} \tag{252}$$

Similarly, the ratio of trajectory probabilities is, to first and second order,

$$\frac{\wp([\mathbf{\Gamma}]|\beta_0, \beta_1)}{\wp([\mathbf{\Gamma}^\ddagger]|\beta_0, \beta_1)} = e^{\beta_1\Delta^0 E_1[\mathbf{\Gamma}]} e^{\beta_1[E_{1\Delta}(\mathbf{\Gamma}_0)+E_{1\Delta}(\mathbf{\Gamma}_f)]} \tag{253}$$

where the adiabatic work done on the trajectory is

$$\Delta^0 E_1[\mathbf{\Gamma}] = \beta_1 \int_0^f \mathrm{d}t\ \dot{E}_1^0(\mathbf{\Gamma}(t|\mathbf{\Gamma}_0)) \tag{254}$$

Note that the argument is the actual trajectory, not the adiabatic trajectory, and in the integrand appears the adiabatic time derivative, not the actual time derivative. The change in the entropy of the reservoirs over a trajectory is $\Delta S_r[\mathbf{\Gamma}]/k_B = \beta_1[\Delta^0 E_1[\mathbf{\Gamma}] - E_1(\mathbf{\Gamma}_f) + E_1(\mathbf{\Gamma}_0)]$, which in conjunction with the above ratio may be used to derive the heat flow version of the fluctuation theorem, Eq. (181) and of the work theorem, Eq. (182).

VIII. MONTE CARLO SIMULATIONS OF HEAT FLOW

The availability of a phase space probability distribution for the steady state means that it is possible to develop a Monte Carlo algorithm for the computer simulation of nonequilibrium systems. The Monte Carlo algorithm that has been developed and applied to heat flow [5] is outlined in this section, following a brief description of the system geometry and atomic potential.

A. System Details

A Lennard-Jones fluid was simulated. All quantities were made dimensionless using the well depth ϵ_{LJ}, the diameter σ_{LJ}, and the time constant

$\tau_{LJ} = \sqrt{m_{LJ}\sigma_{LJ}^2/\epsilon_{LJ}}$, where m_{LJ} is the mass. In addition, Boltzmann's constant was set equal to unity. The pair potential was cut and shifted at $R_{cut} = 2.5$. No tail correction was used. The shift to make the potential zero at the cutoff is necessary for consistency between the Monte Carlo and the molecular dynamics aspects of the computations.

A spatial neighbor table was used with cubic cells of side length ≈ 0.6 [1]. At the beginning of the simulation a list of neighbor cells within the cutoff of each cell was calculated and stored. The neighborhood volume composed of such small neighbor cells can be made to closely approximate the cutoff sphere. In contrast, most conventional neighbor lists are based on cubes of side length equal to the potential cutoff [84]. The advantage of the present small neighbor cells is that they reduce the enveloping neighborhood volume from 27 large cubes (each of size R_{cut}, neighborhood volume $27R_{cut}^3$) to approximately 667 small cubes, giving a neighborhood volume on the order of $(4\pi/3)(R_{cut} + 0.6)^3$. The number of neighbors that enter a force or potential calculation with these small cells is almost a factor of three smaller than for the conventional cells of length R_{cut}.

Both a uniform bulk fluid and an inhomogeneous fluid were simulated. The latter was in the form of a slit pore, terminated in the z-direction by uniform Lennard-Jones walls. The distance between the walls for a given number of atoms was chosen so that the uniform density in the center of the cell was equal to the nominal bulk density. The effective width of the slit pore used to calculate the volume of the subsystem was taken as the region where the density was nonzero. For the bulk fluid in all directions, and for the slit pore in the lateral directions, periodic boundary conditions and the minimum image convention were used.

The energy per atom consists of kinetic energy, singlet, and pair potential terms,

$$\epsilon_i = \frac{1}{2m}\mathbf{p}_i \cdot \mathbf{p}_i + w(q_{iz}) + \frac{1}{2}\sum_{j=1}^{N}{}^{j\neq i}\, u(q_{ij}) \tag{255}$$

where $w(z)$ is the wall potential (if present). In terms of this the zeroth moment is just the total energy,

$$E_0(\mathbf{\Gamma}) = \sum_{i=1}^{N} \epsilon_i \tag{256}$$

and the first moment is

$$E_1(\mathbf{\Gamma}) = \sum_{i=1}^{N} q_{iz}\epsilon_i \tag{257}$$

The adiabatic rate of change of the first energy moment is

$$\dot{E}_1^0(\mathbf{\Gamma}) = \sum_{i=1}^{N} \dot{q}_{iz}\epsilon_i + q_{iz}\dot{\epsilon}_i \tag{258}$$

where the velocity is $\dot{q}_{iz} = p_{iz}/m_{\text{LJ}}$. Using Hamilton's equations, it is readily shown that

$$\dot{\epsilon}_i = \frac{-1}{2}\sum_{j=1}^{N}{}^{j\neq i} u'(q_{ij})\frac{\mathbf{q}_{ij}\cdot[\mathbf{p}_i + \mathbf{p}_j]}{mq_{ij}} \tag{259}$$

This holds whether or not the singlet potential is present. In the case of periodic boundary conditions, it is quite important to use the minimum image convention for all the separations that appear in this expression. This may be rewritten in the convenient form [4]

$$\dot{E}_1^0 = \sum_{i\alpha} \kappa_{i\alpha}p_{i\alpha}/m \tag{260}$$

where

$$\mathbf{\kappa}_i = \frac{\epsilon_i}{m}\hat{\mathbf{z}} - \sum_{j=1}^{N}{}^{j\neq i} u'(q_{ij})\frac{[q_{iz} - q_{jz}]}{2mq_{ij}}\mathbf{q}_{ij} \tag{261}$$

The Monte Carlo algorithms require ΔE_0, ΔE_1, and $\Delta\dot{E}_1^0$. In attempting to move atom n in phase space, the n-dependent contribution to these formulas was identified and only the change in this was calculated for each attempted move.

It was necessary periodically to generate an adiabatic trajectory in order to obtain the odd work and the time correlation functions. In calculating $E_1(t)$ on a trajectory, it is essential to integrate $\dot{E}_1^0(t)$ over the trajectory rather than use the expression for $E_1(\mathbf{\Gamma}(t))$ given earlier. This is because $\dot{E}_1^0$ is insensitive to the periodic boundary conditions, whereas E_1 depends on whether the coordinates of the atom are confined to the central cell, or whether the itinerant coordinate is used, and problems arise in both cases when the atom leaves the central cell on a trajectory.

Because the starting position of each trajectory was taken from a Boltzmann-weighted distribution in $6N$-dimensional phase space, the center of mass velocity of the system (the total linear momentum) was generally nonzero. Prior to commencing each molecular dynamics trajectory, the z-component of the center of mass velocity was zeroed at constant kinetic energy by shifting and rescaling the z-component of the momenta. (Only the z-component of the first energy moment was used.) It was found that a nonzero center of mass velocity made a nonnegligible contribution to the conductivity. Conventional molecular dynamics simulations are performed with zero center of mass velocity, which is of course the most appropriate model of reality. For the bulk case the total z-momentum was conserved at zero along the molecular dynamics trajectory. For the inhomogeneous simulations, the momentum during the adiabatic evolution was not conserved due to collisions with the walls. In this case an additional external force was applied to each atom that was equal and opposite to the net wall force per atom, which had the effect of conserving the z-component of the total linear momentum at zero along the molecular dynamics trajectory.

B. Metropolis Algorithm

Monte Carlo simulations were performed in $6N$-dimensional phase space, where $N = 120$–500 atoms [5]. The Metropolis algorithm was used with umbrella sampling. The weight density was

$$\omega(\mathbf{\Gamma}) = e^{-\beta_0 E_0(\mathbf{\Gamma})} e^{-\beta_1 E_1(\mathbf{\Gamma})} e^{\alpha\beta_1 \dot{E}_1^0(\mathbf{\Gamma})} \tag{262}$$

The umbrella weight used to generate the configurations was corrected by using the exact steady-state probability density, Eq. (247), to calculate the averages (see later). The final exponent obviously approximates the odd work, $\beta_1 E_{1\Delta}$, but is about a factor of 400 faster to evaluate. In the simulations α was fixed at 0.08, although it would be possible to optimize this choice or to determine α on the fly [3].

A trial move of an atom consisted of a small displacement in its position and momentum simultaneously. Step lengths of 0.9 in velocity and 0.09 in position gave an acceptance rate of about 50%. A cycle consisted of one trial move of all the atoms.

Averages were collected after every 50 cycles. For this the required odd work was obtained from the adiabatic Hamiltonian trajectory generated forward and backward in time, starting at the current configuration. A second order integrator was used,

$$\begin{aligned} q_{n\alpha}(t+\Delta_t) &= q_{n\alpha}(t) + \Delta_t \dot{q}_{n\alpha}(t) + \frac{\Delta_t^2}{2m} F_{n\alpha}(t) \\ p_{n\alpha}(t+\Delta_t) &= p_{n\alpha}(t) + \frac{\Delta_t}{2}\left[F_{n\alpha}(t) + F_{n\alpha}(t+\Delta_t)\right] \end{aligned} \tag{263}$$

where n labels the atom, $\alpha = x$, y, or z labels the component, and $F_{n\alpha}(t) \equiv F_{n\alpha}(\mathbf{q}^N(t))$ is the force, which does not depend on the momenta. Obviously one evaluates $F_{n\alpha}(t+\Delta_t)$ after evaluating the new positions and before evaluating the new momenta. Typically, the time step was $\Delta_t = 10^{-3}$. The zeroth energy moment in general increased by less than 1% over the trajectory.

Labeling the current configuration by i, the trajectory is $\mathbf{\Gamma}_0(t|\mathbf{\Gamma}_i)$. This was calculated both forward and backward in time, $-t_f \leq t \leq t_f$. The running integral for $E_{1\Delta}(\mathbf{\Gamma}_i;\tau)$, $0 \leq \tau \leq t_f$, was calculated along the trajectory using both the trapezoidal rule and Simpson's rule, with indistinguishable results. The average flux was calculated as a function of the time interval,

$$\langle \dot{E}_1^0 \rangle_\tau = \frac{\sum_i \dot{E}_1^0(\mathbf{\Gamma}_i) e^{-\alpha\beta_1 \dot{E}_1^0(\mathbf{\Gamma}_i)} e^{\beta_1 E_{1\Delta}(\mathbf{\Gamma}_i;\tau)}}{\sum_i e^{-\alpha\beta_1 \dot{E}_1^0(\mathbf{\Gamma}_i)} e^{\beta_1 E_{1\Delta}(\mathbf{\Gamma}_i;\tau)}} \tag{264}$$

Note how the umbrella weight used in the Metropolis scheme is canceled here. The thermal conductivity was obtained as a function of the time interval, $\lambda(\tau) = \langle \dot{E}_1^0 \rangle_\tau / \beta_1 V k_B T_0^2$. Compared to implementing the steady-state probability directly in the Metropolis algorithm, not only is the umbrella method orders of magnitude faster in generating configurations, but it also allows results as a function of τ to be collected, and it reduces the correlation between consecutive, costly trajectories, by inserting many cheap, umbrella steps. On the order of 50,000 trajectories were generated for each case studied.

C. Nonequilibrium Molecular Dynamics

Perhaps the most common computer simulation method for nonequilibrium systems is the nonequilibrium molecular dynamics (NEMD) method [53, 88]. This typically consists of Hamilton's equations of motion augmented with an artificial force designed to mimic particular nonequilibrium fluxes, and a constraint force or thermostat designed to keep the kinetic energy or temperature constant. Here is given a brief derivation and critique of the main elements of that method.

Following Ref. 84, let the Hamiltonian for the nonequilibrium state be represented by

$$\mathcal{H}_{\mathrm{ne}}(\mathbf{\Gamma}) = \mathcal{H}_0(\mathbf{\Gamma}) + \mathcal{F}(t)\mathcal{A}(\mathbf{\Gamma}) \tag{265}$$

Here the perturbation to the usual Hamiltonian, $\mathcal{F}(t)\mathcal{A}(\mathbf{\Gamma})$, is switched on at a certain time,

$$\mathcal{F}(t) = \begin{cases} \mathcal{F}, & t > 0 \\ 0, & t < 0 \end{cases} \tag{266}$$

In this case the nonequilibrium equations of motion are

$$\dot{q}_{i\alpha} = p_{i\alpha}/m + \mathcal{F}(t)\mathcal{A}_{\mathrm{p}i\alpha} \tag{267}$$

and

$$\dot{p}_{i\alpha} = f_{i\alpha} - \mathcal{F}(t)\mathcal{A}_{qi\alpha} \tag{268}$$

with the subscripts on the nonequilibrium potential denoting a derivative. (A thermostat is generally also added.) The function $\mathcal{A}$ has even parity so that the equations of motion are time reversible. It is straightforward to show that at a given point in phase space, the rate of change of the nonequilibrium potential is the same in the nonequilibrium and in the natural system,

$$\dot{\mathcal{A}}(\mathbf{\Gamma}) = \dot{\mathcal{A}}^0(\mathbf{\Gamma}) \tag{269}$$

The average of an odd function of phase space at a time $t > 0$ after switching on the nonequilibrium perturbation, assuming that the system is initially Boltzmann distributed, is

$$\begin{aligned}
\langle \mathcal{B}(t) \rangle_{\text{ne}} &= \frac{\int d\mathbf{\Gamma}_1 e^{-\beta\mathcal{H}_0(\mathbf{\Gamma}_1)} \mathcal{B}(\mathbf{\Gamma}_{\mathcal{F}}(t|\mathbf{\Gamma}_1))}{\int d\mathbf{\Gamma}_1 e^{-\beta\mathcal{H}_0(\mathbf{\Gamma}_1)}}. \\
&= \frac{\int d\mathbf{\Gamma}_1 e^{-\beta\mathcal{H}_0(\mathbf{\Gamma}_{\mathcal{F}})} e^{-\beta\mathcal{F}[\mathcal{A}(\mathbf{\Gamma}_{\mathcal{F}})-\mathcal{A}(\mathbf{\Gamma}_1)]} \mathcal{B}(\mathbf{\Gamma}_{\mathcal{F}})}{\int d\mathbf{\Gamma}_1 \, e^{-\beta\mathcal{H}_{\mathcal{F}}(\mathbf{\Gamma}_{\mathcal{F}})} e^{\beta\mathcal{F}[\mathcal{A}(\mathbf{\Gamma}_{\mathcal{F}})-\mathcal{A}(\mathbf{\Gamma}_1)]}} \\
&= \frac{-\beta\mathcal{F} \int d\mathbf{\Gamma}_{\mathcal{F}} e^{-\beta\mathcal{H}_0(\mathbf{\Gamma}_{\mathcal{F}})} [\mathcal{A}(\mathbf{\Gamma}_{\mathcal{F}}) - \mathcal{A}(\mathbf{\Gamma}_1)] \mathcal{B}(\mathbf{\Gamma}_{\mathcal{F}})}{\int d\mathbf{\Gamma}_{\mathcal{F}} e^{-\beta\mathcal{H}_{\mathcal{F}}(\mathbf{\Gamma}_{\mathcal{F}})}} \\
&= -\beta\mathcal{F} \langle [\mathcal{A}(\mathbf{\Gamma}_{\mathcal{F}}) - \mathcal{A}(\mathbf{\Gamma}_1)] \mathcal{B}(\mathbf{\Gamma}_{\mathcal{F}}) \rangle_0 \\
&\approx -\beta\mathcal{F} \int_0^t dt' \langle \dot{\mathcal{A}}(t') \mathcal{B}(0) \rangle_0
\end{aligned} \tag{270}$$

where $\mathbf{\Gamma}_{\mathcal{F}} \equiv \mathbf{\Gamma}_{\mathcal{F}}(t|\mathbf{\Gamma}_1)$. This result uses the fact that the nonequilibrium Hamiltonian is a constant on the nonequilibrium trajectory, and that the Jacobian of the transformation along a nonequilibrium trajectory is unity. It also linearizes everything with respect to $\mathcal{F}$ and neglects terms with total odd parity.

There is an approximation implicit in the final line. The subscript zero implies an average for an isolated system (i.e., on an adiabatic or bare Hamiltonian trajectory), whereas the actual trajectory used to obtain this result is the modified one, $\mathbf{\Gamma}_{\mathcal{F}}(t|\mathbf{\Gamma}_1) \neq \mathbf{\Gamma}_0(t|\mathbf{\Gamma}_1)$. In so far as these are the same to leading order, this difference may be neglected.

The choices of $\mathcal{A}$, $\mathcal{B}$, and $\mathcal{F}$ are dictated by the Green–Kubo relation for the particular flow of interest. For heat flow one chooses $\mathcal{B}(\mathbf{\Gamma}) = \dot{E}_1^0(\mathbf{\Gamma})$, $\dot{\mathcal{A}}(\mathbf{\Gamma}) = -\dot{E}_1^0(\mathbf{\Gamma})$, and $\mathcal{F} = T_0/T_1 = \beta_1/\beta_0$.

Depending on the point of view, it is either a strength or a weakness of the NEMD method that it gives a uniform structure for the nonequilibrium system (e.g., for heat flow the subsystem does not acquire the applied temperature gradient, nor does it have gradients in energy or density). On the one hand, such imposed uniformity makes the simulations compatible with periodic boundary conditions, and it does not affect the dynamics in the linear regime. On the other hand, the incorrect structure precludes reliable results for the dynamics in the nonlinear regime when the two are coupled. It is possible to develop NEMD equations that do correctly account for the structure by analyzing the linear response of functions of opposite parity to that used above, as was done at the end of Section IVB.

In the practical implementation of the NEMD method, it is usual to set the momentum derivative of the nonequilibrium potential to zero, $\mathcal{A}_{\mathrm{p}i\alpha} = 0$ [53, 89]. Presumably the reason for imposing this condition is that it preserves the classical relationship between velocity and momentum, $\dot{q}_{i\alpha} = p_{i\alpha}/m$. In view of this condition, the rate of change of the nonequilibrium potential reduces to

$$\dot{\mathcal{A}}(\mathbf{\Gamma}) = \sum_{i\alpha} \mathcal{A}_{\mathrm{q}i\alpha} p_{i\alpha}/m \quad \text{(NEMD)} \tag{271}$$

The Green–Kubo result demands that this be equated to the negative of the natural rate of change of the first energy moment, Eq. (260), which means that

$$\mathcal{A}_{\mathrm{q}i\alpha} = -\kappa_{i\alpha} \quad \text{(NEMD)} \tag{272}$$

However, this leads to the contradiction that $\dot{\mathcal{A}}^0(t) = -\dot{E}_1^0(t)$, but $\mathcal{A}^0(t) \neq -E_1^0(t)$.

The problem arises because one does not have the freedom to make the momentum derivative zero. One can see this from the usual condition on second derivatives,

$$\frac{\partial^2 \mathcal{A}}{\partial q_{i\alpha}\, \partial p_{j\gamma}} = \frac{\partial^2 \mathcal{A}}{\partial p_{j\gamma}\, \partial q_{i\alpha}} \tag{273}$$

From the fact that

$$\frac{\partial \kappa_{i\alpha}}{\partial p_{j\gamma}} = \frac{\delta_{ij} \delta_{\alpha z} p_{j\gamma}}{m} \tag{274}$$

one concludes that the momentum derivative of the nonequilibrium potential must be nonzero. It is in fact equal to

$$\mathcal{A}_{\mathrm{p}i\alpha} = -\frac{\partial E_1}{\partial p_{i\alpha}} = -\frac{q_{iz} p_{i\alpha}}{m} \tag{275}$$

This means that

$$\mathcal{A}_{qi\alpha} = -\kappa_{i\alpha} + q_{iz}f_{i\alpha} \tag{276}$$

where $\mathbf{f}_i$ is the force on atom i. With these, $\dot{\mathcal{A}}^0(\mathbf{\Gamma}) = -\dot{E}_1^0(\mathbf{\Gamma})$ and $\mathcal{A}^0(\mathbf{\Gamma}) = -E_1^0(\mathbf{\Gamma})$. Using these forces, the nonequilibrium trajectory is properly derived from the nonequilibrium Hamiltonian, and the adiabatic incompressibility of phase space is assured, $\nabla \cdot \dot{\mathbf{\Gamma}}_{\mathcal{F}} = 0$ (provided no thermostat is applied).

D. Monte Carlo Results

1. Structure

Figure 3 shows the profiles induced in a bulk system by an applied temperature gradient. These Monte Carlo results [1] were obtained using the static probability distribution, Eq. (246). Clearly, the induced temperature is equal to the applied temperature. Also, the slopes of the induced density and energy profiles can be obtained from the susceptibility, as one might expect since in the linear regime there is a direct correspondence between the slopes and the moments [1].

The energy susceptibility is given in Fig. 4. This was again obtained using the static probability distribution. In this case the susceptibility was obtained directly from the ratio of the induced energy moment to the applied temperature gradient, Eq. (226), and from the fluctuations, Eq. (224), with indistinguishable results. (In the latter formula E_1 was replaced by its departure from equilibrium, $\delta E_1 \equiv E_1 - \langle E_1 \rangle_{\mathrm{st}}$.) The line passing through the points was obtained from bulk properties [1], which shows that the nonequilibrium structure is related directly to that of an equilibrium system.

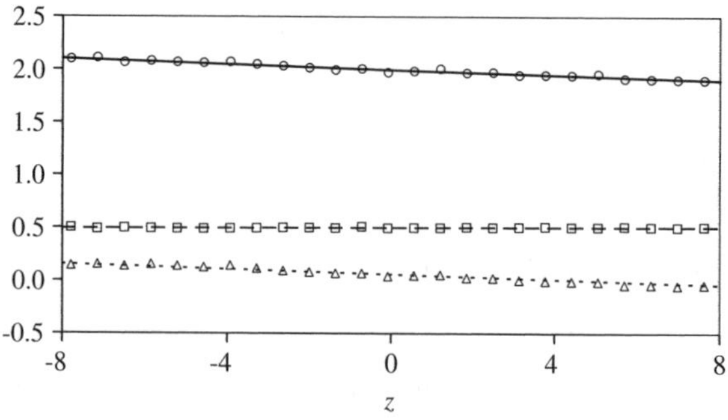

Figure 3. Induced temperature (top), number density (middle), and energy density (bottom) profiles for $\beta_1 = 0.0031$ and $T_0 = 2$, $\rho = 0.5$. The symbols are Monte Carlo results using the static probability distribution, Eq. (246), and the lines are either the applied temperature or the profiles predicted from the simulated susceptibility. (From Ref. 1.)

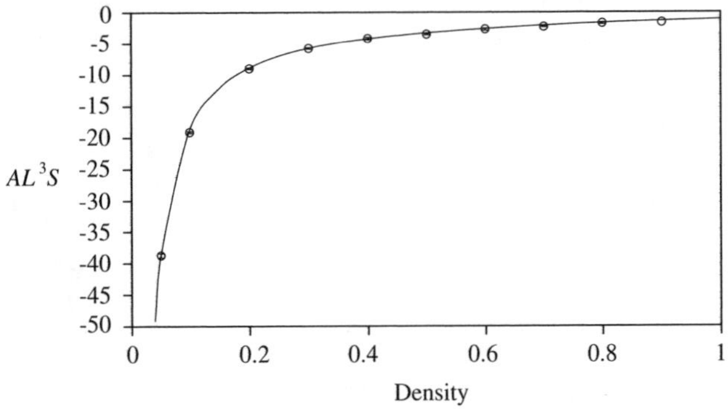

Figure 4. Susceptibility of the energy moment at $T_0 = 2$. The symbols are static Monte Carlo results [1] and the curve is obtained from a local thermodynamic approximation [1] using the bulk susceptibilities from a Lennard-Jones equation of state [90]. (From Ref. 1.)

2. *Dynamics*

Figure 5 shows the decay of the first energy moment following a fluctuation [2]. The fluctuation was induced by sampling the static probability distribution, Eq. (246), which has no preferred direction in time, and the configurations were used as starting points for independent trajectories. The trajectories were generated adiabatically forward and backward in time. The point of the figure is that on short time scales, $t \lesssim \tau_{\text{short}}$, the moment displays a quadratic dependence on time, $E_1(t) \sim t^2$, whereas on long time scales, $t \gtrsim \tau_{\text{short}}$, it decays linearly in

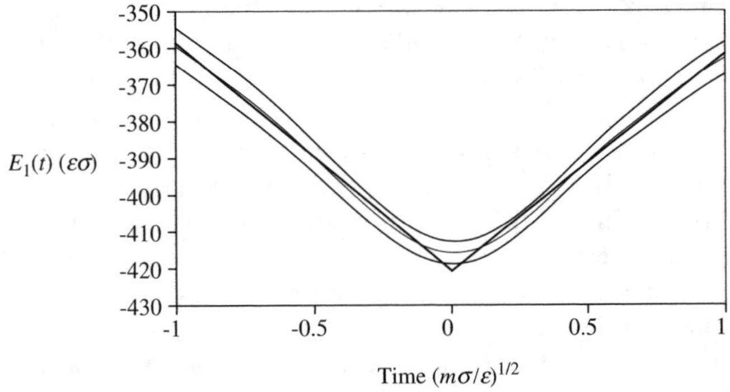

Figure 5. Molecular dynamics simulation of the decay forward and backward in time of the fluctuation of the first energy moment of a Lennard-Jones fluid (the central curve is the average moment, the enveloping curves are estimated standard error, and the lines are best fits). The starting positions of the adiabatic trajectories are obtained from Monte Carlo sampling of the static probability distribution, Eq. (246). The density is 0.80, the temperature is $T_0 = 2$, and the initial imposed thermal gradient is $\beta_1 = 0.02$. (From Ref. 2.)

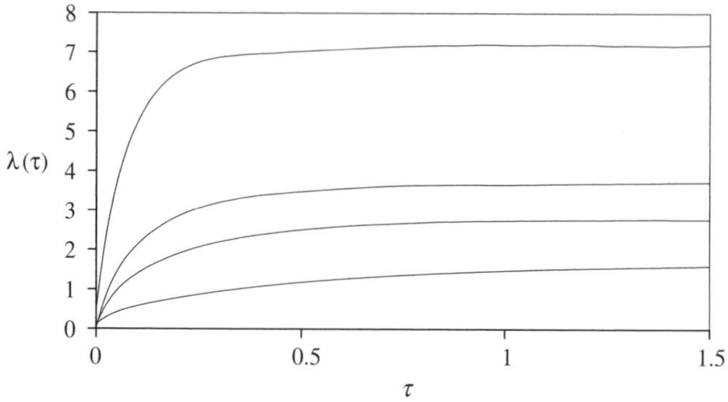

Figure 6. The dependence of the thermal conductivity on the time interval for the odd work, $\tilde{W}_1(\mathbf{\Gamma};\tau)$. The curves are $\lambda(\tau) = \langle \dot{E}_1^0(0)\rangle_\tau / Vk_BT_0^2\beta_1$ for densities of, from bottom to top, 0.3, 0.5, 0.6, and 0.8, and $T_0 = 2$. (From Ref. 5.)

time, $E_1(t) \sim |t|$. This is consistent with conclusions drawn from the second entropy analysis of fluctuations, Section IIC.

Figure 6 tests the dependence of the thermal conductivity on the time interval used to calculate $E_{1\Delta}(\mathbf{\Gamma};\tau)$ [5]. These are Monte Carlo simulations using the Metropolis algorithm, umbrella sampling, and the steady-state probability distribution, Eq. (247). Provided that the limit is not too small, for $\tau \gtrsim 1$, the thermal conductivity is independent of the integration limit used for $E_{1\Delta}$, as can be seen from the figure. This asymptotic or plateau value is "the" thermal conductivity. The values of τ required to reach the respective plateaus here appear comparable to straight Green–Kubo equilibrium calculations [3], but the present steady-state simulations used about one-third the number of trajectories for comparable statistical error.

In the chapter it was assumed that the change in moment over the relevant time scales was negligible, $\tau|\dot{E}_1| \ll |E_1|$. In the case of $\rho = 0.8$ at the largest value of τ in Fig. 5, $\langle E_1\rangle_{ss} = -432$ and $\langle \dot{E}_1\rangle_{ss} = 161$, and so this assumption is valid in this case. Indeed, this assumption was made because on long time scales the moment must return to zero and the rate of change of moment must begin to decrease. There is no evidence of this occurring in any of the cases over the full interval shown in Fig. 6.

Table I shows the values of the relaxation time calculated using Eqs. (249) and (250). Both the inertial time and the long time decrease with increasing density. This is in agreement with the trend of the curves in Fig. 6. Indeed, the actual estimates of the relaxation times in Table I are in semiquantitative agreement with the respective boundaries of the plateaux in Fig. 6. The estimate of τ_{long}, the upper limit on τ that may be used in the present theory, is perhaps a little conservative.

TABLE I
Thermal Conductivity and Relaxation Times for Various Densities[a] at $T_0 = 2$

ρ	λ	τ_{short} Eq. (249)	τ_{long} Eq. (250)
0.3	1.63(8)	0.404(19)	3.22(16)
0.5	2.78(13)	0.233(11)	5.31(34)
0.6	3.76(16)	0.197(9)	3.41(18)
0.8	7.34(18)	0.167(4)	1.36(3)

[a]The standard error of the last few digits is in parentheses. Data from Ref. 5.

Figure 7 compares the thermal conductivity obtained from nonequilibrium Monte Carlo simulations [5] with previous NEMD results [89, 91]. The good agreement between the two approaches validates the present phase space probability distribution. Of course, since the analysis in Section IVB shows that the present steady-state probability gives the Green–Kubo formula, the results in Fig. 7 test the simulation algorithm rather than the probability distribution per se. The number of time steps that we required for an error of about 0.1 was about 3×10^7 (typically 2×10^5 independent trajectories, each of about 75 time steps forward and backward to get into the intermediate regime). This obviously depends on the size of the applied thermal gradient (the statistical error decreases with increasing gradient), but appears comparable to that required by NEMD simulations [89]. No attempt was made to optimize the present algorithm in terms of the number of Monte Carlo cycles between trajectory evaluations or the value of the umbrella parameter.

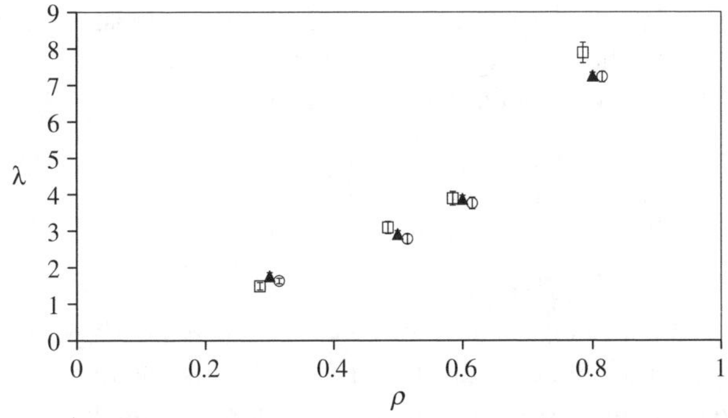

Figure 7. Nonequilibrium Monte Carlo results for the thermal conductivity ($T_0 = 2$). The circles and squares are the present steady-state results for bulk and inhomogeneous systems, respectively (horizontally offset by ±0.015 for clarity), and the triangles are NEMD results [89, 91]. (From Ref. 5.)

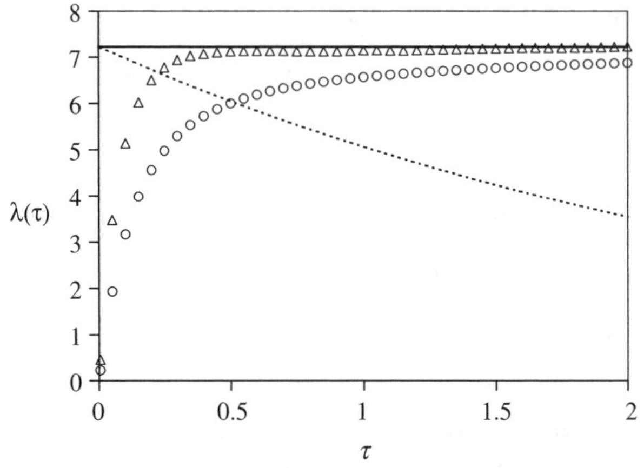

Figure 8. The dimensionless thermal conductivity, $k_B^{-1}\tau_{LJ}\sigma_{LJ}\lambda(\tau)$, at $\rho = 0.8$ and $T_0 = 2$. The symbols are the simulation data, with the triangles using the instantaneous velocity at the end of the interval, $\lambda'(\tau)$, Eq. (277), and the circles using the coarse velocity over the interval, $\lambda''(\tau)$, Eq. (278). The solid line is the second entropy asymptote, essentially Eq. (229), and the dotted curve is the Onsager–Machlup expression $\lambda'_{OM}(\tau)$, Eq. (280). (Data from Ref. 6.)

Figure 7 also shows results for the thermal conductivity obtained for the slit pore, where the simulation cell was terminated by uniform Lennard-Jones walls. The results are consistent with those obtained for a bulk system using periodic boundary conditions. This indicates that the density inhomogeneity induced by the walls has little effect on the thermal conductivity.

Figure 8 shows the τ-dependent thermal conductivity for a Lennard-Jones fluid ($\rho = 0.8$, $T_0 = 2$) [6]. The nonequilibrium Monte Carlo algorithm was used with a sufficiently small imposed temperature gradient to ensure that the simulations were in the linear regime, so that the steady-state averages were equivalent to fluctuation averages of an isolated system.

The conductivity was obtained from the simulations as

$$\lambda'(\tau) = \frac{1}{Vk_B T_0^2 \beta_1} \langle \dot{E}_1^0 \rangle_{ss,\tau} = \frac{-1}{Vk_B T_0^2} \langle E_1(t) \dot{E}_1^0(t+\tau) \rangle_0 \tag{277}$$

Here V is the volume, T_0 is the average applied temperature, β_1 is essentially the applied inverse temperature gradient, and E_1 is the first energy moment. It was also obtained using the coarse velocity over the interval, which is essentially the time integral of the above expression,

$$\lambda''(\tau) = \frac{-1}{Vk_B T_0^2 \tau} \langle E_1(t)[E_1(t+\tau) - E_1(t)] \rangle_0 \tag{278}$$

These two expressions follow from Eq. (229): multiply it by $E_1(t)$, take the isolated system average, and do an integration by parts using the fact that $T_{s1}^{-1} = \partial S/\partial E_1$. The coarse grained expression follows by using it directly, and the terminal velocity expression follows by taking the τ derivative. "The" thermal conductivity λ_∞ is obtained from the simulations as the plateau limit of these.

Onsager and Machlup [32] gave an expression for the probability of a path of a macrostate, $\wp[\mathbf{x}]$. The exponent may be maximized with respect to the path for fixed end points, and what remains is conceptually equivalent to the constrained second entropy used here, although it differs in mathematical detail. The Onsager–Machlup functional predicts a most likely terminal velocity that is exponentially decaying [6, 42]:

$$\overline{\dot{\mathbf{x}}}(\mathbf{x}_1, \tau) = -\hat{\tau} e^{-|\tau| \underline{\underline{\mu}}} \underline{\underline{\mu}} \mathbf{x}_1 \tag{279}$$

Consequently, the time correlation function given by Onsager–Machlup theory is [6]

$$\lambda'_{\mathrm{OM}}(\tau) = \lambda_\infty e^{-\tau\mu} \tag{280}$$

where the inverse time constant is $\mu = S/2a_2$, the first entropy matrix is $S = -k_B/\langle E_1^2 \rangle_0$, and the second entropy matrix is $a_2 = -1/2VT_0^2\lambda_\infty$.

The exponential decay predicted by the Onsager–Machlup theory, and by the Langevin and similar stochastic differential equations, is not consistent with the conductivity data in Fig. 8. This and the earlier figures show a constant value for $\lambda(\tau)$ at larger times, rather than an exponential decay. It may be that if the data were extended to significantly larger time scales it would exhibit exponential decay of the predicted type.

It ought to be stressed that the simulation results for the time-correlation function were obtained in a system with periodic boundary conditions. In particular, $E_1(\tau)$ was obtained by integrating $\dot{E}_1^0(\tau')$ over time. The latter function depends on the separation between pairs of atoms, not their absolute position, and was evaluated using the nearest image convention (see Eq. (258) et seq.). Figure 9a shows data from simulations performed on a system confined between two walls, with periodic boundaries only in the lateral directions, and the energy moment measured parallel to the normal to the walls. It can be seen that as the width of the fluid increases, the extent of the plateau region for the conductivity increases and the eventual rate of decay slows. These data suggest that the extent of the plateau region scales with the size of the system, and that periodic systems are effectively infinite, at least on the time scales used in these simulations.

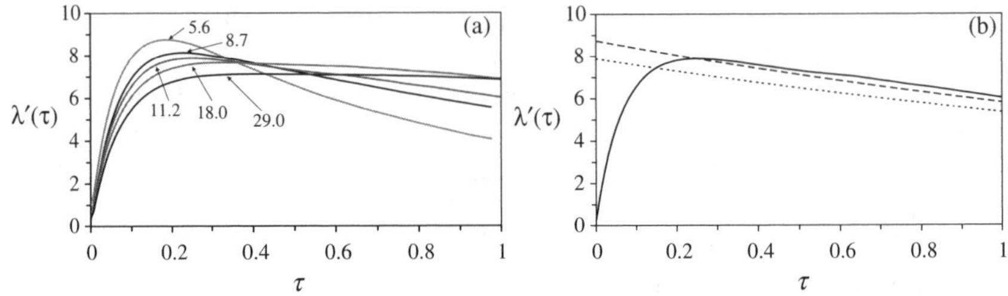

Figure 9. Simulated thermal conductivity $\lambda'(\tau)$ for a Lennard-Jones fluid. The density in the center of the system is $\rho = 0.8$ and the zeroth temperature is $T_0 = 2$. (a) A fluid confined between walls, with the numbers referring to the width of the fluid phase. (From Ref. 6.) (b) The case $L_z = 11.2$ compared to the Markov (dashed) and the Onsager–Machlup (dotted) prediction.

Figure 9b compares the decay of the correlations in a confined fluid case with the present Markov prediction, Eq. (80), and the Onsager–Machlup prediction, Eq. (280). Note that the Markov prediction has been designed to pass through the maximum, $\lambda'(t) = \lambda_\infty(1 - \tau^*\mu)^{(-1+t/\tau^*)}$. Strictly speaking, it is only applicable for $t > \tau^*$, as the Markov approximation is predicated on a constant transport coefficient. It is possible to apply the theory from $\tau = 0$, and without fitting it is still a reasonable approximation for $t > \tau^*$ (not shown). For the state point of the figure, the Onsager–Machlup decay time is $\mu^{-1} = 2.57$, and the Markov decay time is $-\tau^*/\ln(1 - \tau^*\mu) = 2.44$. Evidently there is little to choose between them in this case. The evident failure of the Onsager–Machlup expression for $\tau \lesssim \tau^*$ is undoubtedly due to the same cause as for the Markov theory: using the fixed transport coefficient is not valid on short time scales. The dependence of the Markov decay time on τ^* will give different decay times for the different cases shown in Fig 9a, since the location of the peak varies.

On the basis of the results in Fig. 9 and the discussion in Section IIG, the interpretation of the transport coefficient as the maximal value of the correlation function may be given as follows. After a fluctuation in structure, over time a flux develops in an isolated system, as predicted by the optimized second entropy for each time, and reaches a maximum in the intermediate time regime. In the thermodynamic limit this plateau region may be quite extensive. The end of the plateau region and the decrease in flux occurs when the decrease in the magnitude of the original fluctuation due to the total flux to date becomes relatively significant, because after the transient regime the flux is proportional to the current magnitude of the fluctuation. This interpretation is supported by the success of the Markov analysis in the figure. If a reservoir is connected, the steady-state flux has ample time to develop, but as there is no decrease in the magnitude of the static structure due to replenishment by the reservoir, the flux remains at the same value it would have in the steady-state regime of the isolated system.

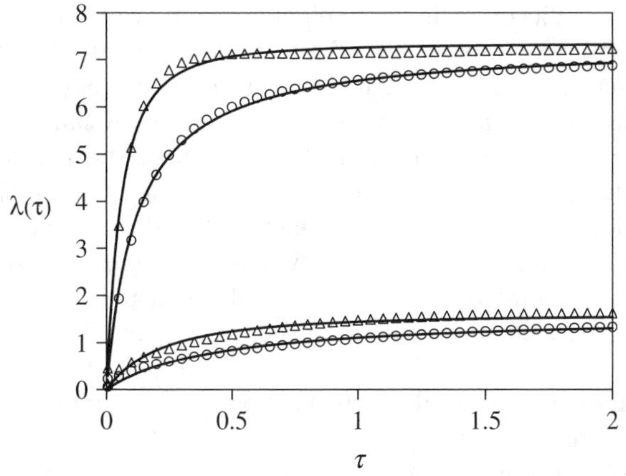

Figure 10. Fitted thermal conductivity, $\lambda'(\tau)$ and $\lambda''(\tau)$, using $\tau^2 A(\tau) = -0.0083 - 0.0680|\tau|$, for $\rho = 0.8$ (upper data and curves), and $\tau^2 A(\tau) = -0.139 - 0.314|\tau|$, for $\rho = 0.3$ (lower data and curves). (From Ref. 6.)

It is also of interest to explore the small-time or transient regime. Using the functional form $\tau^2 A(\tau) = \sigma_i + |\tau| a_2$, the second entropy, Eq. (227), may be extrapolated to smaller time scales. The thermal conductivities become

$$\lambda'(\tau) = \frac{-\tau}{(\sigma_1 + \tau\sigma_2)} + \frac{\tau^2\sigma_2}{(\sigma_1 + \tau\sigma_2)^2} \tag{281}$$

and

$$\lambda''(\tau) = \frac{-\tau}{2(\sigma_1 + \tau\sigma_2)} \tag{282}$$

Figure 10 shows small-time fits to the thermal conductivity using the above functional form. It can be seen that quite good agreement with the simulation data can be obtained with this simple time-dependent transport function. Such a function can be used in the transient regime or to characterize the response to time-varying applied fields.

IX. CONCLUSION

The Second Law of Nonequilibrium Thermodynamics that is described here is very general. Is it too general to provide a basis for detailed quantitative calculations? Is there a need for such a law, particularly given its undeniable, indeed deliberate, similarity, with the traditional equilibrium Second Law?

One of the benefits of having an explicit nonequilibrium Second Law is that it focuses attention on time, and on the fact that the equilibrium Second Law offers no prescription for the speed of time or for the rate of change. Similarly, by introducing the second entropy, which could also be called the transition entropy, it makes the point that any quantitative theory for the time evolution of a system must go beyond the ordinary or first entropy. Almost all theories of the nonequilibrium state to date have been based on the first entropy or its rate of change. It can be argued that the slow rate of progress of nonequilibrium theory itself is due directly to the failure to distinguish between the first and second entropies, a distinction that the nonequilibrium Second Law makes extant.

Despite the significance of the second entropy, the concept is really very easy to grasp. The first entropy gives the weight of molecular configurations associated with the structure that is a macrostate. The second entropy gives the weight of molecular configurations associated with transitions between pairs of macrostates. These transitions may be called dynamic structure, and because they occur over a specified time interval, they are the same as a rate or a flux.

In the equilibrium Second Law, the first entropy increases during spontaneous changes in structure, and when the structure stabilizes (i.e., change ceases), the first entropy is a maximum. This state is called the equilibrium state. Similarly, in the nonequilibrium Second Law, the second entropy increases during spontaneous changes in flux, and when the flux stabilizes, the second entropy is a maximum. This state is called the steady state. The present nonequilibrium Second Law has the potential to provide the same basis for the steady state that Clausius' Second Law has provided for the equilibrium state.

Of course, depending on the system, the optimum state identified by the second entropy may be the state with zero net transitions, which is just the equilibrium state. So in this sense the nonequilibrium Second Law encompasses Clausius' Second Law. The real novelty of the nonequilibrium Second Law is not so much that it deals with the steady state but rather that it invokes the speed of time quantitatively. In this sense it is not restricted to steady-state problems, but can in principle be formulated to include transient and harmonic effects, where the thermodynamic or mechanical driving forces change with time. The concept of transitions in the present law is readily generalized to, for example, transitions between velocity macrostates, which would be called an acceleration, and spontaneous changes in such accelerations would be accompanied by an increase in the corresponding entropy. Even more generally it can be applied to a path of macrostates in time.

Arguably a more practical approach to higher-order nonequilibrium states lies in statistical mechanics rather than in thermodynamics. The time correlation function gives the linear response to a time-varying field, and this appears in computational terms the most useful methodology, even if it may lack the satisfying universality of the Second Law. One reason for focussing on the steady state is the simplicity of many of the analytic formulas that emerge. But as discussed in Sections IIE and IIG, using the time correlation function itself can also be fruitful [92].

For nonequilibrium statistical mechanics, the present development of a phase space probability distribution that properly accounts for exchange with a reservoir, thermal or otherwise, is a significant advance. In the linear limit the probability distribution yielded the Green–Kubo theory. From the computational point of view, the nonequilibrium phase space probability distribution provided the basis for the first nonequilibrium Monte Carlo algorithm, and this proved to be not just feasible but actually efficient. Monte Carlo procedures are inherently more mathematically flexible than molecular dynamics, and the development of such a nonequilibrium algorithm opens up many, previously intractable, systems for study. The transition probabilities that form part of the theory likewise include the influence of the reservoir, and they should provide a fecund basis for future theoretical research. The application of the theory to molecular-level problems answers one of the two questions posed in the first paragraph of this conclusion: the nonequilibrium Second Law does indeed provide a quantitative basis for the detailed analysis of nonequilibrium problems.

The second question concerned the close similarity between the formulations of the two Second Laws. The justification for a distinct nonequilibrium Second Law was given earlier, but the reason for casting it in such a familiar fashion may be discussed. To some extent, there is no choice in the matter, since as a law of nature there are only so many ways it can be expressed without changing its content. Beyond this, there are actually several advantages in the new law being so closely analogous to the old one. Besides the obvious psychological fact that familiarity breeds acceptance, familiarity also enables intuition, and so many nonequilibrium concepts can now be grasped simply by analogy with their equilibrium counterparts. In fact, in setting up a one-to-one correspondence between the two Second Laws, the groundwork has been laid to carry over the known equilibrium results to the nonequilibrium context. That is, all known equilibrium principles, theorems, and relationships ought to have a nonequilibrium version based on the correspondence established by the new Second Law.

The philosophical and conceptual ramifications of the nonequilibrium Second Law are very deep. Having established the credentials of the Law by the detailed analysis outlined earlier, it is worth considering some of these large-scale consequences. Whereas the equilibrium Second Law of Thermodynamics implies that order decreases over time, the nonequilibrium Second Law of Thermodynamics explains how it is possible that order can be induced and how it can increase over time. The question is of course of some relevance to the creation and evolution of life, society, and the environment.

The present analysis shows that when a thermodynamic gradient is first applied to a system, there is a transient regime in which dynamic order is induced and in which the dynamic order increases over time. The driving force for this is the dissipation of first entropy (i.e., reduction in the gradient), and what opposes it is the cost of the dynamic order. The second entropy provides a quantitative expression for these processes. In the nonlinear regime, the fluxes couple to the static structure, and structural order can be induced as well. The nature of this combined order is to dissipate first entropy, and in the transient regime the rate of dissipation increases with the evolution of the system over time.

Prigogine has used the term "dissipative structures" to describe the order induced during nonequilibrium processes [11, 83]. Physical examples include convective rolls, chemical clock reactions, and chaotic flows. These are also suggested to be related to the type of order that living organisms display—a theme that continues to be widely studied [13–18]. While one can argue about the mathematical analysis and physical interpretation of these studies, there is little doubt that the broad picture that emerges is consistent with the nonequilibrium Second Law. At its simplest, the argument is that life exists on the energy gradient provided by the Sun. Living systems also exploit gradients in material resources due to their spatial segregation. The raison d'etre for life is to reduce these gradients, and the role of evolution is to increase the rate at which these gradients are dissipated. Cope's rule, which is the observation that body size increases as species evolve [93], may simply be a manifestation of the extensivity of the second entropy. Changes in environment and society occur on shorter time scales than biological evolution, and the historical record certainly reveals an increasing capacity for entropy production over time. It is sobering to reflect that we evolve in order to hasten our own end.

Acknowledgments

Angus Gray-Weale is thanked for useful discussions on the symmetry and time dependence of the transport coefficients, and the Australian Research Council is thanked for financial support.

References

1. P. Attard, *J. Chem. Phys.* **121**, 7076 (2004).
2. P. Attard, *J. Chem. Phys.* **122** 154101 (2005).
3. P. Attard, *J. Chem. Phys.* **122**, 244105 (2005).
4. P. Attard, *J. Chem. Phys.* **124**, 024109 (2006).
5. P. Attard, *J. Chem. Phys.* **124**, 224103 (2006).
6. P. Attard, *J. Chem. Phys.* **125**, 214502 (2006).
7. P. Attard, *J. Chem. Phys.* **127**, 014503 (2007).
8. P. Attard, *Phys. Chem. Chem. Phys.* **8**, 3585 (2006).

9. P. Attard, *Thermodynamics and Statistical Mechanics: Equilibrium by Entropy Maximisation*, Academic Press, London, 2002.
10. L. Onsager, *Phys. Rev.* **37**, 405 (1931); **38**, 2265 (1931).
11. I. Prigogine, *Introduction to Thermodynamics of Irreversible Processes*, Interscience, New York, 1967.
12. S. R. de Groot and P. Mazur, *Non-equilibrium Thermodynamics*, North-Holland, Amsterdam, 1969.
13. G. W. Paltridge, *Nature* **279**, 630 (1979).
14. R. Swenson and M. T. Turvey, *Ecological Psychol.* **3**, 317 (1991).
15. E. D. Schneider and J. J. Kay, *Math. Comp. Mod.* **19**, 25 (1994).
16. R. Dewar, *J. Phys. Math. Gen.* **36**, 631 (2003).
17. A. Kleidon and R. D. Lorenz (eds.), *Non-equilibrium Thermodynamics and the Production of Entropy: Life, Earth, and Beyond*, Springer, Berlin, 2005.
18. L. M. Martyushev and V. D. Seleznev, *Phys. Rep.* **426**, 1 (2006).
19. M. Paniconi and Y. Oono, *Phys. Rev. E* **55**, 176 (1997).
20. Y. Oono and M. Paniconi, *Prog. Theor. Phys. Suppl.* **130**, 29 (1998).
21. B. C. Eu, *Nonequilibrium Statistical Mechanics*, Kluwer Academic Publishers, Dordrecht, 1998.
22. D. Jou, J. Casa-Vázquez, and G. Lebon, *Rep. Prog. Phys.* **62**, 1035 (1999).
23. J. Casa-Vázquez and D. Jou, *Rep. Prog. Phys.* **66**, 1937 (2003).
24. H. B. G. Casimir, *Rev. Mod. Phys.* **17**, 343 (1945).
25. H. Grabert, R. Graham, and M. S. Green, *Phys. Rev. A* **21**, 2136 (1980).
26. M. S. Green, *J. Chem. Phys.* **22**, 398 (1954).
27. R. Kubo, *Rep. Prog. Phys.* **29**, 255 (1966).
28. J.-P. Hansen and I. R. McDonald, *Theory of Simple Liquids*, Academic Press, London, 1986.
29. J. P. Boon and S. Yip, *Molecular Hydrodynamics*, Dover, New York, 1991.
30. R. Zwanzig, *Non-equilibrium Statistical Mecahnics*, Oxford University Press, Oxford, UK, 2001.
31. H. B. Callen and R. F. Greene, *Phys. Rev.* **86**, 702 (1952). R. F. Greene and H. B. Callen, *Phys. Rev.* **88**, 1387 (1952).
32. L. Onsager and S. Machlup, *Phys. Rev.* **91**, 1505 (1953).
33. H. Risken, *The Fokker–Planck Equation*, Springer-Verlag, Berlin, 1984.
34. R. L. Stratonovich, *Sel. Transl. Math. Stat. Prob.* **10**, 273 (1971).
35. J. Keizer, *Statistical Thermodynamics of Nonequilibrium Processes*, Springer-Verlag, New York, 1987.
36. W. Horsthemke and A. Bach, *Z. Phys. B* **22**, 189 (1975).
37. H. Haken, *Z. Phys. B* **24**, 321 (1976).
38. R. Graham, *Z. Phys. B* **26**, 281 (1977).
39. K. Yasue, *J. Math. Phys.* **19**, 1671 (1978).
40. H. Grabert and M. S. Green, *Phys. Rev. A* **19**, 1747 (1979).
41. B. H. Lavenda, *Foundations Phys.* **9**, 405 (1979).
42. K. L. C. Hunt and J. Ross, *J. Chem. Phys.* **75**, 976 (1981).
43. B. H. Lavenda, *Nonequilibrium Statistical Thermodynamics*, Wiley, Chichester, 1985.

44. G. L. Eyink, *J. Stat. Phys.* **61**, 533 (1990).
45. B. Peng, K. L. C. Hunt, P. M. Hunt, A. Suarez, and J. Ross, *J. Chem. Phys.* **102**, 4548 (1995).
46. P. Ao, *J. Phys. A Math. Gen.* **37**, L25 (2004).
47. N. Hashitsume, *Prog. Theor. Phys.* **8**, 461 (1952).
48. I. Gyarmati, *Z. Phys. Chem. Leipzig* **239**, 133 (1968).
49. I. Gyarmati, *Nonequilibrium Thermodynamics, Field Theory, and Variational Principles*, Springer, Berlin, 1970.
50. G. N. Bochkov and Yu. E. Kuzovlev, *JETP Lett.* **30**, 46 (1980).
51. D. J. Evans and A. Baranyai, *Phys. Rev. Lett.* **67**, 2597 (1991).
52. A. Baranyai, *J. Chem. Phys.* **105**, 7723 (1996).
53. D. J. Evans and G. P. Morriss, *Statistical Mechanics of Nonequilibrium Liquids*, Academic, London, 1990.
54. W. H. Hoover, *Computational Statistical Mechanics*, Elsevier, Amsterdam, 1991.
55. C. Jarzynski, *Phys. Rev. Lett.* **78**, 2690 (1997).
56. D. J. Evans, E. G. D. Cohen, and G. P. Morriss, *Phys. Rev. Lett.* **71**, 2401 (1993).
57. D. J. Evans, *Mol. Phys.* **101**, 1551 (2003).
58. G. N. Bochkov and Yu. E. Kuzovlev, *Sov. Phys. JETP* **45**, 125 (1977).
59. G. N. Bochkov and Yu. E. Kuzovlev, *Sov. Phys. JETP* **49**, 543 (1977).
60. G. N. Bochkov and Yu. E. Kuzovlev, *Physica* **106A**, 443 (1981).
61. G. E. Crooks, *Phys. Rev. E* **60**, 2721 (1999).
62. G. E. Crooks, *Phys. Rev. E* **61**, 2361 (2000).
63. G. M. Wang, E. M. Sevick, E. Mittag, D. J. Searles, and D. J. Evans, *Phys. Rev. Lett.* **89**, 050601 (2002).
64. C. Jarzynski, *Phys. Rev. E* **56**, 5018 (1997).
65. G. E. Crooks, *J. Stat. Phys.* **90**, 1481 (1998).
66. J. Liphardt, S. Dumont, S. B. Smith, I. Tinoco Jr., and C. Bustamante, *Science* **296**, 1832 (2002).
67. R. P. Feynman, *Statistical Mechanics: A Set of Lectures*, Benjamin, Reading, MA, 1972.
68. T. Yamada and K. Kawasaki, *Prog. Theor. Phys.* **38**, 1031 (1967).
69. T. Yamada and K. Kawasaki, *Prog. Theor. Phys.* **53**, 111 (1975).
70. G. P. Morriss and D. J. Evans, *Mol. Phys.* **54**, 629 (1985).
71. G. P. Morriss and D. J. Evans, *Phys. Rev. A* **37**, 3605 (1988).
72. T. Hatano and S. Sasa, *Phys. Rev. Lett.* **86**, 3463 (2001).
73. E. H. Trepagnier, C. Jarzynski, F. Rotort, G. E. Crooks, C. J. Bustamante, and J. Liphardt, *PNAS* **101**, 15038 (2004).
74. O. Mazonka and C. Jarzynski, arXiv:cond-mat/9912121 (1999).
75. R. J. Glauber, *J. Math. Phys.* **4**, 294 (1963).
76. K. Kawasaki, *Phys. Rev.* **145**, 145 (1966).
77. J. S. Langer, *Ann. Phys.* **54**, 258 (1969).
78. H. Metiu, K. Kitahara, and J. Ross, *J. Chem. Phys.* **63**, 5116 (1975).
79. H. C. Anderson, *J. Chem. Phys.* **72**, 2384 (1980).
80. P. Attard, *J. Chem. Phys.* **116**, 9616 (2002).
81. S. Boinepalli and P. Attard, *J. Chem. Phys.* **119**, 12769 (2003).
82. R. F. Fox and G. E. Uhlenbeck, *Phys. Fluids* **13**, 1893 (1970).

83. I. Prigogine, *Time, Structure, and Fluctuations*, Nobel Lecture, 1977.
84. M. P. Allen and D. J. Tildesley, *Computer Simulation of Liquids*, Oxford University Press, Oxford, UK, 1987.
85. B. L. Holian, G. Ciccotti, W. G. Hoover, B. Moran, and H. A. Posch, *Phys. Rev. A* **39**, 5414 (1989).
86. G. P. Morriss, *Phys. Rev. A* **39**, 4811 (1989).
87. D. J. Evans and D. J. Searles, *Phys. Rev. E* **52**, 5839 (1995).
88. D. J. Evans and G. P. Morriss, *Comput. Phys. Rep.* **1**, 297 (1984).
89. P. J. Daivis and D. J. Evans, *Phys. Rev. E* **48**, 1058 (1993).
90. J. K. Johnson, J. A. Zollweg, and K. E. Gubbins, *Mol. Phys.* **78**, 591 (1993).
91. D. J. Evans, *Phys. Rev. A* **34**, 1449 (1986).
92. A. Gray-Weale and P. Attard, *J. Chem. Phys.* **127**, 044503 (2007).
93. W. H. Gould, *Hens Teeth and Horse's Toes: Further Reflections on Natural History*, Norton, New York, 1983.

MULTIPARTICLE COLLISION DYNAMICS: SIMULATION OF COMPLEX SYSTEMS ON MESOSCALES

RAYMOND KAPRAL

University of Toronto, Toronto, Canada

CONTENTS

Advances in Chemical Physics, Volume 140, edited by Stuart A. Rice
Copyright © 2008 John Wiley & Sons, Inc.

I. INTRODUCTION

Complex systems can display complex phenomena that are not easily described at a microscopic level using molecular dynamics simulation methods. Systems may be complex because some or all of their constituents are complex molecular species, such as polymers, large biomolecules, or other molecular aggregates. Such systems can exhibit the formation of patterns arising from segregation of constituents or nonequilibrium effects and molecular shape changes induced by flows and hydrodynamic interactions. Even when the constituents are simple molecular entities, turbulent fluid motions can exist over a range of length and time scales.

Many of these phenomena have their origins in interactions at the molecular level but manifest themselves over mesoscopic and macroscopic space and time scales. These features make the direct simulation of such systems difficult because one must follow the motions of very large numbers of particles over very long times. These considerations have prompted the development of coarse-grain methods that simplify the dynamics or the system in different ways in order to be able to explore longer length and time scales. The use of mesoscopic dynamical descriptions dates from the foundations of nonequilibrium statistical mechanics. The Boltzmann equation [1] provides a field description on times greater than the time of a collision and the Langevin equation [2] replaces a molecular-level treatment of the solvent with a stochastic description of its properties.

The impetus to develop new types of coarse-grain or mesoscopic simulation methods stems from the need to understand and compute the dynamical properties of large complex systems. The method of choice usually depends on the type of information that is desired. If properties that vary on very long distance and time scales are of interest, the nature of the dynamics can be altered, while still preserving essential features to provide a faithful representation of these properties. For example, fluid flows described by the Navier–Stokes equations will result from dynamical schemes that preserve the basic conservation laws of mass, momentum, and energy. The details of the molecular interactions may be unimportant for such applications. Some coarse-grain approaches constructed in this spirit, such as the *lattice Boltzmann method* [3] and *direct simulation Monte Carlo* [4], are based on the Boltzmann equation. In *dissipative particle dynamics* [5, 6], several atoms are grouped into simulation sites whose dynamics is governed by conservative and frictional forces designed to reproduce the thermodynamics and hydrodynamics of the system. *Smoothed particle hydrodynamics* [7] is used to discretize continuum equations by employing weight functions assigned to fictitious particles so that hydrodynamic simulations can be carried out in a molecular dynamics framework. A number of different routes have been taken to construct coarse-grain models for the intermolecular potentials by grouping a number of atoms together to form sites that interact through effective potentials [8–10]. Molecular dynamics simulations may then be carried out more effectively on these coarse-grained entities.

In this chapter we describe another mesoscopic dynamical scheme that is based on a coarse-grain description of molecular collisions. In molecular dynamics a many-body system interacting through an intermolecular potential evolves by Newton's equations of motion. From a kinetic theory perspective, the time evolution of the system is governed by collisional encounters among the molecules. Description of the collisions is difficult since it entails the solution of a many-body scattering problem in a dense phase system to determine the relevant cross sections. In many applications this level of molecular detail is unnecessary or even unwanted. On coarse-grained distance and time scales individual collisional encounters are not important; instead it is the net effect of many collisions that plays a crucial role in determining the system properties. On such scales only the generic features of molecular dynamics are important. These include the conservation laws and the symplectic nature of the dynamics. If these generic features are preserved, then the dynamics will capture many essential features on coarse-grained scales, which do not depend on specific features of the intermolecular potentials. This is the approach adopted here. We construct a fictitious *multiparticle collision* (MPC) dynamics that accounts for the effects of many real collisions and yet preserves the conservation laws and the phase space structure of full molecular dynamics [11].

There are several attractive features of such a mesoscopic description. Because the dynamics is simple, it is both easy and efficient to simulate. The equations of motion are easily written and the techniques of nonequilibriun statistical mechanics can be used to derive macroscopic laws and correlation function expressions for the transport properties. Accurate analytical expressions for the transport coefficient can be derived. The mesoscopic description can be combined with full molecular dynamics in order to describe the properties of solute species, such as polymers or colloids, in solution. Because all of the conservation laws are satisfied, hydrodynamic interactions, which play an important role in the dynamical properties of such systems, are automatically taken into account.

We begin with a description of multiparticle collision dynamics and discuss its important properties. We show how it can be combined with full molecular dynamics (MD) to construct a hybrid MPC–MD method that can be used to simulate complex systems. The equations of motion of MPC dynamics form the starting point for the derivation of macroscopic laws. Discrete-time projection operator methods are used to obtain the macroscopic laws and Green–Kubo expressions are used for the transport properties. The method is then applied to colloidal suspensions and polymer solutions. It is also used to study chemically reacting systems both close to and far from equilibrium. Finally, generalizations of the MPC dynamics are described that allow one to study nonideal and immiscible solutions.

II. MULTIPARTICLE COLLISION DYNAMICS

Consider a system of N particles with masses m in a volume $V = L^3$. Particle i has position $\mathbf{r}_i$ and velocity $\mathbf{v}_i$ and the phase point describing the microscopic state of the system is $\mathbf{x}^N \equiv (\mathbf{r}^N, \mathbf{v}^N) = (\mathbf{r}_1, \mathbf{r}_2, \ldots, \mathbf{r}_N, \mathbf{v}_1, \mathbf{v}_2, \ldots, \mathbf{v}_N)$. We assume that the particles comprising the system undergo collisions that occur at discrete-time intervals τ and free stream between such collisions. If the position of particle i at time t is $\mathbf{r}_i$, its position at time $t + \tau$ is

$$\mathbf{r}_i^* = \mathbf{r}_i + \mathbf{v}_i \tau \tag{1}$$

The collisions that take place at the times τ represent the effects of many real collisions in the system.[1] These effective collisions are carried out as follows.[2] The volume V is divided into N_c cells labeled by cell indices ξ. Each cell is assigned at random a rotation operator $\hat{\omega}_\xi$ chosen from a set Ω of rotation operators. The center of mass velocity of the particles in cell ξ is $\mathbf{V}_\xi = N_\xi^{-1} \sum_{i=1}^{N_\xi} \mathbf{v}_i$, where N_ξ is the instantaneous number of particles in the cell. The postcollision velocities of the particles in the cell are then given by

$$\mathbf{v}_i^* = \mathbf{V}_\xi + \hat{\omega}_\xi (\mathbf{v}_i - \mathbf{V}_\xi) \tag{2}$$

The set of rotations used in MPC dynamics can be chosen in various ways and the specific choice will determine the values of the transport properties of the system, just as the choice of the intermolecular potential will determine the transport properties in a system evolving by full molecular dynamics through Newton's equations of motion. It is often convenient to use rotations about a randomly chosen direction, $\hat{\mathbf{n}}$, by an angle α chosen from a set of angles. The

[1]Without loss of generality, the time τ may be set to unity if only MPC and free streaming determine the dynamics. In hybrid models discussed later that combine molecular and MPC dynamics, its value influences the transport properties of the system. Anticipating such an extension, we allow τ to remain arbitrary here.

[2]The multiparticle collision rule was first introduced in the context of a lattice model with a stochastic streaming rule in Ref. 12.

postcollision velocity $\mathbf{v}_i^*$ of particle i in cell ξ arising from the rotation by angle α is given explicitly by

$$\begin{aligned}\mathbf{v}_i^* &= \mathbf{V}_\xi + \hat{\omega}_\xi(\mathbf{v}_i - \mathbf{V}_\xi) = \mathbf{V}_\xi + \hat{\mathbf{n}}\hat{\mathbf{n}} \cdot (\mathbf{v}_i - \mathbf{V}_\xi) \\ &\quad + (\mathbf{I} - \hat{\mathbf{n}}\hat{\mathbf{n}}) \cdot (\mathbf{v}_i - \mathbf{V}_\xi)\cos\alpha - \hat{\mathbf{n}} \times (\mathbf{v}_i - \mathbf{V}_\xi)\sin\alpha \end{aligned} \tag{3}$$

The unit vector $\hat{\mathbf{n}}$ may be taken to lie on the surface of a sphere and the angles α may be chosen from a set Ω of angles. For instance, for a given α, the set of rotations may be taken to be $\Omega = \{\alpha, -\alpha\}$. This rule satisfies detailed balance. Also, α may be chosen uniformly from the set $\Omega = \{\alpha | 0 \leq \alpha \leq \pi\}$. Other rotation rules can be constructed. The rotation operation can also be carried out using quaternions [13]. The collision rule is illustrated in Fig. 1 for two particles. From this figure it is clear that multiparticle collisions change both the directions and magnitudes of the velocities of the particles.

In multiparticle collisions the same rotation operator is applied to each particle in the cell ξ but every cell in the system is assigned a different rotation operator so collisions in different cells are independent of each other. As a result of free streaming and collision, if the system phase point was $(\mathbf{r}^N, \mathbf{v}^N)$ at time t, it is $(\mathbf{r}^{*N}, \mathbf{v}^{*N})$ at time $t + \tau$.

For consistency we refer to this model as multiparticle collision (MPC) dynamics, but it has also been called stochastic rotation dynamics. The difference in terminology stems from the placement of emphasis on either the multiparticle nature of the collisions or on the fact that the collisions are effected by rotation operators assigned randomly to the collision cells. It is also referred to as real-coded lattice gas dynamics in reference to its lattice version precursor.

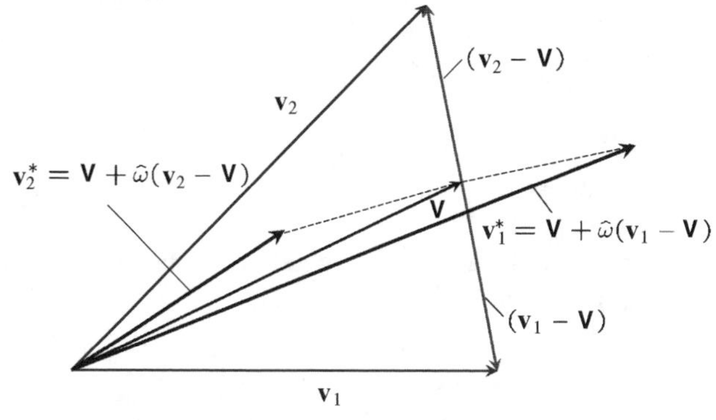

Figure 1. Application of the multiparticle collision rule for two particles with initial velocities $\mathbf{v}_1$ and $\mathbf{v}_2$ leading to postcollision velocities $\mathbf{v}_1^*$ and $\mathbf{v}_2^*$, respectively. Intermediate velocity values in the rule $(\mathbf{v}_1 - \mathbf{V})$ and $(\mathbf{v}_2 - \mathbf{V})$ are shown as thin solid lines while $\hat{\omega}(\mathbf{v}_1 - \mathbf{V})$ and $\hat{\omega}(\mathbf{v}_2 - \mathbf{V})$ are shown as dashed lines.

A. Properties of MPC Dynamics

It is easy to verify that multiparticle collisions conserve mass, momentum, and energy in every cell. Mass conservation is obvious. Momentum and energy conservation are also easily established. For momentum conservation in cell ξ we have

$$\sum_{i=1}^{N_\xi} m\mathbf{v}_i^* = \sum_{i=1}^{N_\xi} m(\mathbf{V} + \hat{\omega}_\xi[\mathbf{v}_i - \mathbf{V}]) = \sum_{i=1}^{N_\xi} m\mathbf{v}_i \tag{4}$$

where we have used the fact that the rotation operator is the same for every particle in the cell. Similarly, energy conservation is established by direct calculation as

$$\sum_{i=1}^{N_\xi} \frac{m}{2}|\mathbf{v}_i^*|^2 = \sum_{i=1}^{N_\xi} \frac{m}{2}|\mathbf{V} + \hat{\omega}_\xi[\mathbf{v}_i - \mathbf{V}]|^2 = \sum_{i=1}^{N_\xi} \frac{m}{2}|\mathbf{v}_i|^2 \tag{5}$$

Phase space volumes are also preserved in MPC dynamics. The phase space volume element $d\mathbf{x}^N$ is invariant with respect to both the streaming and collision steps. Under streaming the volume element transforms according to $d\mathbf{x}^N(t+\tau) = J(\mathbf{x}^N(t+\tau); \mathbf{x}^N(t))d\mathbf{x}^N(t)$, where $J = \det\mathbf{J}$ and the Jacobean matrix has elements $J_{i,j} = \partial\mathbf{x}_i(t+\tau)/\partial\mathbf{x}_j(t)$. Direct calculation shows that $J = 1$ for free streaming. The invariance of the phase space volume element in multiparticle collisions is a consequence of the semidetailed balance condition and the fact that rotations do not change phase space volumes. Letting $p(\hat{\omega}|\mathbf{v}^N)$ be the conditional probability of the rotation $\hat{\omega}$ given $\mathbf{v}^N$, we have

$$\begin{aligned} d\mathbf{x}^{*N} &= d\mathbf{r}^N d\mathbf{v}^{*N} = d\mathbf{r}^N \sum_{\hat{\omega}, \mathbf{v}^N | \hat{\omega}(\mathbf{v}^N) = \mathbf{v}^{*N}} p(\hat{\omega}|\mathbf{v}^N) d\mathbf{v}^N \\ &= d\mathbf{x}^N \sum_{\hat{\omega}, \mathbf{v}^N | \hat{\omega}(\mathbf{v}^N) = \mathbf{v}^{*N}} p(\hat{\omega}|\mathbf{v}^N) = d\mathbf{x}^N \end{aligned} \tag{6}$$

The last equalities follow from the fact that the rotations do not depend on the velocities.

Assuming the MPC dynamics is ergodic, the stationary distribution is microcanonical and is given by

$$\mathsf{P}(\mathbf{x}^N) = AV^{-1}\delta\left(\frac{m}{2}\sum_{i=1}^{N}|\mathbf{v}_i|^2 - \frac{3k_BTN}{2}\right)\delta\left(\sum_{i=1}^{N}[\mathbf{v}_i - \mathbf{u}]\right) \tag{7}$$

where $\mathbf{u}$ is the mean velocity of the system and A is a normalization constant. If this expression is integrated over the coordinates and velocities of particles with

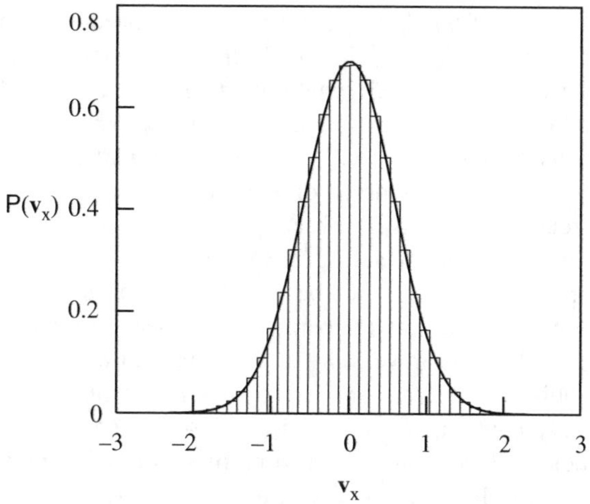

Figure 2. Comparison of the simulated velocity distribution (histogram) with the Maxwell–Boltzmann distribution function (solid line) for $k_BT = \frac{1}{3}$. The system had volume $V = 100^3$ cells of unit length and $N = 10^7$ particles with mass $m = 1$. Rotations $\hat{\omega}_\xi$ were selected from the set $\Omega = \{\pi/2, -\pi/2\}$ about axes whose directions were chosen uniformly on the surface of a sphere.

$i = 2, \ldots, N$, in the limit of large N we obtain the Maxwell–Boltzmann distribution,

$$\mathsf{P}(\mathbf{x}_1) = \frac{N}{V}\left(\frac{m\beta}{2\pi}\right)^{3/2} e^{-\beta m|\mathbf{v}_1-\mathbf{u}|^2/2} \tag{8}$$

where $\beta = 1/k_BT$.

One may also show that MPC dynamics satisfies an H theorem and that any initial velocity distribution will relax to the Maxwell–Boltzmann distribution [11]. Figure 2 shows simulation results for the velocity distribution function that confirm this result. In the simulation, the particles were initially uniformly distributed in the volume and had the same speed $|\mathbf{v}| = 1$ but different random directions. After a relatively short transient the distribution function adopts the Maxwell–Boltzmann form shown in the figure.

B. Galilean Invariance and Grid Shifting

In the description of MPC dynamics, the size of the collision cell was not specified. Given the number density $\bar{n} = N/V$ of the system, the cell size will control how many particles, on average, participate in the multiparticle collision event. This, in turn, controls the level of coarse graining of the system. As originally formulated, it was assumed that on average particles should free stream a distance comparable to or somewhat greater than the cell length in the time τ. Thus if $\bar{v} \sim (k_B T/m)^{1/2}$ is the mean thermal speed at temperature T so that the mean free path is $\Lambda = \bar{v}\tau$, the MPC cell length a should be chosen to be $a \approx \Lambda$. (Alternatively, for unit cell length, the temperature should be chosen so that this condition is satisfied.) Then, on average, particles will travel to neighboring cells where they will undergo multiparticle collisions with molecules they have not encountered on the previous time step. This will lead to rapid decorrelation of collision events. The dimensionless mean free path $\lambda = \Lambda/a$ is an important parameter in MPC dynamics.

In some applications it may not be possible or desirable to satisfy these conditions. For example, at low temperatures the mean free path may be very small so that particles may travel only a small fraction of a cell length. If the multiparticle collision rule were applied to such a system the collision events would be strongly correlated. Ihle and Kroll [14, 15] combined grid shifting with multiparticle collisions in order to improve the mixing properties of the dynamics for small λ. In the grid-shifting algorithm, prior to the application of the collision step, all particles in the system are shifted by a translation vector whose components are chosen from a uniform distribution on the interval $[-a/2, a/2]$. After collision, the particles are shifted back to their original positions. If ξ_i^s is the shifted cell coordinate of particle i, then the postcollision velocity in this modified MPC rule is

$$\mathbf{v}_i^* = \mathbf{V}_{\xi_i^s} + \hat{\omega}_{\xi_i^s}(\mathbf{v}_i - \mathbf{V}_{\xi_i^s}) \tag{9}$$

Thus even if the mean free path is small compared to the cell length, particle (or equivalently grid) shifting will cause particles to collide with molecules in nearby cells, thereby reducing the effects of locally correlated collision events in the same cell.

C. Multicomponent Systems

Multiparticle collision dynamics can be generalized to treat systems with different species. While there are many different ways to introduce multiparticle collisions that distinguish between the different species [16, 17], all such rules should conserve mass, momentum, and energy. We suppose that the N-particle system contains particles of different species $\alpha = A, B, \ldots$ with masses m_α. Different multiparticle collisions can be used to distinguish the interactions among the species. For this purpose we let $\mathbf{V}_\xi^{(\alpha)}$ denote the center of mass velocity of particles of species α in the cell ξ,[3]

$$\mathbf{V}_\xi^{(\alpha)} = \frac{1}{N_\xi{}^{(\alpha)}} \sum_{i=1}^{N_\xi{}^{(\alpha)}} \mathbf{v}_i^\alpha \tag{10}$$

[3]We use the same symbol α for chemical species and the rotation angle when confusion is unlikely.

where $N_\xi^{(\alpha)}$ is the number of particles of species α in cell ξ and $\mathbf{v}_i^\alpha$ is the velocity of particle i of species α. The center of mass velocity of all $N_\xi = \sum_\alpha N_\xi^{(\alpha)}$ particles in the cell ξ is given by

$$\mathbf{V}_\xi = \frac{\sum_\alpha N_\xi^{(\alpha)} m_\alpha \mathbf{V}_\xi^{(\alpha)}}{\sum_\alpha N_\xi^{(\alpha)} m_\alpha} \tag{11}$$

A sequence of multiparticle collisions may then be carried out. The first MPC event involves particles of all species and is analogous to that for a single component system: a rotation operator $\hat{\omega}$ is applied to every particle in a cell. The all-species collision step is

$$\mathbf{v}_i^{\alpha''} = \mathbf{V}_\xi + \hat{\omega}_\xi(\mathbf{v}_i^\alpha - \mathbf{V}_\xi) \tag{12}$$

where $\mathbf{v}_i^{\alpha''}$ is the velocity of particle i of species α after this step. The second set of MPC events involve only particles of the same species. The rotation operator $\hat{\omega}^\alpha$ is applied to each particle of species α in the cell. It changes from cell to cell and from species to species. The species-specific rotation operator

$$\mathbf{v}_i^{\alpha *} = \mathbf{V}_\xi^{(\alpha)''} + \hat{\omega}_\xi^\alpha(\mathbf{v}_i^{\alpha''} - \mathbf{V}_\xi^{''(\alpha)}) \tag{13}$$

is applied. Here $\mathbf{V}_\xi^{(\alpha)''}$ is the center of mass velocity of particles of species α after the all-species collision step. In this rule $\hat{\omega}_\xi$ is applied to all particles in the cell, but the $\hat{\omega}_\xi^\alpha$ are applied only on particles of species α.

The full MPC event consists of the concatenation of these independent steps and its net effect is

$$\mathbf{v}_i^* = \mathbf{V}_\xi + \hat{\omega}_\xi(\mathbf{V}_\xi^{(\alpha)} - \mathbf{V}_\xi) + \hat{\omega}_\xi^\alpha \hat{\omega}_\xi(\mathbf{v}_i^\alpha - \mathbf{V}_\xi^{(\alpha)}) \tag{14}$$

This collision dynamics clearly satisfies the conservation laws and preserves phase space volumes.

III. COLLISION OPERATORS AND EVOLUTION EQUATIONS

The algorithmic description of MPC dynamics given earlier outlined its essential elements and properties and provided a basis for implementations of the dynamics. However, a more formal specification of the evolution is required in order to make a link between the mesoscopic description and macroscopic laws that govern the system on long distance and time scales. This link will also provide us with expressions for the transport coefficients that enter the macroscopic laws. The first step in this analysis is formulation of the evolution equations in a form convenient for analysis.

The state of the entire system at time t is described by the N-particle phase space probability density function, $\mathsf{P}(\mathbf{x}^N, t)$. In MPC dynamics the time evolution of this function is given by the Markov chain,

$$\mathsf{P}(\mathbf{r}^N + \mathbf{v}^N\tau, \mathbf{v}^N, t+\tau) = e^{i\mathcal{L}_0\tau}\mathsf{P}(\mathbf{x}^N, t+\tau) = \hat{\mathcal{C}}\mathsf{P}(\mathbf{x}^N, t) \tag{15}$$

The displaced position on the left-hand side reflects the free streaming between collisions generated by the free streaming Liouville operator,

$$i\mathcal{L}_0 = \mathbf{v}^N \cdot \nabla_{\mathbf{r}^N} \tag{16}$$

while the collision operator $\hat{\mathcal{C}}$ on the far right-hand side is defined by

$$\hat{\mathcal{C}}\mathsf{P}(\mathbf{x}^N, t) = \frac{1}{|\Omega|^{N_c}} \sum_{\Omega^{N_c}} \int d\mathbf{v}'^N \mathsf{P}(\mathbf{r}^N, \mathbf{v}'^N, t) \prod_{i=1}^{N} \delta(\mathbf{v}_i - \mathbf{V}_\xi - \hat{\omega}_\xi[\mathbf{v}'_i - \mathbf{V}_\xi]) \tag{17}$$

In this equation $|\Omega|$ is the number of rotation operators in the set. Equation (15) is the MPC analogue of the Liouville equation for a system obeying Newtonian dynamics.

A similar form for the collision operator applies if the grid-shifting algorithm is employed [15]. Letting $\mathbf{b}$ be an index that specifies a specific choice of origin for the center of a cell, we may label each collision operator by this index: $\hat{\mathcal{C}}_{\mathbf{b}}$. Grid shifting involves a random translation of the grid by a vector distance whose components are drawn uniformly in the interval $[-a/2, a/2]$, as discussed earlier. Since this shift is independent of the system phase point and time, the grid-shift collision operator $\hat{\mathcal{C}}_G$ is a superposition of the $\hat{\mathcal{C}}_{\mathbf{b}}$, the fixed-grid collision operators, and is given by

$$\hat{\mathcal{C}}_G\mathsf{P}(\mathbf{x}^N, t) = \frac{1}{V_c}\int_{V_c} d\mathbf{b}\, \hat{\mathcal{C}}_{\mathbf{b}}\mathsf{P}(\mathbf{x}^N, t) \tag{18}$$

where $V_c = a^3$ is the volume of the cell.

In the discussion of kinetic equations and transport properties it is convenient to write the evolution equation in more compact form as

$$\mathsf{P}(\mathbf{x}^N, t+\tau) = \int d\mathbf{x}'^N \mathsf{L}(\mathbf{x}^N, \mathbf{x}'^N)\mathsf{P}(\mathbf{x}'^N, t) \equiv \hat{\mathsf{L}}\mathsf{P}(\mathbf{x}^N, t) \tag{19}$$

where

$$\hat{\mathsf{L}}\mathsf{P}(\mathbf{x}^N, t) = \frac{1}{|\Omega|^{N_c}} \sum_{\Omega^{N_c}} \int d\mathbf{x}'^N \prod_{i=1}^{N} \delta(\mathbf{v}_i - \mathbf{V}_\xi - \hat{\omega}_\xi[\mathbf{v}'_i - \mathbf{V}_\xi])\delta(\mathbf{r}'_i - (\mathbf{r}_i + \mathbf{v}_i\tau))\mathsf{P}(\mathbf{x}'^N, t) \tag{20}$$

These evolution equations form the starting point for the derivation of macroscopic kinetic equations, which we now consider. They also serve as the starting point for the proof of the H theorem. This proof can be found in Ref. 11.

IV. MACROSCOPIC LAWS AND TRANSPORT COEFFICIENTS

In addition to the fact that MPC dynamics is both simple and efficient to simulate, one of its main advantages is that the transport properties that characterize the behavior of the macroscopic laws may be computed. Furthermore, the macroscopic evolution equations can be derived from the full phase space Markov chain formulation. Such derivations have been carried out to obtain the full set of hydrodynamic equations for a one-component fluid [15, 18] and the reaction-diffusion equation for a reacting mixture [17]. In order to simplify the presentation and yet illustrate the methods that are used to carry out such derivations, we restrict our considerations to the simpler case of the derivation of the diffusion equation for a test particle in the fluid. The methods used to derive this equation and obtain the autocorrelation function expression for the diffusion coefficient are easily generalized to the full set of hydrodynamic equations.

The diffusion equation describes the evolution of the mean test particle density $\bar{n}(\mathbf{r}, t)$ at point $\mathbf{r}$ in the fluid at time t. Denoting the Fourier transform of the local density field by $\bar{n}_{\mathbf{k}}(t)$, in Fourier space the diffusion equation takes the form

$$\frac{\partial}{\partial t}\bar{n}_{\mathbf{k}}(t) = -Dk^2\bar{n}_{\mathbf{k}}(t) \tag{21}$$

where D is the diffusion coefficient. Our goal is to derive this equation from Eq. (19). We are not interested in the evolution of P itself but only the mean local test particle density, which is given by

$$\bar{n}_{\mathbf{k}}(t) = \int d\mathbf{x}^N\, n_{\mathbf{k}}(\mathbf{r}_1)\mathsf{P}(\mathbf{x}^N, t) \tag{22}$$

where the microscopic test particle density is $n_{\mathbf{k}}(\mathbf{r}_1) = e^{-\mathbf{k}\cdot\mathbf{r}_1}$. From Eq. (19) it follows that

$$\bar{n}_{\mathbf{k}}(t+\tau) = \int d\mathbf{x}^N n_{\mathbf{k}}(\mathbf{r}_1)\hat{\mathsf{L}}\mathsf{P}(\mathbf{x}^N, t) \tag{23}$$

This equation does not provide a closed expression for $\bar{n}_{\mathbf{k}}$ and to close it we use projection operator methods. We introduce a projection operator $\mathcal{P}$ defined by

$$\mathcal{P}h(\mathbf{x}^N) = n_{-\mathbf{k}}(\mathbf{r}_1)\mathsf{P}_0(\mathbf{x}^N)\langle n_{\mathbf{k}}(\mathbf{r}_1)n_{-\mathbf{k}}(\mathbf{r}_1)\rangle^{-1}\int d\mathbf{x}^N\, n_{\mathbf{k}}(\mathbf{r}_1)h(\mathbf{x}^N) \tag{24}$$

where $\mathsf{P}_0(\mathbf{x}^N)$ is the equilibrium density and $h(\mathbf{x}^N)$ is an arbitrary function of the phase space coordinates. The equilibrium density is stationary under MPC evolution so that $\mathsf{P}_0(\mathbf{x}^N) = \hat{\mathsf{L}}\mathsf{P}_0(\mathbf{x}^N)$. The angle brackets denote an average over the equilibrium distribution, $\langle\cdots\rangle = \int d\mathbf{x}^N \cdots \mathsf{P}_0(\mathbf{x}^N)$. For the test particle density $\langle n_{\mathbf{k}}(\mathbf{r}_1)n_{-\mathbf{k}}(\mathbf{r}_1)\rangle = 1$, so the projection operator takes the simpler form

$$\mathcal{P}h(\mathbf{x}^N) = n_{-\mathbf{k}}(\mathbf{r}_1)\mathsf{P}_0(\mathbf{x}^N)\int d\mathbf{x}^N\, n_{\mathbf{k}}(\mathbf{r}_1)h(\mathbf{x}^N) \tag{25}$$

which we use in the following derivation. The complement of $\mathcal{P}$ is $\mathcal{Q} = 1 - \mathcal{P}$. One may easily verify by direct computation that $\mathcal{P}$ and $\mathcal{Q}$ are projection operators. We note that $\mathcal{P}\mathsf{P}(\mathbf{x}^N,t) = n_{-\mathbf{k}}(\mathbf{r}_1)\mathsf{P}_0(\mathbf{x}^N)\bar{n}_{\mathbf{k}}(t)$. Using these results we may write Eq. (23) as

$$\bar{n}_{\mathbf{k}}(t+\tau) = \langle n_{\mathbf{k}}(\mathbf{r}_1)\hat{\mathsf{L}}n_{-\mathbf{k}}(\mathbf{r}_1)\rangle\bar{n}_{\mathbf{k}}(t) + \int d\mathbf{x}^N\, n_{\mathbf{k}}(\mathbf{r}_1)\hat{\mathsf{L}}\mathcal{Q}\mathsf{P}(\mathbf{x}^N,t) \tag{26}$$

The equation of motion for $\mathcal{Q}\mathsf{P}$ is

$$\mathcal{Q}\mathsf{P}(\mathbf{x}^N,t+\tau) = \mathcal{Q}\hat{\mathsf{L}}(\mathcal{P}+\mathcal{Q})\mathsf{P}(\mathbf{x}^N,t) \tag{27}$$

which may be solved formally by iteration to give

$$\mathcal{Q}\mathsf{P}(\mathbf{x}^N,t) = (\mathcal{Q}\hat{\mathsf{L}})^t\mathcal{Q}\mathsf{P}(\mathbf{x}^N,0) + \sum_{j=1}^{n}(\mathcal{Q}\hat{\mathsf{L}})^{j-1}\mathcal{Q}(\hat{\mathsf{L}}-1)\mathcal{P}\mathsf{P}(\mathbf{x}^N,t-j\tau) \tag{28}$$

where $t = n\tau$. If the system is prepared in an initial state where the phase space density is perturbed by displacing only the test particle density from its equilibrium value so that $\mathsf{P}(\mathbf{x}^N,0) = n_{\mathbf{k}}(\mathbf{r}_1)\mathsf{P}_0(\mathbf{x}^N)$, then $\mathcal{Q}\mathsf{P}(\mathbf{x}^N,0) = 0$ and we may drop the first term in Eq. (28). Taking this initial condition and substituting this result into Eq. (26) yields a non-Markovian evolution equation for the mean test particle density:

$$\begin{aligned}\bar{n}_{\mathbf{k}}(t+\tau) &= \langle n_{\mathbf{k}}(\mathbf{r}_1)\hat{\mathsf{L}}n_{-\mathbf{k}}(\mathbf{r}_1)\rangle\bar{n}_{\mathbf{k}}(t)\\ &\quad + \sum_{j=1}^{n}\langle n_{\mathbf{k}}(\mathbf{r}_1)(\hat{\mathsf{L}}-1)(\mathcal{Q}\hat{\mathsf{L}})^{j-1}\mathcal{Q}(\hat{\mathsf{L}}-1)n_{-\mathbf{k}}(\mathbf{r}_1)\rangle\bar{n}_{\mathbf{k}}(t-j\tau)\end{aligned} \tag{29}$$

For small wavevectors the test particle density is a nearly conserved variable and will vary slowly in time. The correlation function in the memory term in the above equation involves evolution, where this slow mode is projected out of the dynamics. Consequently it will decay much more rapidly. In this case we may make a Markovian approximation and write the evolution equation as

$$\bar{n}_{\mathbf{k}}(t+\tau) = (1+K(k))\bar{n}_{\mathbf{k}}(t) \tag{30}$$

where

$$\begin{aligned} K(k) &= \langle n_{\mathbf{k}}(\mathbf{r}_1)(\hat{\mathsf{L}}-1)n_{-\mathbf{k}}(\mathbf{r}_1)\rangle \\ &\quad + \sum_{j=1}^{\infty}\langle n_{\mathbf{k}}(\mathbf{r}_1)(\hat{\mathsf{L}}-1)(\mathcal{Q}\hat{\mathsf{L}})^{j-1}\mathcal{Q}(\hat{\mathsf{L}}-1)n_{-\mathbf{k}}(\mathbf{r}_1)\rangle \end{aligned} \tag{31}$$

The last step in the derivation is the calculation of $K(k)$ in the small wavevector limit and the connection of this quantity to the diffusion coefficient. In the Appendix we show that $K(k)$ can be written

$$K(k) = \langle n_{\mathbf{k}}(\mathbf{r}_1)(\hat{\mathsf{L}}-1)n_{-\mathbf{k}}(\mathbf{r}_1)\rangle + \sum_{j=1}^{\infty}\langle f_{\mathbf{k}}(\mathbf{x}_1)\hat{\mathsf{L}}^j\tilde{f}_{-\mathbf{k}}(\mathbf{x}_1)\rangle \tag{32}$$

where

$$\begin{aligned} f_{\mathbf{k}}(\mathbf{x}_1) &= n_{\mathbf{k}}(S(\mathbf{x}_1,\tau)) - \langle n_{\mathbf{k}}(S(\mathbf{x}_1,\tau))n_{-\mathbf{k}}(\mathbf{x}_1)\rangle n_{\mathbf{k}}(\mathbf{x}_1) \\ \tilde{f}_{-\mathbf{k}}(\mathbf{x}_1) &= n_{-\mathbf{k}}(\mathbf{x}_1) - n_{-\mathbf{k}}(S(\mathbf{x}_1,\tau))\langle n_{\mathbf{k}}(S(\mathbf{x}_1,\tau))n_{-\mathbf{k}}(\mathbf{x}_1)\rangle \end{aligned} \tag{33}$$

and $S(\mathbf{x}^N,\tau)$ stands for the phase point at time τ whose value at time zero was $\mathbf{x}^N$. To obtain a more explicit expression for this quantity we may substitute the expression for $n_{\mathbf{k}}$. We find

$$f_{\mathbf{k}}(\mathbf{x}_1) = ik\tau\hat{\mathbf{k}}\cdot\mathbf{v}_1 + \mathcal{O}(k^2), \quad \tilde{f}_{-\mathbf{k}}(\mathbf{x}_1) = ik\tau\hat{\mathbf{k}}\cdot\mathbf{v}_1 + \mathcal{O}(k^2) \tag{34}$$

In addition,

$$\langle n_{\mathbf{k}}(\mathbf{r}_1)(\hat{\mathsf{L}}-1)n_{-\mathbf{k}}(\mathbf{r}_1)\rangle = -\tfrac{1}{2}k^2\tau^2\langle(\hat{\mathbf{k}}\cdot\mathbf{v}_1)^2\rangle + \mathcal{O}(k^4) \tag{35}$$

Thus, to lowest order in k,

$$K(k) = -\tfrac{1}{2}k^2\tau^2\langle(\hat{\mathbf{k}}\cdot\mathbf{v}_1)^2\rangle - \sum_{j=1}^{\infty}k^2\tau^2\langle(\hat{\mathbf{k}}\cdot\mathbf{v}_1)\hat{\mathsf{L}}^j(\hat{\mathbf{k}}\cdot\mathbf{v}_1)\rangle \tag{36}$$

which we see is $\mathcal{O}(k^2)$.

To derive the diffusion equation we return to Eq. (30), which we write as

$$\bar{n}_{\mathbf{k}}(t+\tau) = e^{\tau\partial/\partial t}\bar{n}_{\mathbf{k}}(t) = (1+K(k))\bar{n}_{\mathbf{k}}(t) \tag{37}$$

making use of the time translation operator. This yields the operator identity $e^{\tau\partial/\partial t} = 1 + K(k)$, whose logarithm is $\tau\,\partial/\partial t = \ln(1 + K(k)) = K(k) + \mathcal{O}(k^4)$. Applying this expression for the operator to $\bar{n}_{\mathbf{k}}(t)$, we obtain

$$\frac{\partial}{\partial t}\bar{n}_{\mathbf{k}}(t) = \frac{K(k)}{\tau}\bar{n}_{\mathbf{k}}(t) \tag{38}$$

which has the same form as the diffusion equation (21) if we identify $k^2 D = -K(k)/\tau$. This identification yields the discrete Green–Kubo expression for the diffusion coefficient,

$$D = \frac{1}{2}\tau\langle v_{1z}^2\rangle + \sum_{j=1}^{\infty}\tau\langle v_{1z}\hat{\mathsf{L}}^j v_{1z}\rangle \tag{39}$$

where we have chosen $\hat{\mathbf{k}}$ to lie along the z-direction. The formula has the same structure as a trapezoidal rule approximation to the usual expression for D as the time integral of the velocity correlation function.

A. Calculation of *D*

In order to compute the discrete Green–Kubo expression for D we must evaluate correlation function expressions of the form $\langle v_{1z}\hat{\mathsf{L}}^j v_{1z}\rangle$. Consider

$$r_D = \frac{\langle v_{1z}\hat{\mathsf{L}}v_{1z}\rangle}{\langle v_{1z}v_{1z}\rangle} = \frac{1}{|\Omega|}\sum_{\omega}\sum_{n=1}^{\infty}\frac{\gamma^n}{n!}e^{-\gamma}\int d\mathbf{v}_1\,\frac{v_{1z}v_{1z}^*\phi(\mathbf{v}_1)}{\langle v_{1z}v_{1x}\rangle} \tag{40}$$

where we have used the fact that the particles are Poisson distributed in the cells with $\gamma = \bar{n}a^3$, the average number of particles in a cell for a system with number density $\bar{n}$. Here v_{1z}^* denotes the post collision velocity given in Eq. (2) determined by the rotation operator ω. Using rotations by $\pm\alpha$ about a randomly chosen axis (see Eq. (3)), this integral may be evaluated to give

$$r_D = \frac{1}{3\gamma}(2(1 - \cos\alpha)(1 - e^{-\gamma}) + \gamma(1 + 2\cos\alpha)) \tag{41}$$

Assuming a single relaxation time approximation, the diffusion coefficient takes the form

$$D \approx -\frac{1}{2}\tau\langle v_z v_z\rangle + \tau\langle v_z v_z\rangle\sum_{j=0}^{\infty} r_D{}^j = \tau\frac{\langle v_x v_x\rangle(1 + r_D)}{2(1 - r_D)} \tag{42}$$

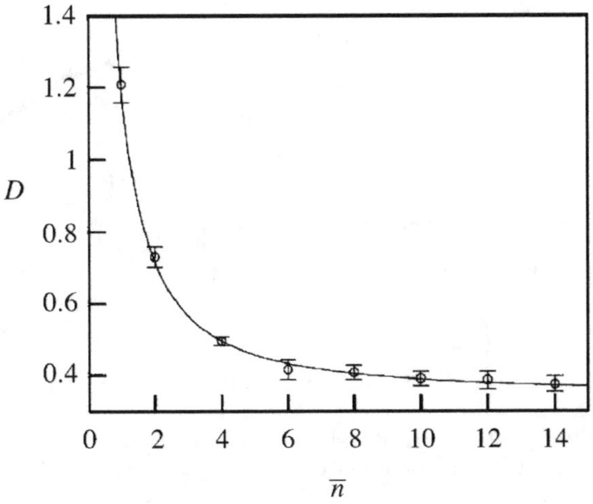

Figure 3. Comparison of the simulated diffusion coefficient (⊙) with the theoretical value D_0 (solid line). The simulation parameters are $\alpha = \pi/2$, $L/a = 100$, $\tau = 1$, $m = 1$, and $k_BT = \frac{1}{3}$.

Substituting r_D into Eq. (42), the diffusion coefficient is given by

$$D = D_0 \equiv \frac{k_BT\tau}{2m}\left(\frac{3\gamma}{(\gamma - 1 + e^{-\gamma})(1 - \cos\alpha)} - 1\right) \tag{43}$$

This analytic formula with $\alpha = \pi/2$ is compared with the simulation results in Fig. 3, where it is seen that it provides an excellent approximation to the simulation results over all of the physically interesting density range.

For the parameters used to obtain the results in Fig. 3, $\lambda \sim 0.6$; so the mean free path is comparable to the cell length. If $\lambda \ll 1$, the correspondence between the analytical expression for D in Eq. (43) and the simulation results breaks down. Figure 4a plots the deviation of the simulated values of D from D_0 as a function of λ. For small λ values there is a strong discrepancy, which may be attributed to correlations that are not accounted for in D_0, which assumes that collisions are uncorrelated in the time τ. For very small mean free paths, there is a high probability that two or more particles will occupy the same collision volume at different time steps, an effect that is not accounted for in the geometric series approximation that leads to D_0. The origins of such corrections have been studied [19–22].

The last issue we address concerns the existence of long-time tails in the discrete-time velocity correlation function. The diffusion coefficient can be written in terms of the velocity correlation function as

$$D = \frac{k_BT}{m}\sum_{j=0}^{\infty}{}' \tau \frac{\langle v_{1z}\hat{\mathcal{L}}^j v_{1z}\rangle}{\langle v_{1z}v_{1z}\rangle} \equiv \frac{k_BT\tau}{m}\sum_{j=0}^{\infty}{}' C_v(j\tau) \tag{44}$$

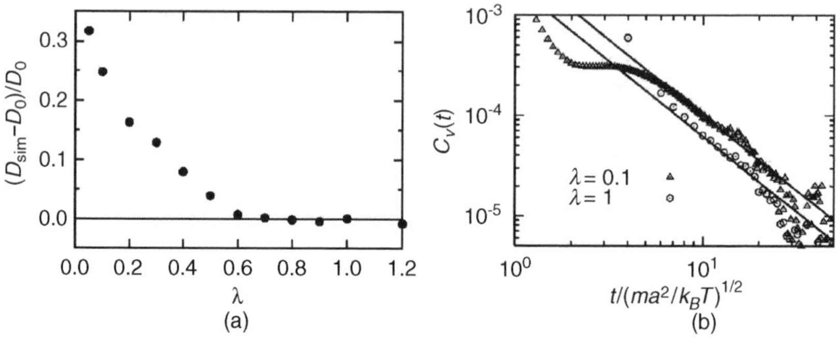

Figure 4. (a) Deviation of the diffusion coefficient from the theoretical prediction in Eq. (43) as a function of λ for $\alpha = 30°$, $\gamma = 5$, and $L/a = 20$. (b) Velocity correlation function versus dimensionless time for $\lambda = 1$ and 0.1. The solid lines are the theoretical prediction of the long-time decay in Eq. (45). The other parameters are the same as in panel (a). (From Ref. 20.)

where the prime on the sum signifies that the $j = 0$ term should be multiplied by a factor of $\frac{1}{2}$ and $C_v(j\tau)$ is the discrete-time velocity correlation function. The test particle density couples to other nearly conserved small wavevector modes in the fluid, leading to collective fluid contributions to the velocity correlation function and diffusion coefficient. The most important of such modes is the fluid velocity field $\mathbf{u}_\mathbf{k}$ and mode coupling to nonlinear products of the form $n_\mathbf{q}\mathbf{u}_{\mathbf{k}-\mathbf{q}}$ leads to an algebraic decay of the velocity correlation function of the form

$$C_v(t) \sim \frac{2}{3\bar{n}}(4\pi(D+\nu)t)^{-3/2} \tag{45}$$

where ν is the kinematic viscosity that characterizes the decay of the transverse fluid velocity field. Since this relation is a consequence of the coupling between the single particle and collective fluid modes, and the existence of conservation laws, one expects that MPC dynamics will also yield such nonanalytic long-time decay. This is indeed the case as is shown in Fig. 4b that graphs $C_v(t)$ versus t on a double logarithmic plot for two values of the dimensionless mean free path [20]. Linear long-time regions with the correct slope are evident in the figure. The departure from linear behavior at very long times is due to finite size effects. Long-time tails in correlation functions in the context of MPC dynamics have been studied in some detail by Ihle and Kroll [23].

V. HYDRODYNAMIC EQUATIONS

The hydrodynamic equations can be derived from the MPC Markov chain dynamics using projection operator methods analogous to those used to obtain the diffusion equation [18, 23–26]. The hydrodynamic equations describe the dynamics of local densities corresponding to the conserved mass, momentum, and energy variables on long distance and time scales. The Fourier transforms of the microscopic variables corresponding to these fields are

$$n_{\mathbf{k}} = \int d\mathbf{r}\, e^{i\mathbf{k}\cdot\mathbf{r}} n(\mathbf{r}) = \sum_{i=1}^{N} e^{i\mathbf{k}\cdot\mathbf{r}_i} \tag{46}$$

$$\mathbf{u}_{\mathbf{k}} = \sum_{\xi} e^{i\mathbf{k}\cdot\xi}\mathbf{u}(\xi) \equiv \sum_{\xi} e^{i\mathbf{k}\cdot\xi} \sum_{i=1}^{N} \mathbf{v}_i \theta(\tfrac{1}{2} - |\mathbf{r}_i - \xi|) \tag{47}$$

$$\epsilon_{\mathbf{k}} = \sum_{\xi} e^{i\mathbf{k}\cdot\xi}\epsilon(\xi) \equiv \sum_{\xi} e^{i\mathbf{k}\cdot\xi} \sum_{i=1}^{N} \frac{m}{2} v_i^2 \theta(\tfrac{1}{2} - |\mathbf{r}_i - \xi|) \tag{48}$$

Here θ is the Heaviside function. The projection operator formalism must be carried out in matrix from and in this connection it is useful to define the orthogonal set of variables, $\{n_{\mathbf{k}}, \mathbf{u}_{\mathbf{k}}, s_{\mathbf{k}}\}$, where the entropy density is $s_{\mathbf{k}} = \epsilon_{\mathbf{k}} - C_v T n_{\mathbf{k}}$ with C_v the specific heat. In terms of these variables the linearized hydrodynamic equations take the form

$$\partial_t n_{\mathbf{k}} = i\mathbf{k} \cdot \mathbf{u}_{\mathbf{k}} \tag{49}$$

$$\partial_t \mathbf{u}_{\mathbf{k}} = i\mathbf{k} \cdot \left[k_B T \rho_{\mathbf{k}} + \frac{s_{\mathbf{k}}}{c_v}\right] - \frac{\eta}{\rho}\left[\mathbf{k}\mathbf{k} - \tfrac{1}{3}k^2\mathbf{1}\right] : \mathbf{u}_{\mathbf{k}} - \frac{\eta_b}{\rho}\mathbf{k}\mathbf{k} : \mathbf{u}_{\mathbf{k}} \tag{50}$$

$$\partial_t s_{\mathbf{k}} = k_B T i\mathbf{k} \cdot \mathbf{u}_{\mathbf{k}} - \frac{\lambda}{\rho} k^2 s_{\mathbf{k}} \tag{51}$$

where $\rho = m\bar{n}$ is the mean mass density and η, η_b, and λ are the shear viscosity, bulk viscosity, and thermal conductivity coefficients, respectively. As in the derivation of the diffusion equation described earlier a projection operator may be constructed that projects the full MPC Markov chain dynamics onto the conserved fields. A Markovian approximation that assumes the conserved fields are slowly decaying functions for small wavenumbers leads to Eq. (51) along with discrete-time Green–Kubo expressions for the transport coefficients [18, 23–26]. Projection operator methods are not the only way to obtain the macroscopic laws and transport coefficients. They have also been obtained through the use of Chapman–Enskog methods [11] and a kinetic theory method based on the computation of moments of the local equilibrium distribution function [27–29].

These derivations yield general expressions for the transport coefficients that may be evaluated by simulating MPC dynamics or approximated to obtain analytical expressions for their values. The shear viscosity is one of the most important transport properties for studies of fluid flow and solute molecule dynamics. The discrete time Green–Kubo expression for the viscosity is the sum of kinetic and collisional contributions [26],

$$\begin{aligned}\eta &= \frac{\rho\tau}{Nk_BT}\sum_{j=0}^{\infty}{}' \langle \sigma_{xy}^{\text{kin}}(0)\sigma_{xy}^{\text{kin}}(j\tau)\rangle + \frac{\rho\tau}{Nk_BT}\sum_{j=0}^{\infty}{}' \langle \sigma_{xy}^{\text{col}}(0)\sigma_{xy}^{\text{col}}(j\tau)\rangle \\ &= \eta^{\text{kin}} + \eta^{\text{col}}\end{aligned} \tag{52}$$

where the kinetic and collisional stress tensors are

$$\sigma_{xy}^{\text{kin}} = -\sum_{i=1}^{N} v_{ix}v_{iy}, \quad \sigma_{xy}^{\text{col}} = -\frac{1}{\tau}\sum_{i=1}^{N} v_{ix}B_{iy} \tag{53}$$

with

$$B_{iy}(j\tau) = \xi_{iy}^{s}((j+1)\tau) - \xi_{iy}^{s}(j\tau) - \tau v_{iy}(j\tau) \tag{54}$$

The collisional contribution arises from grid shifting and accounts for effects on scales where the dimensionless mean free path is small, $\lambda \ll 1$. The discrete Green–Kubo derivation leading to Eq. (52) involves a number of subtle issues that have been discussed by Ihle, Tüzel, and Kroll [26].

For the collision rule using rotations by $\pm\alpha$ about a randomly chosen axis, these expressions may be evaluated approximately to give [26]

$$\eta^{\text{kin}} = \frac{k_BT\tau\rho}{2m}\left(\frac{5\gamma - (\gamma - 1 + e^{-\gamma})(2 - \cos\alpha - \cos 2\alpha)}{(\gamma - 1 + e^{-\gamma})(2 - \cos\alpha - \cos 2\alpha)}\right) \tag{55}$$

and

$$\eta^{\text{col}} = \frac{m}{18a\tau}(\gamma - 1 + e^{-\gamma})(1 - \cos\alpha) \tag{56}$$

Identical results were obtained using the kinetic theory moment method by Kikuchi, Pooley, Ryder, and Yeomans [28, 29].

These expressions for the shear viscosity are compared with simulation results in Fig. 5 for various values of the angle α and the dimensionless mean free path λ. The figure plots the dimensionless quantity $(\nu/\lambda)(\tau/a^2)$ and for fixed γ and α we see that $(\nu^{\text{kin}}/\lambda)(\tau/a^2) \sim \text{const}\,\lambda$ and $(\nu^{\text{col}}/\lambda)(\tau/a^2) \sim \text{const}/\lambda$. Thus we see in Fig. 5b that the kinetic contribution dominates for large λ since particles free stream distances greater than a cell length in the time τ; however, for small λ the collisional contribution dominates since grid shifting is important and is responsible for this contribution to the viscosity.

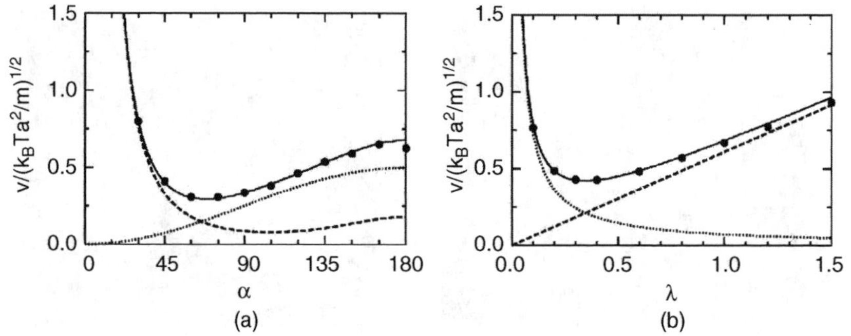

Figure 5. (a) Viscosity as a function of the angle α for $\lambda = 0.2$ and $\gamma = 10$. (b) Viscosity versus λ for $\alpha = 130°$ and $\gamma = 5$. The solid circles are simulation results. The dotted lines are η^{col} and the dashed lines are η^{kin}. The solid lines are the total viscosity $\eta = \eta^{kin} + \eta^{col}$. (From Ref. 20.)

A. Simulation of Hydrodynamic Flows

Since MPC dynamics yields the hydrodynamic equations on long distance and time scales, it provides a mesoscopic simulation algorithm for investigation of fluid flow that complements other mesoscopic methods. Since it is a particle-based scheme it incorporates fluctuations, which are essential in many applications. For macroscopic fluid flow averaging is required to obtain the deterministic flow fields. In spite of the additional averaging that is required the method has the advantage that it is numerically stable, does not suffer from lattice artifacts in the structure of the Navier–Stokes equations, and boundary conditions are easily implemented.

Since hydrodynamic flow fields are described correctly, the method is also useful in applications to rheology. As an example we consider the three-dimensional flow of a fluid between planar walls around a spherical obstacle studied by Allahyarov and Gompper [30]. The flow was generated by imposing a gravitational field of dimensionless strength $g^* = ga/\sqrt{k_BT}$. Figure 6 shows the flow field around the sphere in the middle z-plane of the system for two values of the Reynolds number, $\text{Re} = 2Rv_{max}/\nu$, where R is the sphere radius and v_{max} is the maximum fluid velocity. As the Reynolds number increases, symmetric vortices develop behind the obstacle and the length of the steady wake increases in a manner that agrees with experiment and theoretical predictions. Similar investigations of flow around a cylinder have been carried out [31].

VI. REACTIVE MPC DYNAMICS

Reactive systems form the core of chemistry and most biological functions are based on the operation of complex biochemical reaction networks. In dealing

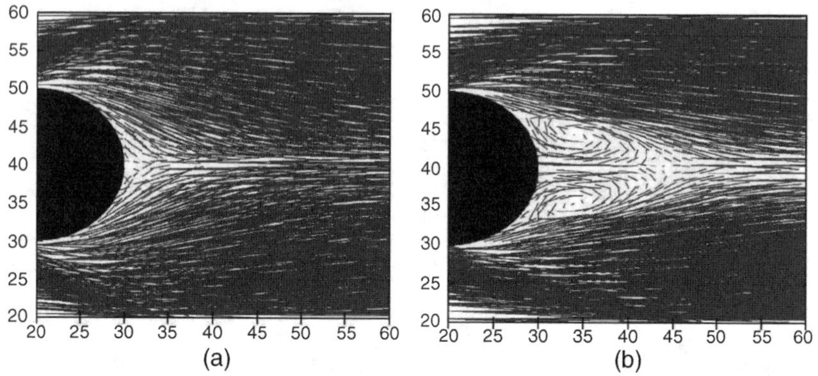

Figure 6. Flow around a sphere. The system size is $50 \times 25 \times 25$ with $\gamma = 8$ particles per cell. The gravitational field strength was $g^* = 0.005$ and the rotation angle for MPC dynamics was $\alpha = \pm\pi/2$. Panel (a) is for a Reynolds number of Re = 24 corresponding to $\lambda = 1.8$ while panel (b) is the flow for Re = 76 and $\lambda = 0.35$. (From Ref. 30.)

with reactions, two features typically come into play: the reactive event itself and the diffusional or other mixing processes that are responsible for bringing the reagents together. There is a large literature on diffusion-influenced reaction dynamics and its description by reaction-diffusion equations of the form

$$\frac{\partial}{\partial t}\mathbf{c}(\mathbf{r}, t) = \mathbf{R}(\mathbf{c}(\mathbf{r}, t)) + \mathbf{D}\nabla^2\mathbf{c}(\mathbf{r}, t) \tag{57}$$

dating from the work of Smoluchowski [32]. Here $\mathbf{c} = (c_1, c_2, \ldots, c_s)$ is a vector of the concentrations of the s chemical species, $\mathbf{R}$ is a vector-valued function of the reaction rates, which is often determined from mass action kinetics, and $\mathbf{D}$ is the matrix of diffusion coefficients. If the system is well mixed, then the simple chemical rate equations,

$$\frac{d}{dt}\mathbf{c}(t) = \mathbf{R}(\mathbf{c}(t)) \tag{58}$$

can be used to describe the evolution of the chemical concentrations. These macroscopic chemical kinetic equations are the analogues of the Navier–Stokes equations for nonreactive fluid flow. (Reaction can also be coupled to fluid flow but we shall not consider this here.)

There are situations where such a macroscopic description of reaction dynamics will break down. For instance, biochemical reactions in the cell may involve only small numbers of molecules of certain species that participate in the mechanism. An example is gene transcription where only tens of free

RNApolymerase molecules are involved in the process. If the system is well stirred so that spatial degrees of freedom play no role, birth–death master equation approaches have been used to describe such reacting systems [33, 34]. The master equation can be simulated efficiently using Gillespie's algorithm [35]. However, if spatial degrees of freedom must be taken into account, then the construction of algorithms is still a matter of active research [36–38].

Spatially distributed reacting systems can be described by a generalization of MPC dynamics that incorporates stochastic birth–death reactive events in the collision step. For simplicity, consider a single reaction among a set of s species X_α, $(\alpha = 1, \ldots, s)$:

$$\nu_1 X_1 + \nu_2 X_2 + \cdots + \nu_s X_s \underset{k_r}{\overset{k_f}{\rightleftharpoons}} \bar{\nu}_1 X_1 + \bar{\nu}_2 X_2 + \cdots + \bar{\nu}_s X_s \tag{59}$$

Here ν_α and $\bar{\nu}_\alpha$ are the stoichiometric coefficients for the reaction. The formulation is easily extended to treat a set of coupled chemical reactions. Reactive MPC dynamics again consists of free streaming and collisions, which take place at discrete times τ. We partition the system into cells in order to carry out the reactive multiparticle collisions. The partition of the multicomponent system into collision cells is shown schematically in Fig. 7. In each cell, independently of the other cells, reactive and nonreactive collisions occur at times τ. The nonreactive collisions can be carried out as described earlier for multi-component systems. The reactive collisions occur by birth–death stochastic rules. Such rules can be constructed to conserve mass, momentum, and energy. This is especially useful for coupling reactions to fluid flow. The reactive collision model can also be applied to far-from-equilibrium situations, where certain species are held fixed by constraints. In this case conservation laws

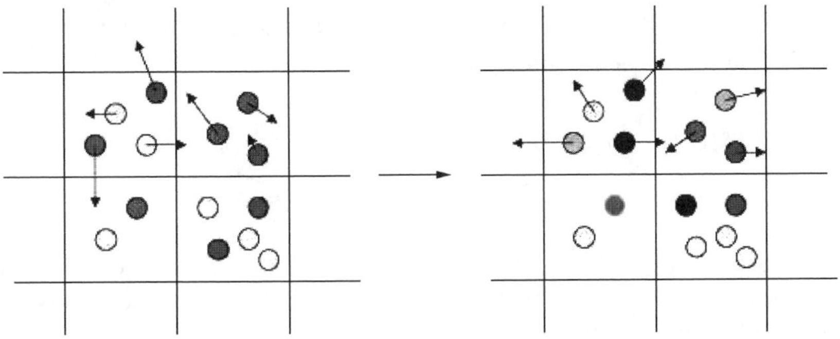

Figure 7. Schematic representation of collision cells for reactive MPC dynamics. Each cell contains various numbers of the different species. The species numbers change in the cells as a result of chemical reactions.

may be violated. Here we describe a simple situation where the reaction is a "coloring" process that is independent of the velocities of the particles.

The reaction transition probabilities in a cell are determined by birth–death probabilistic rules that model the changes in the species particle numbers in the reaction mechanism. Letting $\mathbf{N} = (N^{(1)}, N^{(2)}, \ldots, N^{(s)})$ be the set of all instantaneous cell species numbers, the reaction transition matrix can be written

$$\begin{aligned} W^R(\mathbf{N}|\mathbf{N}') = {} & k_f \prod_{\alpha=1}^{s} \frac{N^{(\alpha)'}!}{(N^{(\alpha)'} - \nu_{(\alpha)})!} \delta_{N^{(\alpha)},N^{(\alpha)'}+\Delta_\alpha} \\ & + k_r \prod_{\alpha=1}^{s} \frac{N^{(\alpha)'}!}{(N^{(\alpha)'} - \bar{\nu}_\alpha)!} \delta_{N^{(\alpha)},N^{(\alpha)'}-\Delta_\alpha} \\ & + (1 - (r_f(\mathbf{N}') + r_r(\mathbf{N}')) \prod_{\alpha=1}^{s} \delta_{N^{(\alpha)},N^{(\alpha)'}} \end{aligned} \tag{60}$$

where

$$r_f(\mathbf{N}) = k_f \prod_{\alpha=1}^{s} \frac{N^{(\alpha)}!}{(N^{(\alpha)} - \nu_\alpha)!}, \quad r_r(\mathbf{N}) = k_r \prod_{\alpha=1}^{s} \frac{N^{(\alpha)}!}{(N^{(\alpha)} - \bar{\nu}_\alpha)!} \tag{61}$$

and $\Delta_\alpha = \bar{\nu}_\alpha - \nu_\alpha$ is the change in the particle number for species α in the reaction. The structure of the reaction transition matrix accounts for the combinatorial choice of reaction partners in the cell at time τ. Reactions are carried out with probabilities determined by the reaction transition matrix. The full collision step then consists of birth–death reaction and velocity changes by multiparticle collisions.

As a simple illustration, consider the irreversible autocatalytic reaction $A + 2B \xrightarrow{k_f} 3B$ [39]. The reaction transition matrix is

$$\begin{aligned} W^R(\mathbf{N}|\mathbf{N}') = {} & k_f N^{(A)'} N^{(B)'} (N^{(B)'} - 1) \delta_{N^{(A)},N^{(A)'}-1} \delta_{N^{(B)},N^{(B)'}+1} \\ & + (1 - k_f N^{(A)'} N^{(B)'} (N^{(B)'} - 1)) \delta_{N^{(A)},N^{(A)'}} \delta_{N^{(B)},N^{(B)'}} \end{aligned} \tag{62}$$

If nonreactive MPC collisions maintain an instantaneous Poissonian distribution of particles in the cells, it is easy to verify that reactive MPC dynamics yields the reaction-diffusion equation,

$$\frac{\partial}{\partial t} \bar{n}_A(\mathbf{r}, t) = -k_f \bar{n}_A \bar{n}_B^2 + D\nabla^2 \bar{n}_A \tag{63}$$

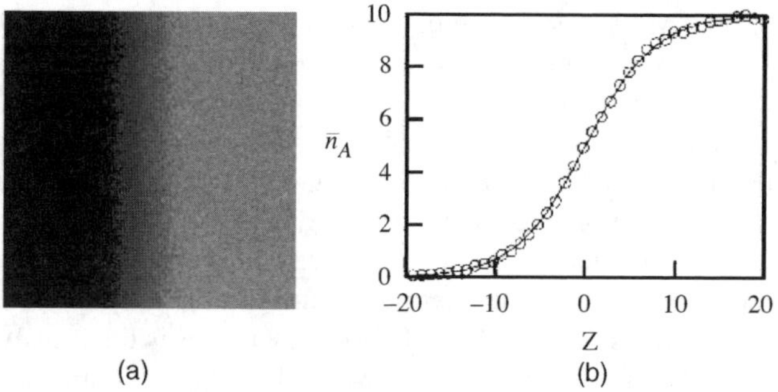

Figure 8. (a) Species density fields. Black denotes high B density and light gray denotes high A density. (b) Comparison of the reactive MPC front profile (circles) with the theoretical profile in Eq. (64) (solid line). The rate constant is $k_f = 0.0005$ and $k_BT = \frac{1}{3}$.

Starting from an initial state where half the system has species A and the other half B, a reaction front will develop as the autocatalyst B consumes the fuel A in the reaction. The front will move with velocity c. The reaction-diffusion equation can be solved in a moving frame, $z = x - ct$, to determine the front profile and front speed,

$$\bar{n}_A(z) = \bar{n}_0(1 + e^{-cz/D})^{-1} \tag{64}$$

where $\bar{n}_A + \bar{n}_B = \bar{n}_0$ and the front speed is $c = (Dk\bar{n}_0^2/2)^{1/2}$. Figure 8 shows the results of front propagation using reactive MPC dynamics for a system where reaction is a very slow process compared to diffusion. In this limit we expect that the reaction-diffusion equation will provide an accurate description of the front. This is indeed the case as the figure shows. In other parameter regimes the reaction-diffusion equation description breaks down.

Reactive MPC dynamics should prove most useful when fluctuations in spatially distributed reactive systems are important, as in biochemical networks in the cell, or in situations where fluctuating reactions are coupled to fluid flow.

VII. HYBRID MPC–MD DYNAMICS

Multiparticle collision dynamics can be combined with full molecular dynamics in order to describe the behavior of solute molecules in solution. Such hybrid MPC–MD schemes are especially useful for treating polymer and colloid dynamics since they incorporate hydrodynamic interactions. They are also useful for describing reactive systems where diffusive coupling among solute species is important.

Hybrid MPC–MD schemes can be constructed in a number of different ways depending on how the solute molecules couple to the fictitious MPC solvent molecules. In one such scheme the solute molecules are assumed to interact with the solvent through an intermolecular potential [18]. More specifically, consider a system with N_s solute molecules and N_b solvent or bath particles. Let $V_s(\mathbf{r}^{N_s})$ be the intermolecular potential among the N_s solute molecules and $V_{sb}(\mathbf{r}^{N_s}, \mathbf{r}^{N_b})$ the interaction potential between the solute and bath particles. There are no bath–bath particle intermolecular forces since these are accounted for by MPC dynamics. The hybrid MPC–MD dynamics is then easily generated by replacing the free streaming step in Eq. (1) by streaming in the intermolecular potential, $V(\mathbf{r}^{N_s}, \mathbf{r}^{N_b}) = V_s(\mathbf{r}^{N_s}) + V_{sb}(\mathbf{r}^{N_s}, \mathbf{r}^{N_b})$, which is generated by the solution of Newton's equations of motion,

$$\dot{\mathbf{r}}_i = \mathbf{v}_i, \quad m_i \dot{\mathbf{v}}_i = -\frac{\partial V}{\partial \mathbf{r}_i} = \mathbf{F}_i \tag{65}$$

Multiparticle collisions are carried out at time intervals τ as described earlier. We can write the equation of motion for the phase space probability density function as a simple generalization of Eq. (15) by replacing the free-streaming operator with streaming in the intermolecular potential. We find

$$e^{i\mathcal{L}\tau} \mathsf{P}(\mathbf{x}^N, t + \tau) = \hat{\mathcal{C}} \mathsf{P}(\mathbf{x}^N, t) \tag{66}$$

The propagator $\exp(i\mathcal{L}\tau)$ on the left-hand side reflects the dynamics generated by the Liouville operator,

$$i\mathcal{L} = \mathbf{v}^N \cdot \nabla_{\mathbf{r}^N} + \mathbf{F} \cdot \nabla_{\mathbf{v}^N} \tag{67}$$

that occurs between multiparticle collisions. This hybrid dynamics satisfies the conservation laws and preserves phase space volumes.

Hybrid MPC–MD schemes may be constructed where the mesoscopic dynamics of the bath is coupled to the molecular dynamics of solute species without introducing explicit solute–bath intermolecular forces. In such a hybrid scheme, between multiparticle collision events at times τ, solute particles propagate by Newton's equations of motion in the absence of solvent forces. In order to couple solute and bath particles, the solute particles are included in the multiparticle collision step [40]. The above equations describe the dynamics provided the interaction potential is replaced by $V_s(\mathbf{r}^{N_s})$ and interactions between solute and bath particles are neglected. This type of hybrid MD–MPC dynamics also satisfies the conservation laws and preserves phase space volumes. Since bath particles can penetrate solute particles, specific structural solute–bath effects cannot be treated by this rule. However, simulations may be more efficient since the solute–solvent forces do not have to be computed.

VIII. SIMULATING REAL SYSTEMS WITH MPC DYNAMICS

Having presented the basic elements of MPC dynamics and the statistical mechanical methods used to derive the macroscopic laws and transport coefficients, we now show how MPC dynamics can be used to study a variety of phenomena. Before doing this we provide some guidelines that can be used to determine the model parameters that are appropriate for simulations. Since MPC dynamics is a self-contained dynamical scheme, the values of transport properties and system conditions can be tuned by varying the parameters that specify the system state and by changing the nature of the collision dynamics. This can be done by changing the cell size a or the mean number of particles per cell γ, the precise form of the collision rule (e.g., the angle α or its distribution) the dimensionless mean free path λ, and so on. Similarly, the macroscopic behavior of real systems is determined by the values of state parameters, such as density and temperature, and the values of transport coefficients that enter into the macroscopic evolution equations. Often what matters most in observing a particular type of behavior, say, fluid turbulence or the swimming motion of a bacterium in water, are the values of these dimensionless numbers. The ability of MPC dynamics to mimic the behavior of real systems hinges on being able to control the values of various dimensionless numbers that are used to characterize the system. A full discussion of such dimensionless numbers for MPC dynamics has been given by Padding and Louis [41] and Hecht et al. [42], which we summarize here.

The Schmidt number $\mathrm{Sc} = \nu/D_0$, where $\nu = \eta/\rho$ is the kinematic viscosity and D_0 is the diffusion coefficient, is the ratio of the rate of diffusive momentum transfer to the rate of diffusive mass transfer. In gases this number is of order unity since momentum transport occurs largely through mass transport. However, in liquids this number is large since collisional effects control momentum transport. Using the results in Eqs. (55), (56), and (43) to compute the Schmidt number, we obtain the estimate $\mathrm{Sc} \approx \frac{1}{3} + 1/18\lambda^2$ if we assume that the cell occupancy is large enough to drop the $\exp(-\gamma)$ terms in the transport coefficients and use the $\alpha = \pi/2$ collision rule. Thus small dimensionless mean free paths are needed to simulate liquid-like regimes of the Schmidt number. For example, for $\lambda = 0.1$ we have $\mathrm{Sc} \approx 6$.

The Reynolds number $\mathrm{Re} = v\ell/\nu$, where v and ℓ are the characteristic velocity and length for the problem, respectively, gauges the relative importance of inertial and viscous forces in the system. Insight into the nature of the Reynolds number for a spherical particle with radius ℓ in a flow with velocity v may be obtained by expressing it in terms of the Stokes time, $\tau_s = \ell/v$, and the kinematic time, $\tau_\nu = \ell^2/\nu$. We have $\mathrm{Re} = \tau_\nu/\tau_s$. The Stokes time measures the time it takes a particle to move a distance equal to its radius while the kinematic time measures the time it takes momentum to diffuse over that distance. For large-scale turbulent flow the Reynolds number is large and inertia dominates. For small particle motion in dense fluids inertial effects are unimportant and the Reynolds number is small. For instance, for swimming bacteria such as *Escherichia coli* the Reynolds number is typically of order 10^{-5} [43]. These low Reynolds numbers result from the small sizes and low velocities of the particles, in conjunction with the fact that they move in a medium with relatively high viscosity. In MPC dynamics low Reynolds numbers can be achieved by considering systems with small dimensionless mean free path λ, but the values that can be obtained are not as low as those quoted above for bacteria.

The Peclet number $\mathrm{Pe} = v\ell/D_c$, where D_c is the diffusion coefficient of a solute particle in the fluid, measures the ratio of convective transport to diffusive transport. The diffusion time $\tau_D = \ell^2/D_c$ is the time it takes a particle with characteristic length ℓ to diffuse a distance comparable to its size. We may then write the Peclet number as $\mathrm{Pe} = \tau_D/\tau_s$, where τ_s is again the Stokes time. For $\mathrm{Pe} > 1$ the particle will move convectively over distances greater than its size. The Peclet number can also be written $\mathrm{Pe} = \mathrm{Re}(\nu/D_c)$, so in MPC simulations the extent to which this number can be tuned depends on the Reynolds number and the ratio of the kinematic viscosity and the particle diffusion coefficient.

IX. FRICTION AND HYDRODYNAMIC INTERACTIONS

Most descriptions of the dynamics of molecular or particle motion in solution require a knowledge of the frictional properties of the system. This is especially true for polymer solutions, colloidal suspensions, molecular transport processes, and biomolecular conformational changes. Particle friction also plays an important role in the calculation of diffusion-influenced reaction rates, which will be discussed later. Solvent multiparticle collision dynamics, in conjunction with molecular dynamics of solute particles, provides a means to study such systems. In this section we show how the frictional properties and hydrodynamic interactions among solute or colloidal particles can be studied using hybrid MPC–MD schemes.

A. Single-Particle Friction and Diffusion

The friction coefficient is one of the essential elements in the Langevin description of Brownian motion. The derivation of the Langevin equation from the microscopic equations of motion provides a Green–Kubo expression for this transport coefficient. Its computation entails a number of subtle features. Consider a Brownian (B) particle with mass M in a bath of N solvent molecules with mass m. The generalized Langevin equation for the momentum $\mathbf{P}$ of the B particle is [44, 45]

$$\frac{d\mathbf{P}(t)}{dt} = -\int_0^t dt' \, \Phi(t')\mathbf{P}(t-t') + \mathbf{f}^+(t) \tag{68}$$

where $\mathbf{f}^+(t)$ is a random force whose time evolution is determined by projected dynamics and the memory kernel $\Phi(t)$ describes the random force correlations,

$$\Phi(t) = \langle \mathbf{f}^+(t) \cdot \mathbf{f} \rangle / \langle P^2 \rangle \tag{69}$$

where the angular brackets denote an equilibrium average. The time-dependent friction coefficient may be defined as the finite-time integral of the projected force autocorrelation function,

$$\zeta(t) = \frac{1}{3k_BT} \int_0^t dt' \, \langle \mathbf{f}^+(t) \cdot \mathbf{f} \rangle. \tag{70}$$

The friction constant is then the infinite-time value of this function: $\zeta = \lim_{t\to\infty} \zeta(t)$.

For a massive B particle ($M \gg m$) and bath relaxation that is rapid compared to the characteristic decay time of the B-particle momentum, Mazur and Oppenheim [45] derived the Langevin equation from the microscopic equations of motion using projection operator methods. If the projected force autocorrelations decay rapidly, one may make a Markovian approximation to obtain the Langevin equation [46],

$$\frac{d\mathbf{P}(t)}{dt} = -\frac{\zeta}{\mu}\mathbf{P}(t) + \mathbf{f}^+(t) \tag{71}$$

where $\mu = \langle P^2 \rangle / 3k_BT$. The conditions under which a Markovian Langevin description is applicable have been given by Tokuyama and Oppenheim [47]. For Langevin dynamics the momentum autocorrelation function decays exponentially and is given by

$$C_P(t) = \langle \mathbf{P}(t) \cdot \mathbf{P} \rangle \langle P^2 \rangle^{-1} = e^{-\zeta t/\mu} \tag{72}$$

Since the diffusion coefficient is the infinite-time integral of the velocity correlation function, we have the Einstein relation, $D = k_BT/\zeta$.

Computer simulations of transport properties using Green–Kubo relations [48] are usually carried out in the microcanonical ensemble. Some of the subtle issues involved in such simulations have been discussed by Español and Zuñiga [49]. From Eq. (70) we see that the time-dependent friction coefficient is given in terms of the force correlation function with projected dynamics. Instead, in MD simulations the time-dependent friction coefficient is computed using ordinary dynamics.

We define the quantity $\zeta_u(t)$, where evolution is by ordinary dynamics, by

$$\zeta_u(t) = \frac{1}{3k_BT}\int_0^t dt' \, \langle \mathbf{f}(t) \cdot \mathbf{f} \rangle \tag{73}$$

Its Laplace transform $\hat{\zeta}_u(z)$ is related to the Laplace transform of $\zeta(t)$, $\hat{\zeta}(z)$, by [48]

$$\hat{\zeta}_u(z) = \frac{\zeta(z)}{z + \zeta(z)/\mu} \tag{74}$$

The form of this equation for small z is $\hat{\zeta}_u(z) \approx \zeta/(z + \zeta/\mu)$. In t-space we have

$$\zeta_u(t) \approx \zeta e^{-\zeta t/\mu} \tag{75}$$

From this expression we see that the friction cannot be determined from the infinite-time integral of the unprojected force correlation function but only from its plateau value if there is time scale separation between the force and momentum correlation functions decay times. The friction may also be estimated from the extrapolation of the long-time decay of the force autocorrelation function to $t = 0$, or from the decay rates of the momentum or force autocorrelation functions using the above formulas.

In the canonical ensemble $\langle P^2 \rangle = 3k_BTM$ and $\mu = M$. In the microcanonical ensemble $\langle P^2 \rangle = 3k_BT\mu = 3k_BTMNm/(M + Nm)$ [49]. If the limit $M \to \infty$ is first taken in the calculation of the force autocorrelation function, then $\mu = Nm$ and the projected and unprojected force correlations are the same in the thermodynamic limit. Since MD simulations are carried out at finite N, the study of the N (and M) dependence of $\zeta_u(t)$ and the estimate of the friction coefficient from either the decay of the momentum or force correlation functions is of interest. Molecular dynamics simulations of the momentum and force autocorrelation functions as a function of N have been carried out [49, 50].

Equation (75) shows that $\zeta_u(t)$ is an exponentially decaying function for long times with a decay constant ζ/μ. For very massive B particles $M \gg mN$ with $M/mN = q = \text{const}$, the decay rate should vary as $1/N$ since $\mu = mNq/(q + 1)$. The time-dependent friction coefficient $\zeta_u(t)$ for a B particle interacting with the mesoscopic solvent molecules through repulsive LJ potentials

$$V_{sb}(r) = 4\epsilon\left(\left(\frac{\sigma}{r}\right)^{12} - \left(\frac{\sigma}{r}\right)^{6} + \frac{1}{4}\right), \quad \text{for } r < 2^{1/6}\sigma \tag{76}$$

and zero for $r \geq 2^{1/6}\sigma$, is shown in Fig. 9 for various values of N [51]. The semilogarithmic plots of $\zeta_u(t)$ for different values of N show the expected linear

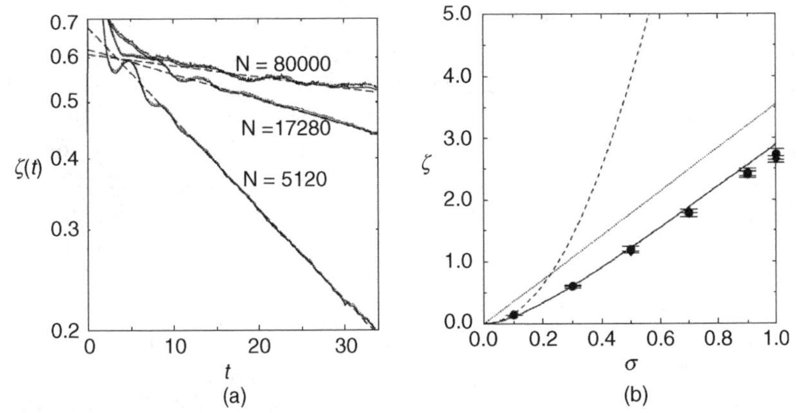

Figure 9. (a) Semilogarithmic plot of time-dependent friction coefficient as function of time for $M/m \approx 20N$ for three values of N. The straight lines show the extrapolation of the exponential long-time decay to $t = 0$ used to determine the value of ζ. (b) Friction constant as a function of σ: circles—ζ from simulation; dotted line—ζ_h; dashed line—ζ_m; and solid line—ζ.

decay at long times. The long-time linear decay may be extrapolated to $t = 0$ to obtain ζ shown in the figure.

In order to examine the nature of the friction coefficient it is useful to consider the various time, space, and mass scales that are important for the dynamics of a B particle. Two important parameters that determine the nature of the Brownian motion are $r_m = (m/M)^{1/2}$, that depends on the ratio of the bath and B particle masses, and $r_\rho = \rho/(3M/4\pi\sigma^3)$, the ratio of the fluid mass density to the mass density of the B particle. The characteristic time scale for B particle momentum decay is $\tau_B = M/\zeta$, from which the characteristic length $\ell_B = (k_BT/M)^{1/2}\tau_B$ can be defined. In derivations of Langevin descriptions, variations of length scales large compared to microscopic length but small compared to ℓ_B are considered. The simplest Markovian behavior is obtained when both $r_m \ll 1$ and $r_\rho \ll 1$, while non-Markovian descriptions of the dynamics are needed when $r_m \ll 1$ and $r_\rho \geq 1$ [47]. The other important times in the problem are $\tau_\nu = \sigma^2/\nu$, the time it takes momentum to diffuse over the B particle radius σ, and $\tau_D = \sigma^2/D_c$, the time it takes the B particle to diffuse over its radius.

The friction coefficient of a large B particle with radius σ in a fluid with viscosity η is well known and is given by the Stokes law, $\zeta_h = 6\pi\eta\sigma$ for stick boundary conditions or $\zeta = 4\pi\eta\sigma$ for slip boundary conditions. For smaller particles, kinetic and mode coupling theories, as well as considerations based on microscopic boundary layers, show that the friction coefficient can be written approximately in terms of microscopic and hydrodynamic contributions as $\zeta^{-1} = \zeta_m^{-1} + \zeta_h^{-1}$. The physical basis of this form can be understood as follows: for a B particle with radius σ a hydrodynamic description of the solvent should be applicable outside a boundary layer, $r > \bar{\sigma}$. Inside the boundary layer the molecular nature of the dynamics can be taken into account. Processes inside the boundary layer contribute a microscopic component to the friction due to collisions between the B and bath particles, which is given by $\zeta_m = \frac{8}{3}\rho\sigma^2(2\pi m k_B T)^{1/2}$. Using a suitable "radiation-like" boundary condition to account for this boundary layer [52], the total friction formula is obtained with ζ_h given by the Stokes law form. Since the microscopic contribution scales as σ^2 while the hydrodynamic contribution scales as σ, the microscopic contribution is important for small particles. The calculation of the friction coefficient with sufficient accuracy for large enough systems to resolve the hydrodynamic and microscopic components for B particles with varying sizes is a problem that can be addressed using MPC dynamics.

Figure 9b shows the friction constant as a function of σ. For large σ the friction coefficient varies linearly with σ in accord with the prediction of the Stokes formula. The figure also shows a plot of ζ_h (slip boundary conditions) versus σ. It lies close to the simulation value for large σ but overestimates the friction for small σ. For small σ, microscopic contributions dominate the friction coefficient as can be seen in the plot of ζ_m. The approximate expression $\zeta^{-1} = \zeta_m^{-1} + \zeta_h^{-1}$ interpolates between the two limiting forms. Cluster friction simulation results have also been interpreted in this way [53]. A discussion of microscopic and hydrodynamic (including sound wave) contributions to the velocity correlation function along with comparisons with MPC simulation results was given by Padding and Louis [41].

B. Hydrodynamic Interactions

The disturbances that solute particles create in the fluid by their motion are transmitted to other parts of the fluid through solvent collective modes, such as the solvent velocity field. These long-range hydrodynamic interactions give rise to a coupling among different solute molecules that influences their motion [54]. If the number density of B particles is n_B, the hydrodynamic screening length within which hydrodynamic interactions become important is $\ell_H = (6\pi n_B \sigma)^{-1/2}$. The time it takes hydrodynamic interactions to become important is $\tau_H = \tau_\nu/\phi$, where $\phi = 4\pi\sigma^3 n_B/3$ is the volume fraction of B particles [55]. It is important to account for such hydrodynamic interactions when dealing with the dynamics of polymers and colloidal suspensions [56, 57].

Brownian motion theory may be generalized to treat systems with many interacting B particles. Such many-particle Langevin equations have been investigated at a molecular level by Deutch and Oppenheim [58]. A simple system in which to study hydrodynamic interactions is two particles fixed in solution at a distance R_{12}. The Langevin equations for the momenta $\mathbf{P}_i$ $(i = 1, 2)$

of Brownian particles with mass M in a solvent take the form

$$\frac{d\mathbf{P}_i(t)}{dt} = -\sum_{j=1}^{2} \mathbf{P}_j(t) \cdot \frac{\zeta^{ij}(\mathbf{R}_{12})}{M} + \mathbf{F}_i(t) + \mathbf{f}_i(t) \tag{77}$$

for times t much greater than the characteristic relaxation time of the bath. Here $\mathbf{F}_i$ is the force on Brownian particle i and $\mathbf{f}_i(t)$ is the random force. The fixed-particle friction tensor is defined as the time integral of the force autocorrelation function,

$$\zeta^{ij}(\mathbf{R}_{12}) = \beta \int_0^\infty dt\, \langle \mathbf{f}_i(0)\mathbf{f}_j(t)\rangle_0 \tag{78}$$

In this expression the time evolution of the random force is given by

$$\mathbf{f}_i(t) = e^{iL_0 t}(\mathbf{F}_i - \langle \mathbf{F}_i \rangle) \equiv e^{iL_0 t}\mathbf{f}_i(0) \tag{79}$$

The Liouvillian $iL_0 \cdot = \{H_0, \cdot\}$, where $\{\cdot, \cdot\}$ is the Poisson bracket, describes the evolution governed by the bath Hamiltonian H_0 in the field of the fixed Brownian particles. The angular brackets signify an average over a canonical equilibrium distribution of the bath particles with the two Brownian particles fixed at positions $\mathbf{R}_1$ and $\mathbf{R}_2$, $\langle \cdots \rangle_0 = Z_0^{-1} \int d\mathbf{r}^N d\mathbf{p}^N e^{-\beta H_0} \cdots$, where Z_0 is the partition function.

If the Brownian particles were macroscopic in size, the solvent could be treated as a viscous continuum, and the particles would couple to the continuum solvent through appropriate boundary conditions. Then the two-particle friction may be calculated by solving the Navier–Stokes equations in the presence of the two fixed particles. The simplest approximation for hydrodynamic interactions is through the Oseen tensor [54],

$$\mathbf{T}_{\alpha\beta}(\mathbf{R}_{12}) = (1 - \delta_{\alpha\beta})\frac{1}{8\pi\eta R_{12}}(\mathbf{1} + \hat{\mathbf{R}}_{12}\hat{\mathbf{R}}_{12}) \tag{80}$$

which is valid when the Brownian particles are separated by distances very large compared to their diameters. Here $\hat{\mathbf{R}}_{12}$ is a unit vector along the inter-particle (z) axis and η is the solvent viscosity. If Oseen interactions are assumed, the friction tensor takes the form

$$\zeta(\mathbf{R}_{12}) = \zeta_0(\mathbf{I} + \zeta_0 \mathbf{T}(\mathbf{R}_{12}))^{-1} \tag{81}$$

where we now use the symbol ζ_0 for the one-particle friction coefficient.

MPC dynamics is able to describe hydrodynamic interactions because it preserves the conservation laws, in particular, momentum conservation, on which these interactions rely. Thus we can test the validity of such an

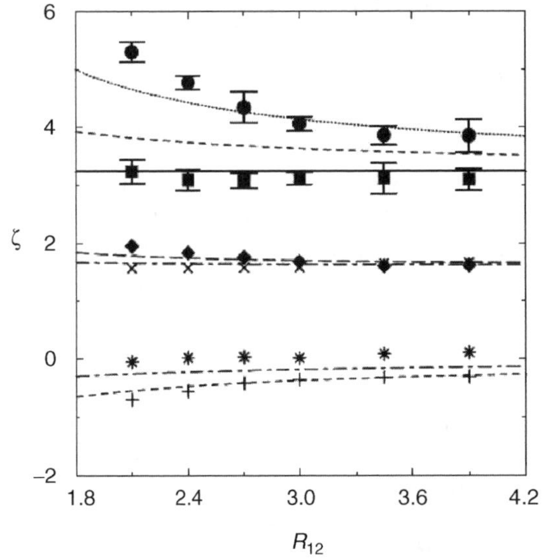

Figure 10. Friction coefficients as a function of R_{12} (units of σ): filled circle—$\zeta_{zz}^{(-)}$; square—$\zeta_{xx}^{(-)}$; blacklozenge—ζ_{zz}^{11}; cross—ζ_{xx}^{11}; asterisk—ζ_{xx}^{12} and plus—ζ_{zz}^{12}, respectively. The solid line indicates twice the single-particle friction, $2\zeta_0$, and the other lines from the top are the analytic results for $\zeta_{zz}^{(-)}$, $\zeta_{xx}^{(-)}$, ζ_{zz}^{11}, ζ_{xx}^{11}, ζ_{xx}^{12}, and ζ_{zz}^{12}, respectively.

approximate macroscopic description of the two-particle friction tensor for particles that are not macroscopically large [59]. The two Brownian particles interact with the bath molecules through repulsive Lennard-Jones intermolecular potentials (Eq. (76)), and their internuclear separation is held fixed by a holonomic constraint on the equations of motion. Bath particle interactions are accounted for by multiparticle collisions. Figure 10 shows the hybrid MD–MPC simulation results for the two-particle friction coefficients for two LJ particles as a function of the interparticle separation, R_{12}. The components of the friction normal to the intermolecular axis, ζ_{xx}^{11} and ζ_{xx}^{12}, are almost independent of R_{12} while the components parallel to this axis, ζ_{zz}^{11} and ζ_{zz}^{12}, increase as the particle separation decreases. There are deviations at small separations from the friction computed using Oseen interactions as might be expected, since this simple hydrodynamic approximation will be inaccurate at small distances. The simulation results for ζ_{xx}^{12} vary much more weakly with internuclear separation than those using the simple hydrodynamic model. The relative friction,

$$\zeta^{(-)}(\mathbf{R}_{12}) = 2(\zeta^{11}(\mathbf{R}_{12}) - \zeta^{12}(\mathbf{R}_{12})) \tag{82}$$

shows this trend clearly. We see that $\zeta_{xx}^{(-)}$ is nearly independent of R_{12} and equal to its asymptotic value of twice the single-particle friction coefficient. The parallel component, $\zeta_{zz}^{(-)}$, increases strongly as R_{12} decreases.

These results show that hydrodynamic interactions and the spatial dependence of the friction tensor can be investigated in regimes where continuum descriptions are questionable. One of the main advantages of MPC dynamics studies of hydrodynamic interactions is that the spatial dependence of the friction tensor need not be specified *a priori* as in Langevin dynamics. Instead, these interactions automatically enter the dynamics from the mesoscopic particle-based description of the bath molecules.

C. Colloidal Suspensions

The methodology discussed previously can be applied to the study of colloidal suspensions where a number of different molecular forces and hydrodynamic effects come into play to determine the dynamics. As an illustration, we briefly describe one example of an MPC simulation of a colloidal suspension of clay-like particles where comparisons between simulation and experiment have been made [42, 60]. Experiments were carried out on a suspension of Al_2O_3 particles. For this system electrostatic repulsive and van der Waals attractive forces are important, as are lubrication and contact forces. All of these forces were included in the simulations. A mapping of the MPC simulation parameters onto the space and time scales of the real system is given in Hecht et al. [42]. The calculations were carried out with an imposed shear field.

The system can exist in a variety of phases depending on the parameters. Figure 11a shows the phase diagram in the ionic strength–pH plane. The pH

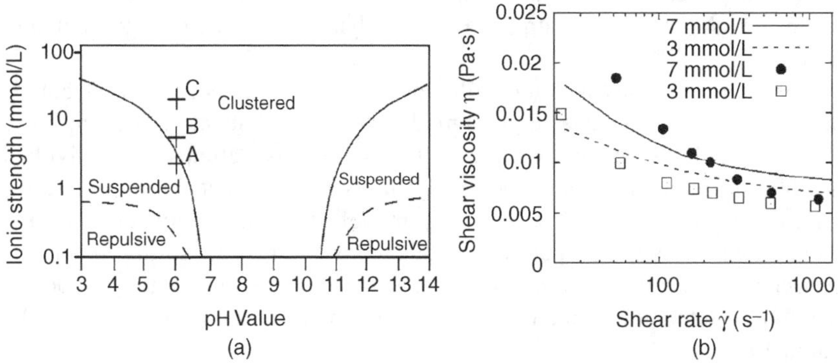

Figure 11. (a) Diagram showing different phases of the suspension. (b) Shear viscosity versus shear rate for the states labelled A and B in the phase diagram. The volume fraction is $\phi = 0.35$. From Hecht, et al., Ref. [60].

controls the surface charge density, which, in turn, influences the electrostatic interactions in the system. Clustering occurs when van der Waals attractions dominate. Stable suspensions are favored when electrostatic repulsion prevents clustering. Repulsion among suspended particles dominates when electrostatic forces are very strong. Figure 11b compares the measured and simulated shear viscosity of the suspension as a function of the shear rate $\dot{\gamma}$ in two regions of the phase diagram. (The shear rate $\dot{\gamma}$ should not be confused with γ, the average number of particles per cell.) When the colloidal particles are suspended (state A) or slightly clustered (state B) shear thinning, where the viscosity decreases with increasing shear rate, is observed. In state B shear thinning is more pronounced. The simulation results are in rough accord with the experimental data. Discrepancies have been attributed to uncertainties in the parameters that enter into the electrostatic effects in the system and how they are modeled, as well as polydispersity and the manner in which lubrication forces are treated.

There have been other MPC dynamics studies of hydrodynamic effects on the transport properties of colloidal suspensions [61–64]. In addition, vesicles that can deform under flow have also been investigated using hybrid MPC–MD schemes [65–69].

X. POLYMERS

It is known that polymer dynamics is strongly influenced by hydrodynamic interactions. When viewed on a microscopic level, a polymer is made from molecular groups with dimensions in the angstrom range. Many of these monomer units are in close proximity both because of the connectivity of the chain and the fact that the polymer may adopt complicated conformations in solution. Polymers are solvated by a large number of solvent molecules whose molecular dimensions are comparable to those of the monomer units. These features make the full treatment of hydrodynamic interactions for polymer solutions very difficult.

For many purposes such a detailed description of the polymer molecule is not necessary. Instead, coarse-grain models of the polymer chain are employed [56,70–72]. In such mesoscopic polymer models, groups of individual neighboring monomers are taken to be units that interact through effective forces. For example, in a bead–spring model of a linear polymer chain the interactions among the polymer beads consist of bead–spring potentials between neighboring beads as well as bead–bead interactions among all beads. Bead–bead interactions among all beads may be taken to be attractive (A) LJ interactions,

$$V_{\rm LJ}(r) = 4\epsilon\left[\left(\frac{\sigma}{r}\right)^{12} - \left(\frac{\sigma}{r}\right)^{6}\right] \tag{83}$$

or repulsive (R) LJ interactions defined in Eq. (76). The non-Hookian bead–spring potential [73, 74] is often described by finitely extensible nonlinear elastic (FENE) interactions,

$$V_{\text{FENE}}(r) = -\frac{\kappa}{2}R_0^2 \ln\left[1 - \left(\frac{r}{R_0}\right)^2\right], \quad r < R_0 \tag{84}$$

where $\kappa = c_1\epsilon/\sigma^2$ and $R_0 = c_2\sigma$, with c_1 and c_2 constants. Bead–spring interactions may even be described by simpler Gaussian or Hookian springs. In Langevin treatments of polymer dynamics hydrodynamic interactions are usually incorporated through Oseen interactions as discussed earlier. While this approximation suffices to capture gross features of hydrodynamic effects, it suffers from the limitations discussed earlier in connection with the two-particle friction tensor.

Hybrid MPC–MD schemes are an appropriate way to describe bead–spring polymer motions in solution because they combine a mesoscopic treatment of the polymer chain with a mesoscopic treatment of the solvent in a way that accounts for all hydrodynamic effects. These methods also allow one to treat polymer dynamics in fluid flows.

A. Polymer Dynamics

The dynamical properties of polymer molecules in solution have been investigated using MPC dynamics [75–77]. Polymer transport properties are strongly influenced by hydrodynamic interactions. These effects manifest themselves in both the center-of-mass diffusion coefficients and the dynamic structure factors of polymer molecules in solution. For example, if hydrodynamic interactions are neglected, the diffusion coefficient scales with the number of monomers as $D \sim D_0/N_b$, where D_0 is the diffusion coefficient of a polymer bead and N_b is the number of beads in the polymer. If hydrodynamic interactions are included, the diffusion coefficient adopts a Stokes–Einstein form $D \sim k_BT/c\pi\eta N_b^{1/2}$, where c is a factor that depends on the polymer chain model. This scaling has been confirmed in MPC simulations of the polymer dynamics [75].

The normal modes (Rouse modes) that characterize the internal dynamics of the polymer can be computed exactly for a Gaussian chain and are given by $\chi_p = (2/N_b)^{1/2} \sum_{i=1}^{N_b} \mathbf{r}_i \cos\left(p\pi(i - \frac{1}{2})/N_b\right)$. The characteristic times scales for the decay of these modes can be estimated from computations of the autocorrelation function of the mode amplitudes, $\langle\chi_p(t)\chi_p(0)\rangle$. Even for non-Gaussian chains these correlation functions provide useful information on the internal dynamics of the chain. Figure 12a plots the autocorrelation functions of two mode amplitudes for a polymer with excluded volume interactions. The figure shows exponential scaling at long times with a characteristic relaxation times τ_p that scale as $\tau_p \sim p^\alpha$ with $\alpha \approx 1.9$. The Zimm theory of polymer

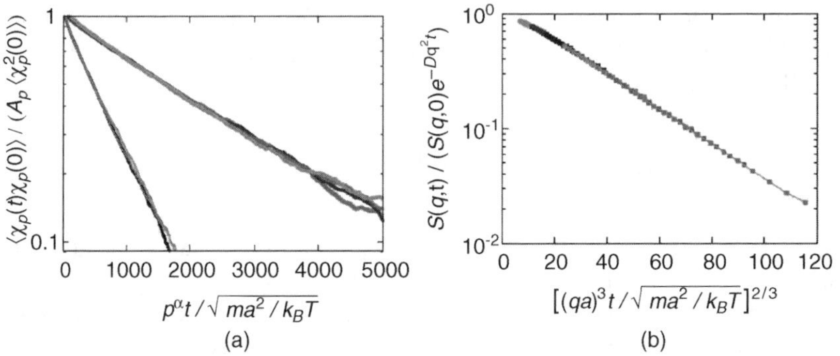

Figure 12. (a) Plots of the Rouse amplitude correlation functions for several modes versus time for polymers with excluded volume interactions and two chain lengths, $N_b = 20$ (lower curves) and $N_b = 40$ (upper curves). (b) Dynamic structure factor versus $q^2t^{2/3}$. (From Ref. 75.)

dynamics, which includes hydrodynamic interactions at the Oseen level of approximation, predicts that $\tau_p \sim p^{-3\nu}$. Since $\nu \approx 0.62$ the simulation results are in close accord with this value.

The dynamic structure factor is $S(\mathbf{q},t) = \langle n_{\mathbf{q}}(t)n_{-\mathbf{q}}(0)\rangle$, where $n_{\mathbf{q}}(t) = \sum_{i=1}^{N_b} e^{i\mathbf{q}\cdot\mathbf{r}_i}$ is the Fourier transform of the total density of the polymer beads. The Zimm model predicts that this function should scale as $S(\mathbf{q},t) = S(\mathbf{q},0)\mathcal{F}(q^\alpha t)$, where $\mathcal{F}$ is a scaling function. The data in Fig. 12b confirm that this scaling form is satisfied. These results show that hydrodynamic effects for polymeric systems can be investigated using MPC dynamics.

B. Collapse Dynamics

The nature of the solvent influences both the structure of the polymer in solution and its dynamics. In good solvents the polymer adopts an expanded configuration and in poor solvents it takes on a compact form. If the polymer solution is suddenly changed from good to poor solvent conditions, polymer collapse from the expanded to compact forms will occur [78]. A number of models have been suggested for the mechanism of the collapse [79–82]. Hydrodynamic interactions are expected to play an important part in the dynamics of the collapse and we show how MPC simulations have been used to investigate this problem. Hybrid MD–MPC simulations of the collapse dynamics have been carried out for systems where bead–solvent interactions are either explicitly included [83] or accounted for implicitly in the multiparticle collision events [84, 85].

Suppose the bead–solvent interactions are described by either repulsive (r) or attractive (a) LJ potentials in the MD–MPC dynamics. The repulsive interactions are given in Eq. (76) while the attractive LJ interactions take the form $cV_{\mathrm{LJ}}(r)$, where c gauges the strength of the bead–solvent potential, and

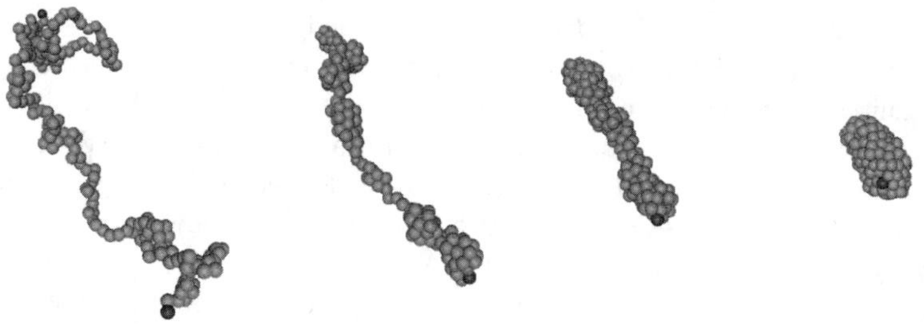

Figure 13. Expanded form of a polymer in a good solvent (far left) and collapsed form in a poor solvent (far right). The end beads in the polymer chain are coded with a different color. Solvent molecules are not shown.

are smoothly truncated to zero at a short distance by a switching function [83]. The bead–bead interactions are either repulsive (R) or attractive (A) LJ potentials. If the system is initially in an expanded configuration with repulsive bead–bead and attractive bead–solvent (Ra) interactions then, following a sudden change to attractive bead–bead and repulsive bead–solvent (Ar) interactions, collapse will ensue. Several polymer configurations during the collapse are shown in Fig. 13. First, "blobs" of polymer beads are formed where portions of the chain are in close proximity. The blobs are separated by segments of the uncollapsed chain. As the collapse progresses, the blobs coalesce to form a thick sausage-shaped structure, which continues to thicken and shrink until the collapsed elongated polymer state is reached.

The time evolution of the radius of gyration Rg(t), where $\mathrm{Rg}^2(t) = N_b^{-1}\sum_{i=1}^{N_b} |(\mathbf{r}_i(t) - \mathbf{r}_{\mathrm{CM}}(t))|^2$, can be used to monitor the collapse dynamics. Here $\mathbf{r}_i$ is the position of bead i and $\mathbf{r}_{\mathrm{CM}}$ is the center of mass of the polymer chain. Figure 14a shows how Rg(t), averaged over several realizations of the

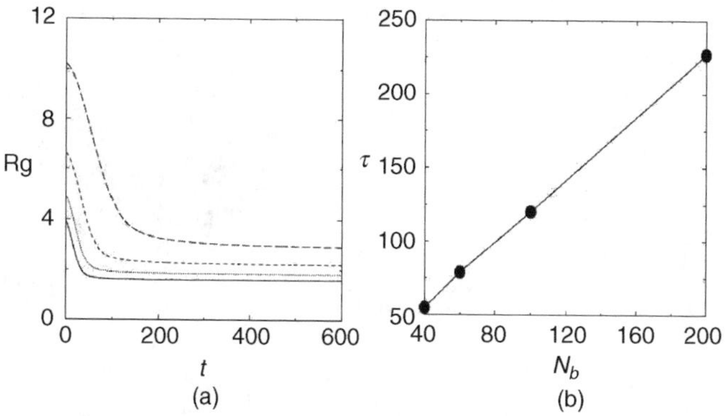

Figure 14. (a) Radius of gyration Rg versus time for several values of N_b, from bottom, $N_b = 40, 60, 100$ and 200; (b) collapse time τ versus N_b for $T = \frac{1}{3}$.

collapse dynamics starting from different configurations of the expanded polymer, varies with time for several values of N_b. The plot shows the decay to a constant average value of the radius of gyration, $\overline{\mathrm{Rg}}$, and the variation of the collapse time with the number of polymer beads (Fig. 14b). The definition of the collapse time is somewhat arbitrary. If it is defined as the time τ for which $\mathrm{Rg}(\tau) = \overline{\mathrm{Rg}} + (\mathrm{Rg}(0) - \overline{\mathrm{Rg}})/20$, the data in Fig. 14b show that τ increases linearly with N_b as $\tau(N_b) = 13.2 + 1.07N_b$.

For (Ar) interactions the collapsed state of the polymer is a tight globule from which solvent is excluded. Figure 15 shows the polymer bead and solvent radial distribution functions relative to the center of mass of the globule,

$$g_{\mathrm{CM}-\nu}(r) = \frac{1}{4\pi r^2 \rho_\nu} \left\langle \sum_i^{N_\nu} \delta(|\mathbf{r}_i - \mathbf{r}_{\mathrm{CM}}| - r) \right\rangle \tag{85}$$

where $\nu = b$ or s labels a polymer or solvent molecule and ρ_ν is the number density of polymer bead or solvent molecules, and the polymer configuration with surrounding solvent molecules. The fact that solvent molecules do not penetrate into the interior of the collapsed polymer can be seen in these results. For other choices of the interaction parameters the solvent may penetrate into the interior of the collapsed polymer.

The effect of hydrodynamic interactions on polymer collapse has also been studied using MPC dynamics, where the polymer beads are included in the multiparticle collision step [28, 84]. Hydrodynamic interactions can be turned off by replacing multiparticle collisions in the cells by sampling of the particle velocities from a Boltzmann distribution. Collapse occurs more rapidly in the

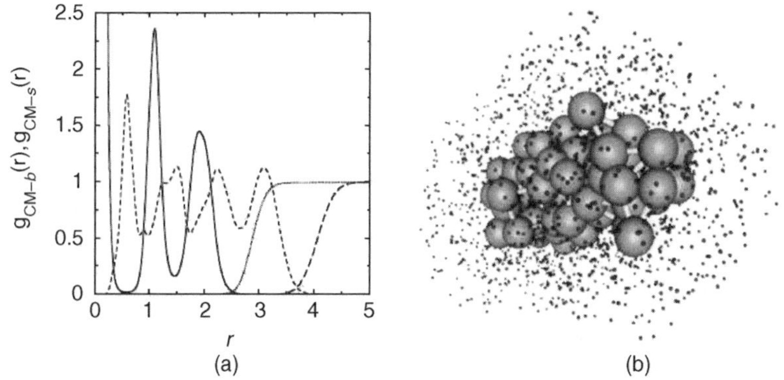

Figure 15. (a) Radial distribution functions $g_{\mathrm{CM}-b}(r)$ (solid line and short dashed line) and $g_{\mathrm{CM}-s}(r)$ (dotted line and long dashed line) versus r for $N_b = 60$ and 200, respectively. (b) Collapsed polymer with $N_b = 60$ and surrounding solvent molecules.

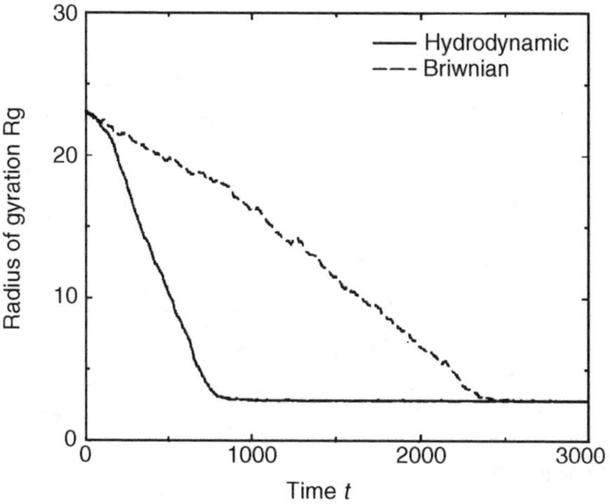

Figure 16. Radius of gyration versus time for MPC dynamics (solid line) that includes hydrodynamic interactions and Boltzmann sampling of velocities (dashed line) without hydrodynamic interactions. System parameters: $N_b = 200$ and $T = 0.8$. From Kikuchi, et al., 2002.

presence of hydrodynamic interactions since these interactions have the effect of reducing the friction of the polymer (Fig. 16). In this model of polymer–solvent interactions, solvent molecules freely penetrate into the interior of the polymer chain. In these studies the collapse time was found to scale as $\tau \sim N_b^{1.40\pm0.08}$ in the presence of hydrodynamic interactions and as $\tau \sim N_b^{1.89\pm0.09}$ in the absence of hydrodynamic interactions.

C. Polymers in Fluid Flows

A fluid flow field can change the conformation or orientation of a polymer in solution and an understanding of such flow effects is important for a number of applications that include microfluidics and flows in biological systems. In order to be able to describe the dynamics of systems of this type, schemes that properly describe both the fluid flow fields and polymer dynamics are required. Since MPC dynamics satisfies both of these criteria, it has been used to investigate linear [86–88] and branched [89] polymers in various flow fields. In addition, translocation [90] of polymers and polymer packing [91, 92] have also been investigated using this mesoscopic simulation method.

As an example of such applications, consider the dynamics of a flexible polymer under a shear flow [88]. A shear flow may be imposed by using Lees–Edwards boundary conditions to produce a steady shear flow $\dot{\gamma} = u/L_y$, where L_y is the length of the system along y and u is the magnitude of the velocities of the boundary planes along the x-direction. An important parameter in these studies is the Weissenberg number, $\mathrm{Wi} = \tau_1\dot{\gamma}$, the product of the longest

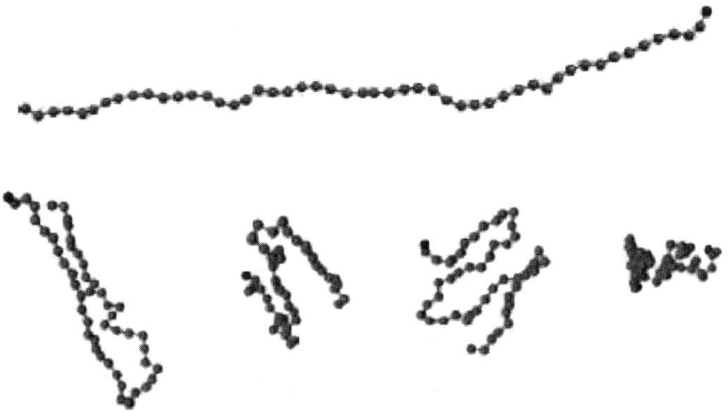

Figure 17. Configurations of an excluded volume polymer chain with $N_b = 50$ for Wi > 1. The top configuration is for a polymer fully extended in the flow direction. The bottom configurations show other configurations that the polymer adopts in the shear flow. (Adapted from Fig. 2 of Ref. 88.)

relaxation time of the polymer and the shear rate. Examples of the shapes the polymer adopts in a strong shear flow are shown in Fig. 17. The changes in the polymer shape can be characterized quantitatively by computing the average of the radius of gyration tensor.

XI. REACTIVE HYBRID MPC–MD DYNAMICS

In studies of reactions in nanomaterials, biochemical reactions within the cell, and other systems with small length scales, it is necessary to deal with reactive dynamics on a mesoscale level that incorporates the effects of molecular fluctuations. In such systems mean field kinetic approaches may lose their validity. In this section we show how hybrid MPC–MD schemes can be generalized to treat chemical reactions.

We begin by considering a reactive system with M finite-sized catalytic spherical particles C, and a total of $N = N_A + N_B$ A and B point particles in a volume V [17]. The C particles catalyze the interconversion between A and B particles according to the reactions

$$A + C \underset{k_r}{\overset{k_f}{\rightleftharpoons}} B + C \tag{86}$$

The macroscopic mass action rate law, which holds for a well-mixed system on sufficiently long time scales, may be written

$$\frac{d}{dt}\bar{n}_A(t) = -k_f n_C \bar{n}_A(t) + k_r n_C \bar{n}_B(t) \tag{87}$$

where $\bar{n}_A(t)$ and $\bar{n}_B(t)$ are the mean number densities of A and B particles, respectively, and n_C is the fixed number density of the catalytic C particles. We pose the question: How do the rate constants depend on the density of the catalytic spheres? If the C density is very low, the catalysts will act independently and the rate constants will not depend on n_C; however, if the density is high, correlated reactive events will lead to a dependence on the concentration of the catalyst.

We first consider the structure of the rate constant for low catalyst densities and, for simplicity, suppose the A particles are converted irreversibly to B upon collision with C (see Fig. 18a). The catalytic particles are assumed to be spherical with radius σ. The chemical rate law takes the form $d\bar{n}_A(t)/dt = -k_f(t)n_C\bar{n}_A(t)$, where $k_f(t)$ is the time-dependent rate coefficient. For long times, $k_f(t)$ reduces to the phenomenological forward rate constant, k_f. If the dynamics of the A density field may be described by a diffusion equation, we have the well known partially absorbing sink problem considered by Smoluchowski [32]. To determine the rate constant we must solve the diffusion equation

$$\frac{\partial}{\partial t} n_A(\mathbf{r}, t) = D_A n_A(\mathbf{r}, t) \tag{88}$$

subject to the boundary condition [93]

$$4\pi D\bar{\sigma}^2 \hat{\mathbf{r}} \cdot (\nabla n_A)(\hat{\mathbf{r}}\bar{\sigma}, t) = k_{0f} n_A(\hat{\mathbf{r}}\sigma, t) \tag{89}$$

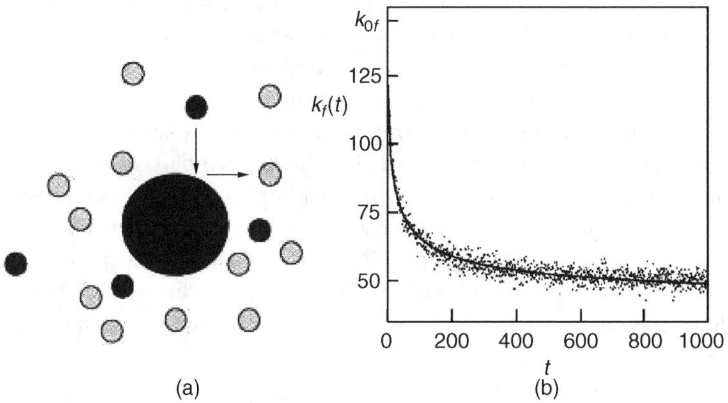

Figure 18. (a) Schematic diagram showing a reactive event $A \rightarrow B$. (b) Plot of the time-dependent rate constant $k_f(t)$ versus t for $\sigma = 10$ and $p_R = 0.5$. The solid line is the theoretical value of $k_f(t)$ using Eq. (90) and $\bar{\sigma} = \sigma + 1$.

This formulation assumes that the continuum diffusion equation is valid up to a distance $\bar{\sigma} > \sigma$, which accounts for the presence of a boundary layer in the vicinity of the catalytic particle where the continuum description no longer applies. The rate constant k_{0f} characterizes the reactive process in the boundary layer. If it approximated by binary reactive collisions of A with the catalytic sphere, it is given by $k_{0f} = p_R \sigma_C^2 (8\pi k_B T/m)^{1/2}$, where p_R is the probability of reaction on collision.

The solution of this problem yields the time-dependent rate coefficient [94]

$$k_f(t) = \frac{k_{0f} k_D}{k_{0f} + k_D} + \frac{k_{0f}^2}{k_{0f} + k_D} \exp\left[\left(1 + \frac{k_{0f}}{k_D}\right)^2 \frac{D}{\bar{\sigma}^2} t\right] \times \operatorname{erfc}\left[\left(1 + \frac{k_{0f}}{k_D}\right)\left(\frac{Dt}{\bar{\sigma}^2}\right)^{1/2}\right] \tag{90}$$

Here $k_D = 4\pi\bar{\sigma}D$ is the rate constant for a diffusion-controlled reaction (Smoluchowski rate constant) for a perfectly absorbing sphere. For long times, $k_f(t)$ approaches its asymptotic constant value $k_f = k_{0f}k_D/(k_{0f} + k_D)$ as

$$k_f(t) \sim k_f \left(1 + \frac{k_{0f}}{k_{0f} + k_D} \frac{\bar{\sigma}}{(\pi D t)^{1/2}}\right) \tag{91}$$

This power law decay is captured in MPC dynamics simulations of the reacting system. The rate coefficient $k_f(t)$ can be computed from $-(d\bar{n}_A(t)/dt)/\bar{n}_A(t)$, which can be determined directly from the simulation. Figure 18 plots $k_f(t)$ versus t and confirms the power law decay arising from diffusive dynamics [17]. Comparison with the theoretical estimate shows that the diffusion equation approach with the radiation boundary condition provides a good approximation to the simulation results.

If the volume fraction $\phi = 4\pi\sigma^3 M/3V$ of catalytic particles is high, reactions at one catalytic particle will alter the A and B particle density fields in the vicinities of other catalytic particles, leading to a many-body contribution to the reaction rate. This coupling has a long range and is analogous to the long-range interactions that determine hydrodynamic contributions to the friction coefficient discussed earlier. Theoretical predictions of the volume fraction dependence of the rate constant have been made and take the form [95, 96]

$$k_f(\phi) = k_f \left[1 + \left(\frac{(k_{0f})^3}{(k_{0f} + k_D)^3} 3\phi\right)^{1/2} + \cdots\right] \tag{92}$$

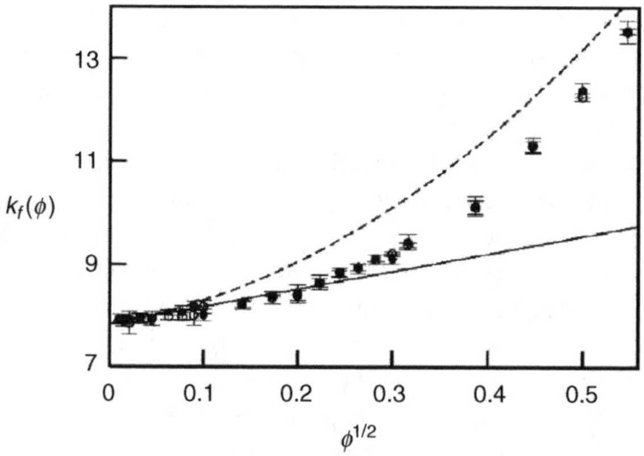

Figure 19. (•) Rate coefficient $k_f(\phi)$ as a function of the square root of the volume fraction $\phi^{1/2}$ for $\sigma = 3$ and $k_B T = \frac{1}{3}$. The solid line is determined using Eq. (92), while the dashed line is obtained using a higher-order approximation to the volume fraction dependence.

The first density correction to the rate constant depends on the square root of the volume fraction and arises from the fact that the diffusion Green's function acts like a screened Coulomb potential coupling the diffusion fields around the catalytic spheres.

MPC dynamics follows the motions of all of the reacting species and their interactions with the catalytic spheres; therefore collective effects are naturally incorporated in the dynamics. The results of MPC dynamics simulations of the volume fraction dependence of the rate constant are shown in Fig. 19 [17]. The MPC simulation results confirm the existence of a $\phi^{1/2}$ dependence on the volume fraction for small volume fractions. For larger volume fractions the results deviate from the predictions of Eq. (92) and the rate constant depends strongly on the volume fraction. An expression for rate constant that includes higher-order corrections has been derived [95]. The dashed line in Fig. 19 is the value of $k_f(\phi)$ given by this higher-order approximation and this formula describes the departure from the $\phi^{1/2}$ behavior that is seen in Fig. 19. The deviation from the $\phi^{1/2}$ form occurs at smaller ϕ values than indicated by the simulation results and is not quantitatively accurate. The MPC results are difficult to obtain by other means.

A. Crowded Environments

The interior of the living cell is occupied by structural elements such as microtubules and filaments, organelles, and a variety of other macromolecular species making it an environment with special characteristics [97]. These systems are crowded since collectively the macromolecular species occupy a large volume fraction of the cell [98, 99]. Crowding can influence both the equilibrium and transport properties of the system [100–102]; for example, it can decrease the diffusion coefficients of macromolecules [103, 104], influence diffusion-controlled reaction rates [105], and lead to shifts in chemical equilibria [106].

We focus on the effects of crowding on small molecule reactive dynamics and consider again the irreversible catalytic reaction $A + C \rightarrow B + C$ as in the previous subsection, except now a volume fraction ϕ_o of the total volume is occupied by obstacles (see Fig. 20). The A and B particles diffuse in this crowded environment before encountering the catalytic sphere where reaction takes place. Crowding influences both the diffusion and reaction dynamics, leading to nontrivial volume fraction dependence of the rate coefficient $k_f(\phi)$ for a single catalytic sphere. This dependence is shown in Fig. 21a. The rate constant has the form discussed earlier,

$$k_f(\phi) = \frac{k_{0f}(\phi)k_D(\phi)}{k_{0f}(\phi) + k_D(\phi)} \tag{93}$$

but now both the microscopic rate constant $k_{0f}(\phi)$ and the Smoluchowski rate constant $k_D(\phi)$ acquire ϕ dependence due to the existence of obstacles in the system.

The volume fraction dependence of $k_{0f}(\phi)$ is plotted in Fig. 21b and shows that it increases strongly with ϕ. Recall that this rate coefficient is independent of ϕ if simple binary collision dynamics is assumed to govern the boundary layer region. The observed increase arises from the obstacle distribution in the vicinity of the catalytic sphere surface. When obstacles are present, a reactive

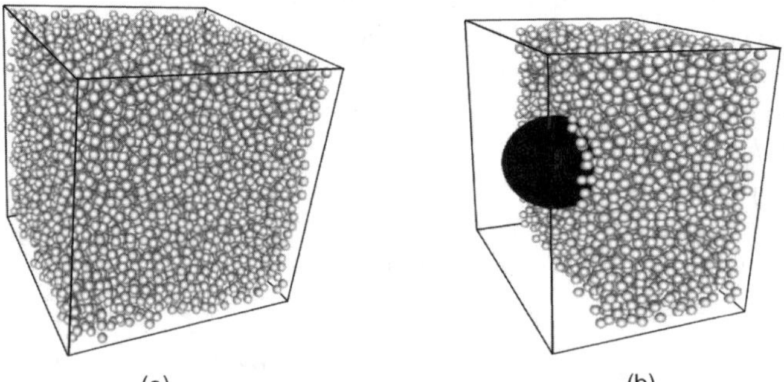

Figure 20. (a) Volume containing spherical obstacles with volume fraction $\phi = 0.15$. (b) The large catalytic sphere (black) with radius $\sigma = 10$ surrounded by obstacle spheres (light grey) with radius $\sigma = 1$ for a volume fraction $\phi = 0.15$. Obstacle spheres in half of the volume are shown in order to see the embedded catalytic sphere. None of the small A or B molecules are shown. They fill the interstices between the obstacles.

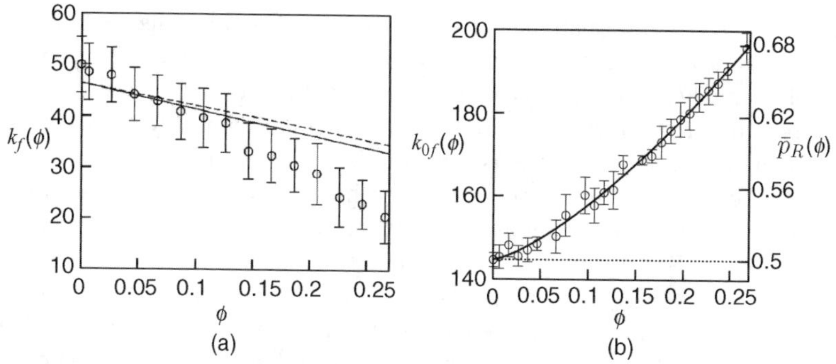

Figure 21. (a) Plot of the asymptotic value of $k_f(\phi)$ as a function of the obstacle volume fraction ϕ. The solid and dashed lines are the theoretical estimates discussed in the text. (b) Microscopic rate constant $k_{0f}(\phi)$ versus ϕ. The dotted line is the predicted value of k_{0f} when no obstacles are present. The figure also shows the effective reaction probability $\bar{p}_R(\phi)$ (see text).

small particle may collide with the catalyst more than once in unit time. Also, due to the obstacle structural ordering near the catalyst for high volume fractions, the local density of A particles is higher near the catalytic sphere than in the bulk of the system, leading to a larger initial rate. These effects can be interpreted as a ϕ dependence of the reaction probability p_R. By writing $k_{0f}(\phi)$ as $k_{0f}(\phi) = \bar{p}_R(\phi)\sigma_C^2(8\pi k_B T/m)^{1/2}$, the effective reaction probability $\bar{p}_R(\phi)$ can be determined and is plotted in Fig. 21b (solid line, right ordinate axis).

The ϕ dependence of $k_D(\phi)$ arises from the variation of the diffusion coefficient $D(\phi)$ with the volume fraction. The diffusion coefficient may be determined as a function of ϕ by carrying out simulations in a system with obstacles but no catalytic sphere (see Fig. 20). The resulting values of $D(\phi)$ can be used to compute $k_D(\phi)$. Using the simulation values of $k_{0f}(\phi)$ and $k_D(\phi)$ determined previously, this estimate for $k_f(\phi)$ is plotted in Fig. 21 (solid line). The estimate given by the dashed line neglects the ϕ dependence of k_{0f}. This has only a small effect on $k_f(\phi)$ since the smaller k_D contribution dominates. We see that crowding produces nontrivial effects on the reaction dynamics that can be explored using MPC dynamics.

XII. SELF-PROPELLED OBJECTS

The description of the motions of swimming bacteria like *E. coli* or molecular motors such as kinesin requires a knowledge of their propulsion mechanisms and the nature of the interactions between these micron and nanoscale objects and the surrounding fluid in which they move. These are only two examples of a large class of small self-propelled objects that one finds in biology. Propulsion occurs by a variety of mechanisms that usually involve the conversion of chemical energy to effect conformational or other changes that drive the motion. Two features characterize such motion: these motors operate in the regime of low Reynolds numbers where inertia is unimportant, and fluctuations may be important because of their small size. As a result, their dynamics differs considerably from that of the self-motion of large objects [43].

Synthetic nanoscale self-propelled objects have been constructed and studied recently [107–109]. The motors are made from bimetallic Pt-Au nanorods immersed in a H_2O_2 solution, which supplies the chemical energy to drive the motion. The catalytic reaction $2H_2O_2(\ell) \rightarrow O_2(g) + 2H_2O(\ell)$ occurs at the Pt end of the rod and is the power source for the motion. Suggested mechanisms for the motion include the surface tension gradient due to O_2 adsorption on the nonreactive Au end or nanobubble formation, although the full mechanism is still a matter of debate. Dreyfus et al. [110] constructed an artificial swimmer that is composed of a red blood cell attached to a filament made from superparamagnetic colloids connected to each other with DNA. Magnetic fields are used to induce oscillatory motion of the filament that drives the swimmer.

A number of different model motor systems have been proposed and studied theoretically in order to gain insight into the nature and origin of the motor motion. There is a large literature on models for biological and Brownian motors [111, 112]. Other motor models include coupled objects with asymmetric shapes residing in heat baths with different temperatures [113], swimmers composed of linked beads that undergo irreversible cyclic conformational changes [114], and motors that utilize an asymmetric distribution of reaction products in combination with phoretic forces to effect propulsion [115].

Multiparticle collision dynamics provides an ideal way to simulate the motion of small self-propelled objects since the interaction between the solvent and the motor can be specified and hydrodynamic effects are taken into account automatically. It has been used to investigate the self-propelled motion of swimmers composed of linked beads that undergo non-time-reversible cyclic motion [116] and chemically powered nanodimers [117]. The chemically powered nanodimers can serve as models for the motions of the bimetallic nanodimers discussed earlier. The nanodimers are made from two spheres separated by a fixed distance R dissolved in a solvent of A and B molecules. One dimer sphere (C) catalyzes the irreversible reaction $A + C \rightarrow B + C$, while nonreactive interactions occur with the noncatalytic sphere (N). The nanodimer and reactive events are shown in Fig. 22. The A and B species interact with the nanodimer spheres through repulsive Lennard-Jones (LJ) potentials in Eq. (76). The MPC simulations assume that the potentials satisfy $V_{CA} = V_{CB} = V_{NA}$, with ϵ_A and V_{NB} with ϵ_B. The A molecules react to form B molecules when they approach the catalytic sphere within the interaction distance $r < r_c$. The B molecules produced in the reaction interact differently with the catalytic and noncatalytic spheres.

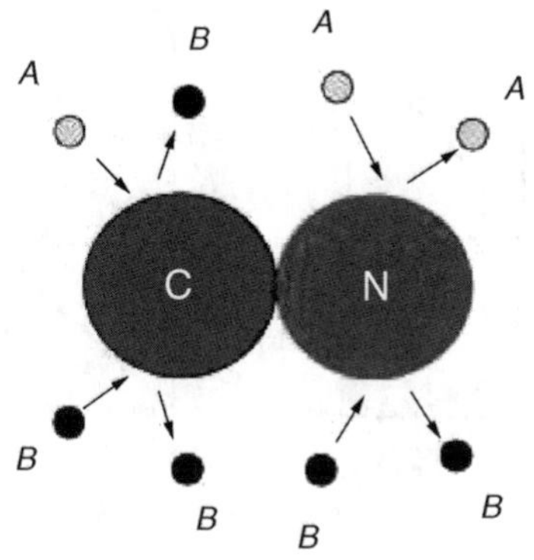

Figure 22. Catalytic (C) and noncatalytic (N) dimer spheres and the collision events that occur on interaction of the A and B species with each sphere.

The velocity of the dimer along its internuclear z-axis can be determined in the steady state where the force due to the reaction is balanced by the frictional force: $\zeta V_z = \langle \hat{\mathbf{z}} \cdot \mathbf{F} \rangle$. Since the diffusion coefficient is related to the friction by $D = k_B T/\zeta = 1/\beta\zeta$, we have

$$V_z = D\langle \hat{\mathbf{z}} \cdot \beta \mathbf{F} \rangle \tag{94}$$

where the average z-component of the force is

$$\begin{aligned}\langle \hat{\mathbf{z}} \cdot \mathbf{F} \rangle = &-\sum_{\alpha=A}^{B} \int d\mathbf{r} \rho_\alpha(\mathbf{r})(\hat{\mathbf{z}} \cdot \hat{\mathbf{r}}) \frac{dV_{C\alpha}(r)}{dr} \\ &-\sum_{\alpha=A}^{B} \int d\mathbf{r}' \rho_\alpha(\mathbf{r}' + \mathbf{R})(\hat{\mathbf{z}} \cdot \hat{\mathbf{r}}') \frac{dV_{N\alpha}(r')}{dr'}\end{aligned} \tag{95}$$

The first and second integrals have their coordinate systems centered on the catalytic C and noncatalytic N spheres, respectively. The local nonequilibrium average microscopic density field for species α is $\rho_\alpha(\mathbf{r}) = \left\langle \sum_{i=1}^{N_\alpha} \delta(\mathbf{r} - \mathbf{r}_{i\alpha}) \right\rangle$. The solution of the diffusion equation can be used to estimate this nonequilibrium density, and thus the velocity of the nanodimer can be computed. The simple model yields results in qualitative accord with the MPC dynamics simulations and shows how the nonequilibrium density field produced by reaction, in combination with the different interactions of the B particles with the noncatalytic sphere, leads to directed motion [117].

Since hydrodynamic interactions are included in MPC dynamics, the collective motion of many self-propelled objects can be studied using this mesoscopic simulation method.

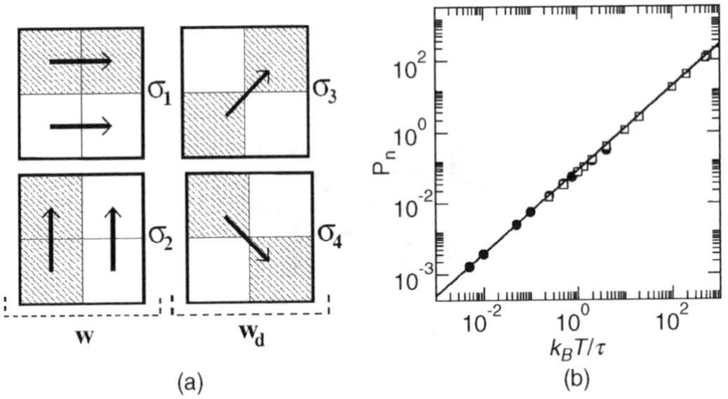

Figure 23. (a) Collision cell structure for MPC collisions that account for excluded volume in a two-dimensional system. Momentum exchange occurs between the pairs of cells in the directions indicated by the arrows. (b) Nonideal pressure as a function of k_BT/τ. (From Ref. 118.)

XIII. GENERALIZATIONS OF MPC DYNAMICS

A. Nonideal Fluids

Multiparticle collision dynamics as formulated earlier has an ideal gas equation of state. Ihle, Tüzel, and Kroll [118] have generalized the collision rule to account for excluded volume effects akin to those in hard sphere fluids. Incorporation of this effect in the dynamics leads to a nonideal equation of state, which is desirable for many applications.

Recall that in MPC dynamics the system is divided into cells with linear dimension a. The excluded volume effect is accounted for by introducing another grid that defines supercells with linear dimension $2a$ within which the multiparticle collisions are carried out. The cell structure is sketched in Fig. 23a for a two-dimensional system. Grid shifting is performed as described earlier but on the interval $[-a, a]$ to account for the doubled size of the supercells. Two cells are randomly selected from every supercell in the system. Collisions are carried out on two double cells in each supercell so that all particles in the system have an opportunity to collide. From the figure we see that there are three possible choices for collisions in the directions indicated on the cubic lattice[4]: horizontal, vertical, and along the diagonals. The directions in which momentum exchange takes place are defined by the unit vectors $\hat{\boldsymbol{\sigma}}_\alpha$ $(\alpha = 1, \ldots, 4)$. The center of mass velocity in a cell ξ is again denoted by $\mathbf{V}_\xi$. If the selected cells have indices 1 and 2, the projection of the center of mass velocity difference on the direction $\hat{\boldsymbol{\sigma}}_i$ is $\Delta\mathbf{V} = \hat{\boldsymbol{\sigma}}_i \cdot (\mathbf{V}_1 - \mathbf{V}_2)$. If $\Delta\mathbf{V} < 0$ no

[4]The introduction of a cubic lattice grid leads to anisotropy in the equations at the Burnett and higher levels.

collision takes place. If $\Delta\mathbf{V} \geq 0$ collision occurs with a probability that depends on $\Delta\mathbf{V}$ and the numbers of particles in the two cells, N_1 and N_2. One choice for this probability is $p = \theta(\Delta\mathbf{V})\tanh(A\,\Delta\,\mathbf{V}N_1N_2)$, where A is a parameter that is used to tune the equation of state.

Given that a collision takes place, the nature of the momentum transfer between the cells must be specified. This should be done in such a way that the total momentum and kinetic energy on the double cell are conserved. There are many ways to do this. A multiparticle collision event may be carried out on all particles in the pair of cells. Alternatively, a hard sphere collision can be mimicked by exchanging the component of the mean velocities of the two cells along $\hat{\boldsymbol{\sigma}}_\alpha$,

$$\hat{\boldsymbol{\sigma}}_\alpha \cdot (\mathbf{v}_i^* - \mathbf{V}) = -\hat{\boldsymbol{\sigma}}_\alpha \cdot (\mathbf{v}_i - \mathbf{V}) \tag{96}$$

Here $\mathbf{V} = (N_1\mathbf{V}_1 + N_2\mathbf{V}_2)/(N_1 + N_2)$ is the mean velocity of the pair of cells. By summing over the particles in cell 1 it is easy to verify that $\hat{\boldsymbol{\sigma}}_\alpha \cdot \mathbf{V}_1^* = -\hat{\boldsymbol{\sigma}}_\alpha \cdot \mathbf{V}_1$, with a similar expression for $\hat{\boldsymbol{\sigma}}_\alpha \cdot \mathbf{V}_2^*$ obtained by interchanging the indices 1 and 2. The components of the velocities normal to $\hat{\boldsymbol{\sigma}}_\alpha$ remain unchanged. This collision rule treats the particles in the two cells as groups that undergo elastic-like collisions.

Although this collision rule conserves momentum and energy, in contrast to the original version of MPC dynamics, phase space volumes are not preserved. This feature arises from the fact that the collision probability depends on $\Delta\mathbf{V}$ so that different system states are mapped onto the same state. Consequently, it is important to check the consistency of the results in numerical simulations to ensure that this does not lead to artifacts.

The pressure can be computed by calculating the average momentum transfer across a fixed plane per unit area and time. The result is $P = \bar{n}k_BT + P_n$, where P_n is the nonideal contribution to the pressure which, for small A, takes the form

$$P_n = \left(\frac{1}{2\sqrt{2}} - \frac{1}{4}\right)\frac{Aa^3}{2}\frac{k_BT}{\tau}\bar{n}^2 + \mathcal{O}(A^3T^2) \tag{97}$$

Since the internal energy is that of an ideal gas, the pressure must be linear in the temperature. We see that this is true provided A is sufficiently small. The nonideal contribution to the pressure is plotted in Fig. 23 as a function of the temperature. Linear scaling is observed over a wide range of temperatures, even when the nonideal contribution to the pressure dominates the ideal contribution. For high enough pressures an ordering transition is observed and solid phase with cubic symmetry is obtained [118]. This collision model can be generalized to binary mixtures and phase segregation can be studied.

B. Immiscible Fluids

Immiscible fluids can be simulated using MPC dynamics by generalizing the collision rule to describe attractive and repulsive interactions among the different species. One method for constructing such a rule has its antecedents in the Rothman–Keller model [119] for immiscible lattice gases. The immiscible lattice-gas model accounted for cohesion in real fluids that arises from short-range attractive intermolecular forces by allowing particles in neighboring sites to influence the configuration of particles at a chosen site. Suppose the two species in a binary mixture are denoted red and blue. The Rothman–Keller model defines a color flux and a color field and constructs collision rules where the "work" performed by the color flux against the color field is a minimum.

In a similar spirit, Inoue et al. [120] and Hashimoto et al. [121] generalized MPC dynamics so that the collision operator reflects the species compositions in the neighborhood of a chosen cell. More specifically, consider a binary mixture of particles with different colors. The color of particle i is denoted by c_i. The color flux of particles with color c in cell ξ is defined as

$$\mathbf{q}_c(\xi) = \sum_{i=1}^{N_\xi} \delta_{cc_i}(\mathbf{v}_i - \mathbf{V}_\xi) \tag{98}$$

that is, just the sum of the velocities, relative to the center of mass velocity in cell ξ, of all particles with color c in cell ξ. The color field is a color gradient arising from the color differences in neighboring cells. It is defined as

$$\mathbf{f}_c(\xi) = \sum_{\{j|\mathbf{r}_j \in \xi'\}} \kappa_{cc_j} \frac{[\mathbf{r}_j] - [\mathbf{r}_i]}{|[\mathbf{r}_j] - [\mathbf{r}_i]|^d} \delta(1 - ([\mathbf{r}_j] - [\mathbf{r}_i])) \tag{99}$$

where ξ' lies in the neighborhood $\mathcal{N}(\xi)$ of the cell ξ and d is the dimension. The parameter $\kappa_{cc'}$ specifies the magnitude and type of interaction: it is positive for attractive interactions and negative for repulsive interactions. An interaction energy is associated to these fields and is minimized to obtain the rotation angle in MPC dynamics.

As an example, consider the red–blue binary mixture, where $c = r$ or b. For this case $\mathbf{f}_r(\xi) = -\mathbf{f}_b(\xi)$ and a potential

$$U(\xi) = -\mathbf{q}_r(\xi) \cdot \mathbf{f}_r(\xi) - \mathbf{q}_b(\xi) \cdot \mathbf{f}_b(\xi) = -\mathbf{q}(\xi) \cdot \mathbf{f}_r(\xi) \tag{100}$$

may be defined where $\mathbf{q}(\xi) = \mathbf{q}_r(\xi) - \mathbf{q}_b(\xi)$. The potential U is at a minimum when $\mathbf{q}(\xi)$ is parallel to $\mathbf{f}_r(\xi)$. The rotation operator $\hat{\omega}_\xi$ for multiparticle collisions in a cell ξ is constructed so that the rotated color flux lies along the color field, $\hat{\mathbf{f}}_r = \hat{\omega}_\xi \hat{\mathbf{q}}$. Let $\hat{\mathbf{h}} = \hat{\mathbf{f}}_r \times \hat{\mathbf{q}}$ be a unit vector that is normal to both the color flux and color field vectors. The angle between $\hat{\mathbf{q}}$ and $\hat{\mathbf{f}}_r$ is θ and $\cos\theta = \hat{\mathbf{f}}_r \cdot \hat{\mathbf{q}}$. Then the rotation operator $\hat{\omega}_\xi$, which effects a rotation of $\hat{\mathbf{q}}$ by θ about $\hat{\mathbf{h}}$, has the same form as Eq. (3),

$$\mathbf{v}_i^* = \mathbf{V}_\xi + \hat{\omega}_\xi(\mathbf{v}_i - \mathbf{V}_\xi) = \mathbf{V}_\xi + \hat{\mathbf{h}}\hat{\mathbf{h}} \cdot (\mathbf{v}_i - \mathbf{V}_\xi) \\ + (\mathbf{I} - \hat{\mathbf{h}}\hat{\mathbf{h}}) \cdot (\mathbf{v}_i - \mathbf{V}_\xi)\cos\theta - \hat{\mathbf{h}} \times (\mathbf{v}_i - \mathbf{V}_\xi)\sin\theta \qquad (101)$$

Immiscible fluid MPC dynamics has been used to investigate microemulsions [122] and droplets in a bifurcating channel [123].

XIV. SUMMARY

Multiparticle collision dynamics describes the interactions in a many-body system in terms of effective collisions that occur at discrete time intervals. Although the dynamics is a simplified representation of real dynamics, it conserves mass, momentum, and energy and preserves phase space volumes. Consequently, it retains many of the basic characteristics of classical Newtonian dynamics. The statistical mechanical basis of multiparticle collision dynamics is well established. Starting with the specification of the dynamics and the collision model, one may verify its dynamical properties, derive macroscopic laws, and, perhaps most importantly, obtain expressions for the transport coefficients. These features distinguish MPC dynamics from a number of other mesoscopic schemes. In order to describe solute motion in solution, MPC dynamics may be combined with molecular dynamics to construct hybrid schemes that can be used to explore a variety of phenomena. The fact that hydrodynamic interactions are properly accounted for in hybrid MPC–MD dynamics makes it a useful tool for the investigation of polymer and colloid dynamics. Since it is a particle-based scheme it incorporates fluctuations so that the reactive and nonreactive dynamics in small systems where such effects are important can be studied.

The dynamical regimes that may be explored using this method have been described by considering the range of dimensionless numbers, such as the Reynolds number, Schmidt number, Peclet number, and the dimensionless mean free path, which are accessible in simulations. With such knowledge one may map MPC dynamics onto the dynamics of real systems or explore systems with similar characteristics. The applications of MPC dynamics to studies of fluid flow and polymeric, colloidal, and reacting systems have confirmed its utility.

The basic model has already been extended to treat more complex phenomena such as phase separating and immiscible mixtures. These developments are still at an early stage, both in terms of the theoretical underpinnings of the models and the applications that can be considered. Further research along such lines will provide even more powerful mesoscopic simulation tools for the study of complex systems.

APPENDIX A: STRUCTURE OF *K(k)*

In order to cast the expression for $K(k)$ into a form that is convenient for its evaluation, it is useful to establish several relations first. To this end we let $S(\mathbf{x}^N, \tau)$ denote the value of the phase point at time τ whose value at time zero was $\mathbf{x}^N$. We then have

$$\begin{aligned}\int d\mathbf{x}^N h(\mathbf{x}^N)\hat{\mathsf{L}}g(\mathbf{x}^N) &= \int d\mathbf{x}^N\, h(\mathbf{x}^N) \int d\mathbf{x}'^N\, \mathsf{L}(\mathbf{x}^N, \mathbf{x}'^N) g(\mathbf{x}'^N)\\ \int d\mathbf{x}'^N \left[\int d\mathbf{x}^N\, h(\mathbf{x}^N)\mathsf{L}(\mathbf{x}^N, \mathbf{x}'^N)\right] g(\mathbf{x}'^N) &= \int d\mathbf{x}'^N\, h(S(\mathbf{x}'^N, \tau)) g(\mathbf{x}'^N)\end{aligned} \tag{A.1}$$

for any phase space functions h and g. Furthermore, we consider integrals of the form

$$\begin{aligned}\int d\mathbf{x}^N h(\mathbf{x}^N)\mathcal{Q}\hat{\mathsf{L}}\mathcal{Q}g(\mathbf{x}^N) &= \int d\mathbf{x}^N\, h(\mathbf{x}^N)\hat{\mathsf{L}}\mathcal{Q}g(\mathbf{x}^N) - \langle h(\mathbf{x}^N) n_{-\mathbf{k}}(\mathbf{r}_1)\rangle\\ &\quad \times \int d\mathbf{x}'^N\, n_{\mathbf{k}}(S(\mathbf{x}'_1, \tau))\mathcal{Q}g(\mathbf{x}'^N)\\ &= \int d\mathbf{x}^N\, h(\mathbf{x}^N)\hat{\mathsf{L}}\mathcal{Q}g(\mathbf{x}^N) + \mathcal{O}(k)\end{aligned} \tag{A.2}$$

The last equality follows from the fact that $n_{\mathbf{k}}(S(\mathbf{x}'_1, \tau)) = e^{i\mathbf{k}\cdot(\mathbf{r}'_i + \mathbf{v}'_i\tau)}$ and $n_{\mathbf{k}}(\mathbf{x}_1)$ is orthogonal to $\mathcal{Q}g(\mathbf{x}^N)$. This result implies that the projected evolution may be replaced by ordinary MPC evolution for small wavevectors in the evaluation of the kinetic coefficients. Finally, we consider averages of the type

$$\begin{aligned}\langle h_{\mathbf{k}}(\mathbf{x}^N)\mathcal{Q}(\hat{\mathsf{L}} - 1) n_{-\mathbf{k}}(\mathbf{x}_1)\rangle &= \int d\mathbf{x}^N\, h(\mathbf{x}^N)\mathcal{Q}(\hat{\mathbf{L}} - 1) n_{-\mathbf{k}}(\mathbf{x}_1)\mathsf{P}_0(\mathbf{x}^N)\\ &= \int d\mathbf{x}^N\, h(\mathbf{x}^N)\mathcal{Q}(\hat{\mathsf{L}} - 1) n_{-\mathbf{k}}(\mathbf{x}_1)\hat{\mathsf{L}}\mathsf{P}_0(\mathbf{x}^N)\end{aligned} \tag{A.3}$$

where in the last line we made use of the invariance of P_0 under the dynamics. Making repeated use of Eq. (A.2) we find

$$\begin{aligned}&\int d\mathbf{x}^N\, h(\mathbf{x}^N)\mathcal{Q}(\hat{\mathsf{L}} - 1) n_{-\mathbf{k}}(\mathbf{x}_1)\hat{\mathsf{L}}\mathsf{P}_0(\mathbf{x}^N)\\ &= \langle h(S(\mathbf{x}^N, 2\tau))[n_{-\mathbf{k}}(S(\mathbf{x}_1, \tau)) - n_{-\mathbf{k}}(S(\mathbf{x}_1, 2\tau))\langle n_{\mathbf{k}}(S(\mathbf{x}_1, \tau)) n_{-\mathbf{k}}(\mathbf{x}_1)\rangle]\rangle\\ &\quad \times \langle h(S(\mathbf{x}^N, \tau))[n_{-\mathbf{k}}(\mathbf{x}_1) - n_{-\mathbf{k}}(S(\mathbf{x}_1, \tau))\langle n_{\mathbf{k}}(S(\mathbf{x}_1, \tau)) n_{-\mathbf{k}}(\mathbf{x}_1)\rangle]\rangle\end{aligned} \tag{A.4}$$

We may now use these results to write the expression for $K(k)$ in a useful form. We define

$$f_{\mathbf{k}}(\mathbf{x}_1) \equiv n_{\mathbf{k}}(S(\mathbf{x}_1,\tau)) - \langle n_{\mathbf{k}}(S(\mathbf{x}_1,\tau))n_{-\mathbf{k}}(\mathbf{x}_1)\rangle n_{\mathbf{k}}(\mathbf{x}_1) \tag{A.5}$$

which is orthogonal to $n_{-\mathbf{k}}(\mathbf{x}_1)$, and

$$\tilde{f}_{-\mathbf{k}}(\mathbf{x}_1) \equiv n_{-\mathbf{k}}(\mathbf{x}_1) - n_{-\mathbf{k}}(S(\mathbf{x}_1,\tau))\langle n_{\mathbf{k}}(S(\mathbf{x}_1,\tau))n_{-\mathbf{k}}(\mathbf{x}_1)\rangle \tag{A.6}$$

which is orthogonal to $n_{\mathbf{k}}(S(\mathbf{x}_1,\tau))$. Making use of the identities established earlier, we have

$$\langle n_{\mathbf{k}}(\mathbf{r}_1)(\hat{\mathsf{L}}-1)(\mathcal{Q}\hat{\mathsf{L}})^{m-1}\mathcal{Q}(\hat{\mathsf{L}}-1)n_{-\mathbf{k}}(\mathbf{r}_1)\rangle = \langle f_{\mathbf{k}}(\mathbf{x}_1)\mathsf{L}^m\tilde{f}_{-\mathbf{k}}(\mathbf{x}_1)\rangle \tag{A.7}$$

Thus we may write $K(k)$ as

$$K(k) = \langle n_{\mathbf{k}}(\mathbf{r}_1)(\hat{\mathsf{L}}-1)n_{-\mathbf{k}}(\mathbf{r}_1)\rangle + \sum_{m=1}^{\infty}\langle f_{\mathbf{k}}(\mathbf{x}_1)\mathsf{L}^m\tilde{f}_{-\mathbf{k}}(\mathbf{x}_1)\rangle \tag{A.8}$$

References

1. L. Boltzmann, *Lectures on Gas Theory*, University of California Press, Berkeley, 1964.
2. P. Langevin, Sur la théorie du mouvement brownian, *Comptes Rendus* **146**, 530 (1908).
3. S. Succi, *The Lattice Boltzmann Equation for Fluid Dynamics and Beyond*, Clarendon Press, New York, 2001.
4. G. A. Bird, *Molecular Gas Dynamics*, Clarendon Press, Oxford, UK, 1976.
5. P. J. Hoogerbrugge and J. M. V. A. Koelman, Simulating microscopic hydrodynamic phenomena with dissipative particle dynamics, *Europhys. Lett.* **19**, 155 (1992).
6. P. Español and P. Warren, Statistical mechanics of dissipative particle dynamics, *Europhys. Lett.* **30**, 191 (1995).
7. J. J. Monaghan, Smoothed particle hydrodynamics, *Ann. Rev. Astron. Astrophys.* **30**, 543 (1992).
8. M. J. Kotelyanskii and D. N. Theodorou (eds.), *Simulation Methods for Polymers*, Marcel Dekker, New York, 2004.
9. M. Karttunen, I. Vattulainen, and A. Lukkarinen (eds.), *Novel Methods in Soft Matter Simulations*, Springer-Verlag, Berlin, 2004.
10. S. O. Nielsen, C. F. Lopez, G. Srinivas, and M. L. Klein, Coarse-grain models and the simulation of soft materials, *J. Phys. Condens. Matter* **16**, R481 (2004).
11. A. Malevanets and R. Kapral, Mesoscopic model for solvent dynamics, *J. Chem. Phys.* **110**, 8605 (1999).
12. A. Malevanets and R. Kapral, Continuous-velocity lattice-gas model for fluid flow, *Europhys. Lett.* **44**, 552 (1998).

13. A. Malevanets and R. Kapral, Mesoscopic multi-particle collision model for fluid flow and molecular dynamics, in *Novel Methods in Soft Matter Simulations*, M. Karttunen, I. Vattulainen, and A. Lukkarinen (eds.), Springer-Verlag, Berlin, 2003, p. 113.

14. T. Ihle and D. M. Kroll, Stochastic rotation dynamics: a Galilean-invariant mesoscopic model for fluid flow, *Phys. Rev. E* **63**, 020201(R) (2001).

15. T. Ihle and D. M. Kroll, Stochastic rotation dynamics. I. Formalism, Galilean invariance, and Green–Kubo relations, *Phys. Rev. E* **67**, 066705 (2003).

16. A. Malevanets and J. M. Yeomans, A particle-based algorithm for the hydrodynamics of binary fluid mixtures, *Comput. Phys. Commun.* **129**, 282 (2000).

17. K. Tucci and R. Kapral, Mesoscopic model for diffusion-influenced reaction dynamics, *J. Chem. Phys.* **120**, 8262 (2004).

18. A. Malevanets and R. Kapral, Solute dynamics in a mesoscale solvent, *J. Chem. Phys.* **112**, 7260 (2000).

19. M. Ripoll, K. Mussawisade, R. G. Winkler, and G. Gompper, Low-Reynolds-number hydrodynamics of complex fluids by multi-particle-collision dynamics, *Europhys. Lett.* **68**, 106 (2004).

20. M. Ripoll, K. Mussawisade, R. G. Winkler, and G. Gompper, Dynamic regimes of fluids simulated by multiparticle-collision dynamics, *Phys. Rev. E* **72**, 016701 (2005).

21. R. G. Winkler, M. Ripoll, K. Mussawisade, and G. Gompper, Simulation of complex fluids by multi-particle-collision dynamics, *Comput. Phys. Commun.* **169**, 326 (2005).

22. E. Tüzel, T. Ihle, and D. M. Kroll, Dynamic correlations in stochastic rotation dynamics, *Phys. Rev. E* **74**, 056702 (2006).

23. T. Ihle and D. M. Kroll, Stochastic rotation dynamics. II. Transport coefficients, numerics, and long-time tails, *Phys. Rev. E* **67**, 066706 (2003).

24. E. Tüzel, M. Strauss, T. Ihle, and D. M. Kroll, Transport coefficients for stochastic rotation dynamics in three dimensions, *Phys. Rev. E* **68**, 036701 (2003).

25. T. Ihle, E. Tüzel, and D. M. Kroll, Resummed Green–Kubo relations for a fluctuating fluid-particle model, *Phys. Rev. E* **70**, 035701(R) (2004).

26. T. Ihle, E. Tüzel, and D. M. Kroll, Equilibrium calculation of transport coefficients for a fluid-particle model, *Phys. Rev. E* **72**, 046707 (2005).

27. J. M. Yeomans, Mesoscale simulations: lattice Boltzmann and particle algorithms, *Physica A* **369**, 159 (2006).

28. N. Kikuchi, C. M. Pooley, J. F. Ryder, and J. M. Yeomans, Transport coefficients of a mesoscopic fluid dynamics model, *J. Chem. Phys.* **119**, 6388 (2003).

29. C. M. Pooley and J. M. Yeomans, Kinetic theory derivation of the transport coefficients of stochastic rotation dynamics, *J. Phys. Chem. B* **109**, 6505 (2005).

30. E. Allahyarov and G. Gompper, Mesoscopic solvent simulations: multiparticle-collision dynamics of three-dimensional flows, *Phys. Rev. E* **66**, 036702 (2002).

31. A. Lamura and G. Gompper, Numerical study of the flow around a cylinder using multi-particle collision dynamics, *Eur. Phys. J.* **9**, 477 (2002).

32. M. von Smoluchowski, Über Brownsche Molekularbewegung unter Einwirkung äusserer Kräfte und deren Zusammenhang mit der verallgemeinerten Diffusionsgleichung, *Ann. Phys.* **48**, 1103 (1915); Zusammenfassende bearbeitungen, *Phys. Z.* **17**, 557 (1916).

33. G. Nicolis and I. Prigogine, *Self-Organization in Non-Equilibrium Systems*, Wiley, Hoboken, NJ, 1977.

34. C. W. Gardiner, *Handbook of Stochastic Processes*, Springer-Verlag, New York, 1985.

35. D. T. Gillespie, Exact stochastic simulation of coupled chemical reactions, *J. Phys. Chem.* **81**, 2340 (1977).

36. J. S. van Zon and P. R. ten Wolde, Simulating biochemical networks at the particle level and in time and space: Green's function reaction dynamics, *Phys. Rev. Lett.* **94**, 128103 (2005).

37. J. V. Rodríguez, J. A. Kaandorp, M. Dobrzyński, and J. K. Blom, Spatial stochastic modelling of the phosphoenolpyruvate-dependent phosphotransferase (PTS) pathway in *Escherichia coli*, *Bioinformatics* **22**, 1895 (2006).

38. R. Grima and S. Schnell, A systematic investigation of the rate laws valid in intracellular environments, *Biophys. Chem.* **124**, 1 (2006).

39. K. Tucci and R. Kapral, Mesoscopic multiparticle collision dynamics of reaction-diffusion fronts, *J. Phys. Chem. B* **109**, 21300 (2005).

40. A. Malevanets and J. M. Yeomans, Dynamics of short polymer chains in solution, *Europhys. Lett.* **52**, 231 (2000).

41. J. T. Padding and A. A. Louis, Hydrodynamic interactions and Brownian forces in colloidal suspensions: coarse-graining over time and length scales, *Phys. Rev. E* **74**, 031402 (2006).

42. M. Hecht, J. Harting, T. Ihle, and H. J. Herrmann, Simulation of claylike colloids, *Phys. Rev. E* **72**, 011408 (2005).

43. E. Purcell, Life at low Reynolds number, *Am. J. Phys.* **45**, 3 (1977).

44. H. Mori, Transport collective motion and Brownian motion, *Prog. Theor. Phys.* **33**, 423 (1965).

45. P. Mazur and I. Oppenheim, Molecular theory of Brownian motion, *Physica* **50**, 241 (1970).

46. S. Chandrasekhar, Stochastic problems in physics and astronomy, *Rev. Mod. Phys.* **15**, 1 (1943).

47. M. Tokuyama and I. Oppenheim, Statistical-mechanical theory of Brownian motion—translational motion in an equilibrium fluid, *Physica A* **94**, 501 (1978).

48. R. Kubo, Statistical-mechanical theory of irreversible processes. 1. General theory and simple applications to magnetic and conduction problems, *J. Phys. Soc. Japan* **12**, 570 (1957); R. Kubo, The fluctuation-dissipation theorem, *Rep. Prog. Phys.* **29**, 255 (1966).

49. P. Español and I. Zúñiga, Force autocorrelation function in Brownian motion theory, *J. Chem. Phys.* **98**, 574 (1993).

50. F. Ould-Kaddour and D. Levesque, Determination of the friction coefficient of a Brownian particle by molecular-dynamics simulation, *J. Chem. Phys.* **118**, 7888 (2003).

51. S. H. Lee and R. Kapral, Friction and diffusion of a Brownian particle in a mesoscopic solvent, *J. Chem. Phys.* **22**, 11163 (2004).

52. J. T. Hynes, R. Kapral, and M. Weinberg, Molecular theory of translational diffusion: microscopic generalization of the normal velocity boundary condition, *J. Chem. Phys.* **70**, 1456 (1970).

53. S. H. Lee and R. Kapral, Cluster structure and dynamics in a mesoscopic solvent, *Physica A* **298**, 56 (2001).

54. J. Happel and H. Brenner, *Low Reynolds Number Hydrodynamics*, Prentice Hall, Englewood Cliffs, NJ, 1965.

55. M. Tokuyama and I. Oppenheim, On the theory of concentrated hard-sphere suspensions, *Physica A* **216**, 85 (1995).

56. H. Yamakawa, *Modern Theory of Polymer Solutions*, Harper & Row, New York, 1971.

57. J. K. G. Dhont, *An Introduction to Dynamics of Colloids*, Elsevier, Amsterdam, 1996.

58. J. M. Deutch and I. Oppenheim, Molecular theory of Brownian motion for several particles, *J. Chem. Phys.* **54**, 3547 (1971).

59. S. H. Lee and R. Kapral, Two-particle friction in a mesoscopic solvent, *J. Chem. Phys.* **122**, 2149016 (2005).
60. M. Hecht, J. Harting, M. Bier, J. Reinshagen, and H. J. Herrmann, Shear viscosity of claylike colloids in computer simulations and experiments, *Phys. Rev. E* **74**, 021403 (2006).
61. Y. Sakazaki, S. Masuda, J. Onishi, Y. Chen, and H. Ohashi, The modeling of colloidal fluids by the real-coded lattice gas, *Math. Comput. Simulation* **72**, 184 (2006).
62. J. T. Padding and A. A. Louis, Hydrodynamic and Brownian fluctuations in sedimenting suspensions, *Phys. Rev. Lett.* **93**, 220601 (2004).
63. E. Falck, J. M. Lahtinen, I. Vattulainen, and T. Ala-Nissila, Influence of hydrodynamics on many-particle diffusion in 2D colloidal suspensions, *Eur. Phys. J. E* **13**, 267 (2004).
64. Y. Inoue, Y. Chen, and H. Ohashi, Development of a simulation model for solid objects suspended in a fluctuating fluid, *J. Stat. Phys.* **107**, 85 (2002).
65. H. Noguchi and G. Gompper, Dynamics of vesicle self-assembly and dissolution, *J. Chem. Phys.* **125**, 164908 (2006).
66. H. Noguchi and G. Gompper, Meshless membrane model based on the moving least-squares method, *Phys. Rev. E* **73**, 021903 (2006).
67. H. Noguchi and G. Gompper, Shape transitions of fluid vesicles and red blood cells in capillary flows, *Proc. Natl. Acad. Sci. USA* **102**, 14159 (2005).
68. H. Noguchi and G. Gompper, Dynamics of fluid vesicles in shear flow: effect of membrane viscosity and thermal fluctuations, *Phys. Rev. E* **72**, 011901 (2005).
69. H. Noguchi and G. Gompper, Fluid vesicles with viscous membranes in shear flow, *Phys. Rev. Lett.* **93**, 258102 (2004).
70. P.-G. de Gennes, *Scaling Concepts in Polymer Physics*, Cornell University Press, Ithaca, 1979.
71. M. Doi and S. F. Edwards, *Theory of Polymer Dynamics*, Clarendon, Oxford, 1989.
72. A. Yu. Grosberg and A. R. Kohkhlov, *Statistical Mechanics of Macromolecules*, AIP Press, New York, 1994.
73. R. B. Bird, C. F. Curtiss, R. C. Armstrong, and O. Hassager, *Dynamics of Polymeric Liquids*, Wiley, Hoboken, NJ, 1987.
74. K. Kremer and G. S. Grest, Dynamics of entangled linear polymer melts: a molecular-dynamics simulation, *J. Chem. Phys.* **92**, 5057 (1990).
75. K. Mussawisade, M. Ripoll, R. G. Winkler, and G. Gompper, Dynamics of polymers in a particle-based mesoscopic solvent, *J. Chem. Phys.* **123**, 144905 (2005).
76. O. Punkkinen, E. Falck, and I. Vattulainen, Dynamics and scaling of polymers in a dilute solution: analytical treatment in two and higher dimensions, *J. Chem. Phys.* **122**, 094904 (2005).
77. E. Falck, O. Punkkinen, I. Vattulainen, and T. Ala-Nissila, Dynamics and scaling of two-dimensional polymers in a dilute solution, *Phys. Rev. E* **68**, 050102 (2003).
78. W. H. Stockmayer, Problems of the statistical thermodynamics of dilute polymer solutions, *Makromol. Chem.* **35**, 54 (1960).
79. P.-G. de Gennes, Kinetics of collapse for a flexible coil, *J. Phys. Lett.* **46**, L639 (1985).
80. A. Yu. Grosberg, S. K. Nechaev, and E. I. Shakhnovich, The role of topological constraints in the kinetics of collapse of macromolecules, *J. Phys. France* **49**, 2095 (1988).
81. A. Halperin and P. M. Goldbart, Early stages of homopolymer collapse, *Phys. Rev. E* **61**, 565 (2000).
82. L. I. Klushin, Kinetics of a homopolymer collapse: beyond the Rouse–Zimm scaling, *J. Chem. Phys.* **108**, 7917 (1998).

83. S. H. Lee and R. Kapral, Mesoscopic description of solvent effects on polymer dynamics, *J. Chem. Phys.* **124**, 214901 (2006).
84. N. Kikuchi, J. F. Ryder, C. M. Pooley, and J. M. Yeomans, Kinetics of the polymer collapse transition: the role of hydrodynamics, *Phys. Rev. E* **71**, 061804 (2005).
85. N. Kikuchi, A. Gent, and J. M. Yeomans, Polymer collapse in the presence of hydrodynamic interactions, *Eur. Phys. J. E* **9**, 63 (2002).
86. R. G. Winkler, K. Mussawisade, M. Ripoll, and G. Gompper, Rod-like colloids and polymers in shear flow: a multi-particle-collision dynamics study, *J. Phys. Condens. Matter* **16**, S3941 (2004).
87. M. A. Webster and J. M. Yeomans, Modeling a tethered polymer in Poiseuille flow, *J. Chem. Phys.* **122**, 164903 (2005).
88. J. F. Ryder and J. M. Yeomans, Shear thinning in dilute polymer solutions, *J. Chem. Phys.* **125**, 194906 (2006).
89. M. Ripoll, R. G. Winkler, and G. Gompper, Star polymers in shear flow, *Phys. Rev. Lett.* **96**, 188302 (2006).
90. I. Ali and J. M. Yeomans, Polymer translocation: the effect of backflow, *J. Chem. Phys.* **123**, 234903 (2005).
91. I. Ali, D. Marenduzzo, and J. M. Yeomans, Polymer packaging and ejection in viral capsids: shape matters, *Phys. Rev. Lett.* **96**, 208102 (2006).
92. I. Ali, D. Marenduzzo, and J. M. Yeomans, Dynamics of polymer packaging, *J. Chem. Phys.* **121**, 8635 (2004).
93. F. C. Collins and G. E. Kimball, Diffusion-controlled reaction rates, *J. Colloid Sci.* **4**, 425 (1949).
94. R. Kapral, Kinetic theory of chemical reactions in liquids, *Adv. Chem. Phys.* **48**, 71 (1981).
95. B. U. Felderhof and J. M. Deutch, Concentration dependence of the rate of diffusion-controlled reactions, *J. Chem. Phys.* **64**, 4551 (1976).
96. J. Lebenhaft and R. Kapral, Diffusion-controlled processes among partially absorbing stationary sinks, *J. Stat. Phys.* **20**, 25 (1979).
97. D. S. Goodsell, Inside a living cell, *Trends Biochem. Sci.* **16**, 203 (1991).
98. A. B. Fulton, How crowded is the cytoplasm? *Cell* **30**, 345 (1982).
99. M. T. Record, Jr., E. S. Courtenay, S. Caley, and H. J. Guttman, Biophysical compensation mechanisms buffering *E. coli* protein–nucleic acid interactions against changing environments, *Trends Biochem. Sci.* **23**, 190 (1998).
100. T. C. Laurent, An early look at macromolecular crowding, *Biophys. Chem.* **57**, 7 (1995).
101. S. B. Zimmerman and A. P. Minton, Macromolecular crowding: biochemical, biophysical, and physiological consequences, *Annu. Rev. Biophys. Biomol. Struct.* **22**, 27 (1993).
102. A. P. Minton, The influence of macromolecular crowding and macromolecular confinement on biochemical reactions in physiological media, *J. Biol. Chem.* **276**, 10577 (2001).
103. K. Luby-Phelps, P. E. Castle, D. L. Taylor, and F. Lanni, Hindered diffusion of inert tracer particles in the cytoplasm of mouse 3T3 cells, *Proc. Natl. Acad. Sci. USA* **84**, 4910 (1987).
104. N. D. Gersohn, K. R. Porter, and B. L. Trus, The cytoplasmic matrix: its volume and surface area and the diffusion of molecules through it, *Proc. Natl. Acad. Sci. USA* **82**, 5030 (1985).
105. S. Schnell and T. E. Turner, Reaction kinetics in intracellular environments with macromolecular crowding: simulations and rate laws, *Prog. Biophys. Mol. Biol.* **85**, 235 (2004).
106. D. Hall and A. P. Minton, Macromolecular crowding: qualitative and semiquantitative successes, quantitative challenges, *Biochim. Biophys. Acta (Proteins Proteomics)* **1649**, 127 (2003).

107. W. F. Paxton, K. C. Kistler, C. C. Olmeda, A. Sen, S. K. St. Angelo, Y. Cao, T. E. Mallouk, P. E. Lammert, and V. H. Crespi, Catalytic nanomotors: autonomous movement of striped nanorods, *J. Am. Chem. Soc.* **126**, 13424 (2004).

108. W. F. Paxton, A. Sen, and T. E. Mallouk, Motility of catalytic nanoparticles through self-generated forces, *Chem. Eur. J.* **11**, 6462 (2005).

109. A. Fournier-Bidoz, A. C. Arsenault, I. Manners, and G. A. Ozin, Synthetic self-propelled nanomotors, *Chem. Commun.* 441 (2005).

110. R. Dreyfus, J. Baudry, M. L. Roper, M. Femigier, H. A. Stone, and J. Bibette, Microscopic artificial swimmers, *Nature* **437**, 862 (2005).

111. J. Howard, *Mechanics of Motor Proteins and the Cytoskeleton*, Sinauer, New York, 2000.

112. P. Reimann, Brownian motors: noisy transport far from equilibrium, *Phys. Rep.* **361**, 57 (2002).

113. C. Van den Broeck, R. Kawai, and P. Meurs, Microscopic analysis of a thermal Brownian motor, *Phys. Rev. Lett.* **93**, 090601 (2004).

114. A. Najafi and R. Golestanian, Simple swimmer at low Reynolds number: three linked spheres, *Phys. Rev. E* **69**, 062901 (2004).

115. R. Golestanian, T. B. Liverpool, and A. Ajdari, Propulsion of a molecular machine by asymmetric distribution of reaction products, *Phys. Rev. Lett.* **94**, 220801 (2005).

116. D. J. Earl, C. M. Pooley, J. F. Ryder, I Bredberg, and J. M. Yeomans, Modeling microscopic swimmers at low Reynolds number, *J. Chem. Phys.* **126**, 064703 (2007).

117. G. Rückner and R. Kapral, Chemically powered nanodimers, *Phys. Rev. Lett.* **98**, 150603 (2007).

118. T. Ihle, E. Tüzel, and D. M. Kroll, Consistent particle-based algorithm with a non-ideal equation of state, *Europhys. Lett.* **73**, 664 (2006).

119. D. H. Rothman and J. M. Keller, Immiscible cellular-automaton fluids, *J. Stat. Phys.* **52**, 1119 (1988).

120. Y. Inoue, Y. Chen, and H. Ohashi, A mesoscopic simulation model for immiscible multiphase fluids, *J. Comput. Phys.* **201**, 191 (2004).

121. Y. Hashimoto, Y. Chen, and H. Ohashi, Immiscible real-coded lattice gas, *Comput. Phys. Commun*, **129**, 56 (2000).

122. T. Sakai, Y. Chen, and H. Ohashi, Real-coded lattice gas model for ternary amphiphilic fluids, *Phys. Rev. E* **65**, 031503 (2002).

123. Y. Inoue, S. Takagi, and Y. Matsumoto, A mesoscopic simulation study of distributions of droplets in a bifurcating channel, *Comp. Fluids* **35**, 971 (2006).

TWO-PATHWAY EXCITATION AS A COHERENCE SPECTROSCOPY

TAMAR SEIDEMAN

Department of Chemistry, Northwestern University, Evanston, Illinois 60208

ROBERT J. GORDON

*Department of Chemistry, University of Illinois at Chicago
Chicago, Illinois 60607*

CONTENTS

Advances in Chemical Physics, Volume 140, edited by Stuart A. Rice
Copyright © 2008 John Wiley & Sons, Inc.

I. BACKGROUND AND MOTIVATION: WHY CHANNEL PHASES?

Coherent control has matured during the past decade from an interesting concept in fundamental research into a major tool in science and technology. Coherent approaches are being applied to manipulate processes ranging from quantum transitions in atoms [1] and the structure and dynamics of electronic wavepackets in atomic and ionic media [2, 3], through propagation of photonic pulses in matter [4, 5] and electron spin dynamics in quantum dots [6], to fragmentation channels in complex molecules [7, 8], reactions in biological systems [9], selective excitation in chromophores [10, 11], and energy flow in photosynthetic bacteria [12]. Potential applications range from quantum information processing [4, 6, 13–16] and measurement technology [17], through enhancement of nonlinear optical processes [5, 18], multiplexed generation of tailored terahertz signals [19], and the development of controlled X-ray sources [3, 20, 21], to photodynamic therapy [10] and new approaches to fast optical switches [22, 23]. For reviews of the theory, experimental realization, and applications of coherent control, we refer the reader to Refs. 24–28.

Much less thoroughly explored, but of similar fundamental value and increasing interest, is the related topic of coherence spectroscopies [29–33]. Underlying this concept is the anticipation that a spectroscopy that exploits the phase properties of light would provide new insights into material properties, beyond what is available from conventional spectroscopies, which utilize only the energy resolution of lasers. Particularly inviting is the possibility of extracting information with regard to the phase properties of matter.

References 29–33 introduce the notion of coherence spectroscopy in the context of two-pathway excitation coherent control. Within the energy domain, two-pathway approach to coherent control [25, 34–36], a material system is simultaneously subjected to two laser fields of equal energy and controllable relative phase, to produce a degenerate continuum state in which the relative phase of the laser fields is imprinted. The probability of the continuum state to evolve into a given product, labeled S, is readily shown (*vide infra*) to vary sinusoidally with the relative phase of the two laser fields ϕ,

$$p^S = A^S + B^S \cos(\phi + \delta^S) \tag{1}$$

where the three constants are elaborated on later. A familiar realization of the two-pathway method is simultaneous excitation through m- and n-photon processes with frequencies satisfying $m\omega_n = n\omega_m$, most commonly one- versus three-photon experiments [34–36] with $3\omega_1 = \omega_3$ (see Ref. 37). In this context, the phase δ^S in Eq. (1) was shown to provide an interesting probe of the material system. In the molecular beam environment [29, 33, 35, 38–46], where coherence is fully maintained on relevant time scales, δ^S probes solely the (ionization and/or dissociation) continua of the isolated molecule, averaged over the spatial coherence length of the laser. In dense environments [47, 48], such as gas cells, interfaces, solutions, and matrices, the phase factor contains much richer information, pertaining to both the molecular continua and the interaction of the system with the environment. These qualitative statements are quantified in the following sections.

From the previous discussion, the reader anticipates that similar information regarding the phase properties of scattering continua and bath–system interactions would be contained in the observables of other coherent control experiments, including pump–dump scenarios, optimal control scenarios, and strong field scenarios [24–27], in all of which a product state is accessed via more than a single pathway, introducing a phase-sensitive component into the observable. Recent research clearly confirms this anticipation. In particular, several publications have illustrated the possibility of utilizing the outcome of optimal control experiments to unravel reaction pathways and mechanisms [7, 11, 12, 21, 28]. Optimal control experiments were shown to provide interesting insights also into the interaction of molecules with a dissipative environment [49–51]. Similar in concept are time-domain pump–probe experiments that likewise use the phase relation between the components of a laser pulse to control multiphoton processes [52, 53] and to gain information about matter wave interference phenomena [54]. Closely related and clearly intriguing is the challenge of developing a true inversion scheme based on coherent control concepts.

The present chapter has no ambition to cover all these topics. We focus solely on the information content of the two-pathway coherent control approach, where the energy-domain, single quantum states approach to the control problem simplifies the phase information and allows analysis at the most fundamental level. We regret having to limit the scope of this chapter and thus exclude much of the relevant literature. We hope, however, that this contribution will entice the reader to explore related literature of relevance.

In the next section we discuss the qualitative physics underlying the channel phase and explore the origin of its information content. Section III surveys the experimental method of quantifying the channel phase, and in Section IV we outline the theoretical framework for its formulation. Here we attempt to make connection with the discussions of Sections II and III, while providing some detail of the underlying scattering theory. To that end we stress simple limits, making use of familiar analogues where possible. Section V provides several examples, starting with the most elementary processes in atomic systems and progressing through diatomic and polyatomic molecules to extended systems and molecules interacting with dissipative environments. The final section concludes with an outlook to future research in this and related areas.

II. QUALITATIVE PHYSICS: HOW DOES IT COME ABOUT?

Envision a bound-free (dissociation or ionization) experiment with a conventional, energy-resolved source, where a bound state $|g\rangle$ is projected onto a dissociation or ionization continuum $|E^-\rangle$. Here g denotes the set of quantum numbers required to fully define the initial state, $E = E_g + \hbar\omega$ is the total energy, E_g being the initial and $\hbar\omega$ the photon energy, and the superscript $-$ indicates incoming wave boundary conditions. In the weak field limit, the observed signal is proportional to the squared modulus of the bound-free matrix element $|\langle E^-|D|g\rangle|^2$, where $D = \vec{\mu} \cdot \vec{\varepsilon}$, $\vec{\mu}$ is the dipole vector operator, and $\vec{\varepsilon}$ is the electromagnetic field. This signal is sensitive to the probability density of the continuum eigenstate in the region of space where the bound state $|g\rangle$ has appreciable amplitude. It probes the *modulus* of the scattering wavefunction but is insensitive to its *phase*. The phase information is lost upon formation of the squared modulus.

Consider, by contrast, a two-color experiment where the continuum is accessed by two laser fields with a well defined relative phase, ε_a and ε_b. A schematic illustration of the experiment envisioned is provided in Fig. 1a, where we consider the specific case of excitation with one- and three-photon fields of

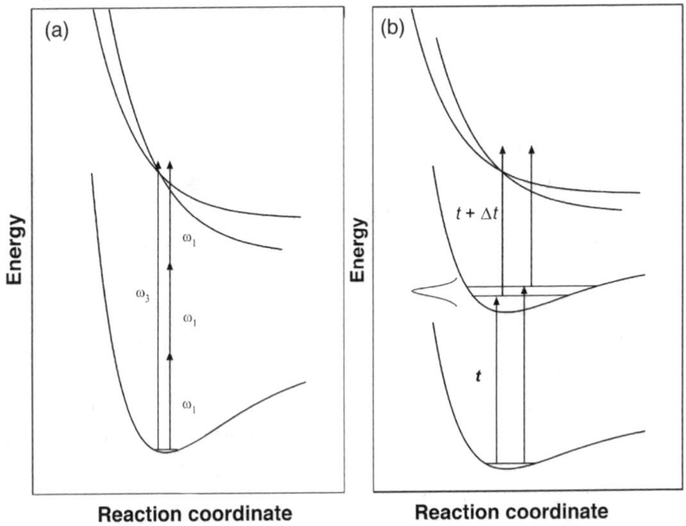

Figure 1. Schematic illustration of two-pathway control in the (a) frequency and (b) time domains. In case (a) the ground state is excited to a coupled continuum by either one photon of frequency ω_3 or three photons of frequency ω_1. Control is achieved by introducing a phase lag between the two fields. In case (b) a two-pulse sequence has sufficient bandwidth to excite a superposition of two intermediate states. Control is achieved by introducing a delay, Δt, between the pulses, resulting in a phase difference of $\omega\,\Delta t$.

equal total photon energy. In the weak field limit, the observable signal is of the form of Eq. (1),

$$\begin{aligned} p &= |\langle E^- | D_a | g\rangle + \langle E^- | D_b | g\rangle|^2 \\ &= p_a + p_b + p_{ab} \end{aligned} \tag{2}$$

where

$$p_\alpha = |\langle E^- | D_\alpha | g\rangle|^2, \quad \alpha = a, b \tag{3}$$

$$\begin{aligned} p_{ab} &= 2\mathrm{Re}\,\{\langle g | D_a | E^-\rangle\langle E^- | D_b | g\rangle\} \\ &= 2\mathrm{Re}\,\{e^{i\phi}\langle g | \bar{D}_a | E^-\rangle\langle E^- | \bar{D}_b | g\rangle\} \\ &= 2\sqrt{p_a p_b}\cos[\phi + \delta(E)] \end{aligned} \tag{4}$$

$\bar{D}_\alpha = |D_\alpha|$, ϕ is the relative phase of the two laser fields, $\phi = \arg\{D_a^* D_b\}$, and δ is the relative phase of the two matrix elements,

$$\delta(E) = \arg\{\langle g | \bar{D}_a | E^-\rangle\langle E^- | \bar{D}_b | g\rangle\}. \tag{5}$$

The former phase, ϕ, serves as an external control tool that can be tuned to vary the interference term and hence the reaction outcome. The latter phase, $\delta(E)$, serves as an analytical tool that provides a route to the phases of the scattering wavefunctions.

Often overlooked, the phase of continuum wavefunctions contains valuable information. It is conveniently illustrated by consideration of the form of the wavefunction within the quasiclassical (WKB) approximation [55],

$$\langle x | E^\pm\rangle = \psi_E^\pm(x) \approx \frac{1}{\sqrt{k(x)}}\exp[\pm i \int^x k(x')dx'], \tag{6}$$

where

$$\lambda(x) = \frac{2\pi}{k(x)} = \frac{h}{\sqrt{2\mu[E - V(x)]}} \tag{7}$$

is the deBroglie wavelength associated with the system, $k(x)$ is the wavenumber, μ is the reduced mass, and $V(x)$ is the potential energy. The prefactor in Eq. (6) is a slowly varying function that is often neglected in zero-order estimates. The *phase* carries the information regarding the complete potential energy curve.

Scattering theory formulates the observables of collisions and bound-free experiments in terms of the partial wave phase shifts, which distinguish the asymptotic form of the exact wavefunction from that of a plane wave and contain all the information about the scattering dynamics. In the case of a central (spherically symmetric) potential, for instance,

$$\psi_E(r,\theta) \rightarrow \sum_{J=0}^{\infty} (2J+1) i^J e^{i\delta_J} \frac{\sin(kr - J\pi/2 + \delta_J)}{(2\pi)^{3/2} kr} P_J(\cos\theta) \tag{8}$$

where $k = \sqrt{2\mu E}/\hbar$ and P_J are Legendre polynomials. Equation (8) differs from a plane wave,

$$e^{ikr\cos\theta} \rightarrow \sum_{J=0}^{\infty} (2J+1) i^J \frac{\sin(kr - J\pi/2)}{kr} P_J(\cos\theta) \tag{9}$$

only by the presence of the phases δ_J. The latter take a particularly transparent form in the semiclassical approximation,

$$\delta_J \approx \lim_{r\to\infty} \left\{ \int^r k_J(r) dr - kr + (J + \tfrac{1}{2}) \frac{\pi}{2} \right\}, \tag{10}$$

where $k_J(r)$, defined by analogy to the one-dimensional (1D) analogue in Eq. (7), depends on J through the centrifugal part of the potential. For a repulsive potential δ_J is negative: the potential pushes the wave out as compared to the free wave. For an attractive potential δ_J is positive: the potential pulls the wave in as compared to the free wave analogue [56].

In Section IV we quantify the relation of the information-rich phase of the scattering wavefunction to the observable $\delta(E)$ of Eq. (5). Here we proceed by connecting the two-pathway method with several other phase-sensitive experiments. Consider first excitation from $|g\rangle$ into an electronically excited bound state with a sufficiently broad pulse to span two levels, E_a and E_b,

$$\varepsilon_{\text{ex}}(\omega) = 2^{-3/2} \varepsilon_{\text{ex}} \tau_{\text{ex}} e^{-i(\omega_{\text{ex}} - \omega) t_{\text{ex}}} e^{-\tau_{\text{ex}}^2 (\omega_{\text{ex}} - \omega)^2 / 4}, \quad 2\pi/\tau_{\text{ex}} \sim E_b - E_a \tag{11}$$

We imagine allowing the system to evolve for a period Δt and then using a second pulse to project the superposition onto the scattering eigenstate $|E^-\rangle$. The second pulse, ε_{de}, is given by a similar expression to Eq. (11), with $t_{\text{de}} = t_{\text{ex}} + \Delta t$ and τ_{de} of order $2\pi/(E_b - E_a)$. A schematic illustration of this experiment is provided in Fig. 1b. The signal will have a form similar to Eq. (2), where now the two simultaneous pathways from $|g\rangle$ to $|E^-\rangle$ are distinguished by the two distinct discrete levels that mediate the bound-free process,

$$p = p_a + p_b + p_{ab} \tag{12}$$

with

$$p_\alpha = |\langle E^-|\mu_{21}|E_\alpha\rangle \varepsilon_{\rm de}(\omega_{EE_\alpha})\langle E_\alpha|\mu_{10}|g\rangle \varepsilon_{\rm ex}(\omega_{E_\alpha E_g})|^2, \quad \alpha = a, b \tag{13}$$

and

$$p_{ab} = 2\mathrm{Re}\,\{\langle g|\mu_{01}|E_a\rangle \varepsilon^*_{\rm ex}(\omega_{E_a E_g})\langle E_a|\mu_{12}|E^-\rangle \varepsilon^*_{\rm de}(\omega_{EE_a})\langle E^-|\mu_{21}|E_b\rangle \varepsilon_{\rm de}(\omega_{EE_b}) \times \langle E_b|\mu_{10}|g\rangle \varepsilon_{\rm ex}(\omega_{E_b E_g})\}. \tag{14}$$

Up to a sign, we find

$$p_{ab} = 2\sqrt{p_a p_b}\cos[\omega_{ab}\,\Delta t + \delta(E)], \tag{15}$$

where $\omega_{ab} = (E_b - E_a)/\hbar$, and

$$\delta(E) = \arg\{\langle E^-|\mu_{21}|E_a\rangle\langle E_b|\mu_{12}|E^-\rangle\}. \tag{16}$$

The extension of Eqs. (12)–(16) to a standard pump–probe experiment, where the final state is a (dissociation or ionization) continuum is straightforward. It requires only that we replace the pulse that spans two vibrational eigenstates by a shorter one that spans several eigenstates.

Equations (12)–(16) explain the observations of Ref. 57, described as "one of the most intriguing and as yet unexplained observations of our study," where the time-resolved signal was found to be time delayed from the expected origin to a degree that depended on the probe energy. The above results also correct an error in several publications on pump–probe studies, where the phase factor arising from the complex nature of the final state was disregarded, and the interference term was taken to be of the form $C\cos(\omega\,\Delta t)$. More interestingly, Eqs. (12)–(16) establish the conceptual equivalence of the energy-domain (two pathway) and time-domain (pump–dump) approaches to both coherent control and coherence spectroscopies. To clarify this argument, we note the equivalence of the tunable parameter (or experimental control knob) in Eq. (15) to that in Eq. (4). The former phase, $\omega_{ab}\,\Delta t$, signifies the difference in the accumulated phase during a certain evolution period between two matter waves. The latter, ϕ, is the difference in accumulated phase during a certain evolution period between two light waves (e.g., see Section III for the one-versus three-photon example). More generally, the (relative) phase of scattering eigenstates of the Schrödinger equation, like that of bound eigenstates and that of light waves, can be interpreted directly in terms of a time delay. Bound–bound pump–probe spectroscopies probe the relative phases of bound eigenstates, whereas energy-domain coherence spectroscopy, like differential cross sections, probe the relative phases of scattering eigenstates. Conceptually these experiments are similar. The discussion of such coherence spectroscopies in terms of an evolution period (or a time delay) becomes more appropriate as the superposition becomes broader in energy space and better defined in time, approaching the classical limit.

Finally, we note that the transform-limited pulse of Eq. (11) establishes a very specific phase relation among the components of a wavepacket prior to its evolution, transferring to the superposition of matter waves the phase information imprinted in the superposition of light waves and thereby defining the initial state and determining its free evolution. Specification of the initial phase is analogous to solving an initial-value trajectory problem, in the sense that the initial position and momentum are defined, and the duration of the evolution determines the final state of the trajectory. The case where one defines the initial and final states (analogous to "double ended" trajectory studies) is experimentally realizable with current technology, which allows the preparation of laser pulses with essentially arbitrary phase relations among their frequency components. In fact, feedback control experiments modify iteratively the spectral composition of the laser pulse to guide the material system from a given initial state to a desired final state. Taken in reverse, the time evolution of this spectral composition contains interesting phase information regarding the material system [7, 11, 12, 21, 28, 49–51].

III. EXPERIMENTAL METHODS: HOW TO MEASURE IT?

The main components of energy-domain, two-pathway coherent control experiments are a pair of phase-locked sources of radiation of equal total photon energy, a material target, and a product detector [58]. The radiation fields are most commonly produced with lasers of frequencies ω_n and ω_1, where $\omega_n = n\omega_1$. Phase locking is achieved by using a nonlinear medium to generate the nth harmonic of the fundamental frequency, ω_1. Typically, ω_1 is produced by a tunable dye laser, but frequency-doubled Nd:YAG lasers and CO_2 lasers have also proved useful. A nonlinear crystal is used to double the frequency, and an atomic or molecular gas is used to triple it. The nature of the harmonic generation process guarantees that the two fields have a definite phase relationship.

The electric field of a focused Gaussian beam in TEM_{00} at cylindrical coordinates z and r is given by [59],

$$E(r,z) = E_0 \frac{w_0}{w(z)} \exp\left\{-i(\phi_0 + kz - \eta(z)) - r^2\left[\frac{1}{w(z)^2} + \frac{ik}{2z\zeta(z)}\right]\right\}, \quad (17)$$

where

$$\zeta(z) = 1 + z^2/z_R^2 \tag{18}$$

Here, $w_0 = \lambda_0 f/\pi NW$ is the radius of the field at the focal point ($z = 0$) after being focused by a lens of focal length f and radius W, λ_0 is the wavelength, $k = 2\pi/\lambda_0$ is the wavenumber, N is the refractive index of the medium, and ϕ_0 is a constant phase. The radius of the field at axial distance z is given by

$$w(z) = w_0\zeta(z), \tag{19}$$

where

$$z_R = \pi w_0^2 N/\lambda_0 \tag{20}$$

is the Rayleigh range, and

$$\eta = \tan^{-1}(z/z_R) \tag{21}$$

is the Gouy phase. The Gouy phase shift arises from the increased phase velocity along the curve defined by $w(z)$, as compared with the phase along the geometric path that passes through the focal point [60]. From Eq. (17) we find that the relative phase for m photons of frequency ω_n and n photons of frequency ω_m is given by

$$\phi = (m\phi_n - n\phi_m) + (mk_n - nk_m)\left(z + \frac{r^2}{2z\zeta(z)}\right) + (n - m)\tan^{-1}\left(\frac{z}{z_R}\right) \tag{22}$$

The first term in ϕ is the constant phase of $E(r, z)$ that arises from the refractive index of the medium through which the beams propagate. The second term vanishes because of conservation of momentum in the harmonic generating medium. The third term is the Gouy phase, which changes by $(n - m)\pi$ as the beams pass through a focal point.

Equations (17) and (22) provide the experimentalist with three knobs (see Fig. 2). First, by varying the optical density of the transmitting medium, one may alter the constant term in ϕ (the "refractive phase"), as given by [61]

$$(m\phi_n - n\phi_m) = mn(N_n^0 - N_m^0)\omega_1(D/D^0)l/c \tag{23}$$

where N_m^0 is the refractive index at frequency ω_m under standard conditions, D is the density of the refractive medium, D^0 is its density under standard conditions,

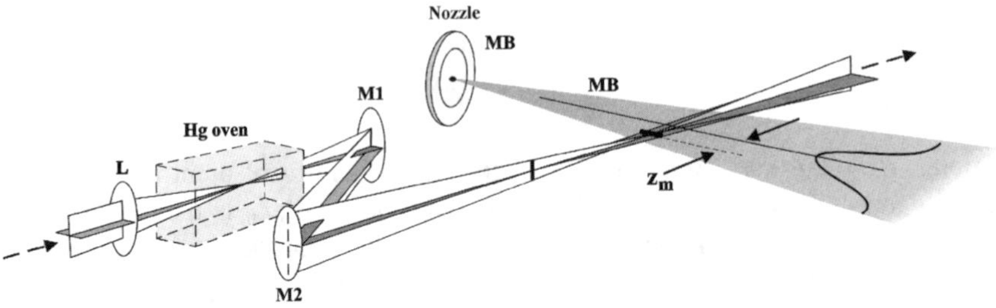

Figure 2. Schematic drawing of a control apparatus. A laser of frequency ω_1 is focused by lens L into a cell containing a tripling medium such as Hg vapor. Mirrors M1 and M2 are mounted inside a phase tuning cell (not shown) containing a refractive medium such as H_2. The folded mirror geometry produces a pair of elliptical astigmatic foci, one of which overlaps the molecular beam (MB) at a distance z_m from the beam axis. (Reproduced with permission from Ref. 62, Copyright 2006 American Physical Society.)

l is the path length, and c is the speed of light in a vacuum. Two options for manipulating this phase are to pass the beams through a phase tuning cell and vary the temperature and/or pressure of the cell or to pass them through a transparent plate and vary the path length by tilting the plate.

The second knob is the Gouy phase, which may be manipulated by scanning the focus of the laser beams across the field of view [62, 63]. The laser is usually focused onto the sample by a pair of curved mirrors. (A lens is avoided because its focal length is wavelength dependent.) Because the detector typically samples products along the axis of the lasers, the effective Gouy phase is a spatial average over the full field of view. This spatial average samples regions of varying intensity along the optical axis. It is not uncommon for some reaction channels to require the absorption of additional photons. For example, dissociation of vinyl chloride requires the absorption of either one ultraviolet or three visible photons, whereas ionization of this molecule requires the absorption of two additional visible photons. (See Fig. 12 in Section VB.) Because the absorption of five photons requires a greater intensity than do three photons, the ionization reaction is confined to a smaller volume near the laser focus, thereby reducing the contribution from the Gouy phase for that channel. A further complication arises from the fact that the mirrors are positioned in a folded configuration (Fig. 2), which shifts the optical axis and creates a pair of astigmatic elliptical foci [59, 64]. Despite these complications, useful insight into the effects of the Gouy phase may be obtained from a simple analytic model in which the laser beam is assumed to have a single circular focus, and the field of view is assumed to have a uniform rectangular profile of width $2d$. (The distance d for example, may be, the radius of the molecular beam.) Assuming that an additional l photons of frequency ω_1 are required to promote a particular reaction channel following the absorption of $m\omega_n$ and $n\omega_m$ photons, the spatially averaged contribution from the Gouy phase is given by

$$\tan \delta^{G}_{lmn}(z_m) = \frac{2J_{(m+n)/2+l}}{2I_{(m+n)/2+l} - I_{(m+n)/2+l-1}} \tag{24}$$

where z_m is the distance of focal point from the center of the field of view, and the definite integrals

$$I_n = \int_{(-d+z_m)/z_R}^{(d+z_m)/z_R} \frac{dz}{\zeta^n} \tag{25}$$

and

$$J_n = \int_{(-d+z_m)/z_R}^{(d+z_m)/z_R} \frac{(z/z_R)dz}{\zeta^n} \tag{26}$$

have simple closed form expressions (see Eq. (18)).

The third experimental knob is the relative amplitudes of the two laser fields. These are typically chosen to maximize the modulation depth without affecting the phase of the interference term.

The other components of an energy-domain, two-pathway coherent control experiment are the target and detector. The most common and versatile configuration is a molecular beam aligned perpendicular to the laser beams and to the axis of a time-of-flight (ToF) mass spectrometer, although in some cases a gas cell or solid target was employed together with a means of detecting either ion density or electron current. If conventional (e.g., Wiley–McLaren [65]) ion optics are used by the ToF detector, the observed quantity is proportional to the total reaction cross section for each product channel. In this case, the matrix element product in Eq. (4) is averaged over all scattering angles, and the mth and nth harmonics must both be either even or odd in order to obtain a nonzero channel phase (see Section IV). More detailed information may be obtained by using a position-sensitive detector to measure the angular distribution of the products. Using this configuration, the relative phases of the outgoing waves in different channels may be obtained, in addition to the relative phases of the transition matrix elements, and the parity restriction is lifted (see Section VA).

In a typical experiment, the laser is focused at some fixed position z_m and the refractive phase is varied by changing the optical density of the tuning medium. The product signal displays interference of the form of Eq. (4). If two product channels, A and B, are observed simultaneously, their modulation curves display a phase lag, which contains contributions from both the channel phase,

$\delta^A(E) - \delta^B(E)$, and the Gouy phase, $\delta^G_{l_A,mn}(z_m) - \delta^G_{l_B,mn}(z_m)$. The contributions from individual channel phases may be determined by measuring the phase lag with respect to a reference molecule that has a known channel phase [32, 66]. The Gouy contribution to the phase lag is sensitive to z_m and the number of additional photons l_A and l_B required to reach the product continua. Calculations for the typical case of $m = 1, n = 3, l_A = 2$, and $l_B = 0$ are shown in Fig. 3 for a circular laser focus and various relative values of the Rayleigh range. The Gouy phase lag is negligible for z_R much greater or much smaller than the molecular beam (or detector slit) width, but for intermediate values of d/z_R its contribution may be as large as $\pm 40°$. For a circular focus, the Gouy phase lag vanishes at $z_m = 0$, but for an astigmatic focus it could in principle be large because of the contribution of the second elliptical focus [63].

Additional information may by obtained from the modulation depth of the signal. For the simple case of one-dimensional, angle-resolved scattering into a single, uncoupled continuum (see Section IVC), the modulation depth is given by (see Eq. (4))

$$M = \frac{\sqrt{p_a p_b}}{\frac{1}{2}(p_a + p_b)} \tag{27}$$

If the amplitudes of the two electric fields are adjusted such that $p_a = p_b$, the limiting value of $M = 1$ is achieved. A variety of factors, however, may

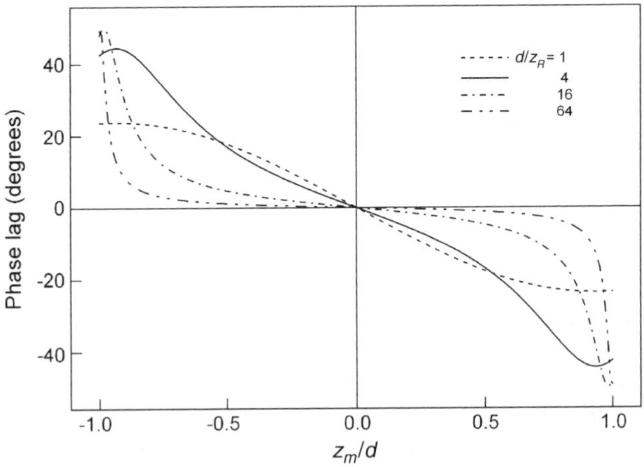

Figure 3. The phase lag produced by the Gouy phase, calculated using the analytic model described in the text for $\omega_3 + 3\omega_1$ excitation, with two additional ω_1 photons in one of the channels. The calculations are performed for various ratios of the molecular beam radius d to the Rayleigh range z_R.

contribute to reduce M. First, the divergence of the laser beam makes it impossible for p_a and p_b to be equal everywhere along the laser axis. If these probabilities are equated at $z_m = 0$, the modulation depth for a circular focus is given by [63]

$$M(z_m) = \frac{R\{[2I_{(m+n)/2+l} - I_{(m+n)/2+l-1}]^2 + 4J^2_{(m+n)/2+1}\}^{1/2}}{\frac{1}{2}(I_{m+l-1} + I_{n+l-1})} \tag{28}$$

where R is the ratio of matrix elements, $p_{ab}/(p_a p_b)^{1/2}$. For $m = 1, n = 3, l_A = 2$, and $l_B = 0$, the modulation depth at $z_m = 0$ is shown by the dashed curve in Fig. 4 to fall off rapidly with d/z_R. If, however, the field amplitudes are set to maximize M, the arithmetic mean in the denominator of Eq. (28) is replaced by the geometric mean, and the modulation depth is depicted by the solid curve in Fig. 4. The falloff of M with d/z_R can explain why modulation depths greater than $\sim 50\%$ have never been reported even for bound-to-bound transitions [58].

Once instrumental effects on M have been accounted for, useful information about the physical system may be deduced from the modulation depth. For isolated molecules, averaging over scattering angles and summing over continuum indices will reduce the ratio R. Further loss of modulation depth may be caused by decoherence in dissipative systems (*vide infra*), making this quantity a potentially useful observable for deducing structural and dynamical effects.

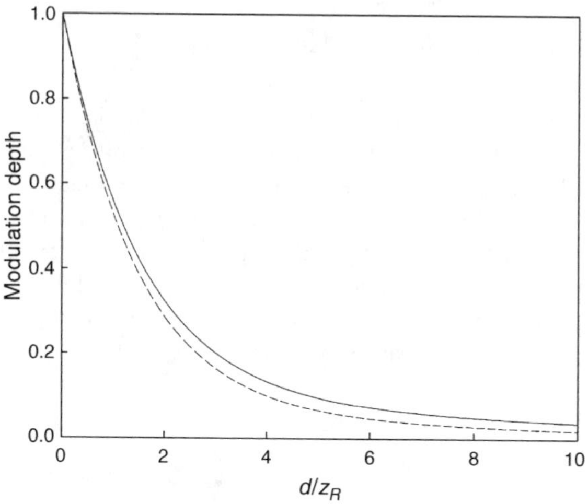

Figure 4. Calculation of the modulation depth using the analytic model described in the text, with the lasers focused on the axis of the molecular beam ($z_m = 0$). The dashed curve was obtained by setting $p_a = p_b$ at $z_m = 0$. The solid curve is the maximum possible modulation depth, obtained by setting $p_a = p_b$ at an optimum location.

IV. THEORY: HOW TO FORMULATE IT?

In this section we first (Section IV A) derive a formal expression for the channel phase, applicable to a general, isolated molecule experiment. Of particular interest are bound-free experiments where the continuum can be accessed via both a direct and a resonance-mediated process, since these scenarios give rise to rich structure of $\delta(E)$, and since they have been the topic of most experiments on the phase problem. In Section IVB we focus specifically on the case considered in Section III, where the two excitation pathways are one- and three-photon fields of equal total photon energy. We note the form of $\delta(E) = \delta_{13}(E)$ in this case and reformulate it in terms of physical parameters. Section IVC considers several limiting cases of δ_{13} that allow useful insight into the physical processes that determine its energy dependence. In the concluding subsection of Section V we note briefly the modifications of the theory that are introduced in the presence of a dissipative environment.

A. Disentangling Resonance from Long-Range Effects

The problem of unimolecular decomposition into one or several continua via simultaneous direct and resonance-mediated routes is conveniently formulated within Feshbach's partitioning framework [56]. Following Refs. 29 and 31, we partition the scattering eigenstate into its bound and continuum projections as

$$|ES\hat{k}^-\rangle = (Q + P)|ES\hat{k}^-\rangle, \tag{29}$$

where $\hat{k}$ is a unit vector in the direction of the scattering state momentum, $P = \Sigma_S P_S, P_S$ projects onto the S continuum, Q projects onto the bound manifold, and $P + Q = I$. The label S has been used in much of the previous literature to distinguish chemically different arrangement channels but is somewhat generalized here to imply a single collective index specifying the continuum state, in order to simplify the notation. We confine attention to the weak field limit of the formalism of Ref. 29, relevant to the experiments of Section III, and describe the dipole coupling within the Golden Rule approximation. Thus

$$Q|ES\hat{k}^-\rangle = QGQH_M P|ES\hat{k}_1^-\rangle \tag{30}$$

and

$$P|ES\hat{k}^-\rangle = P|ES\hat{k}_1^-\rangle + (E^- - PH_M P)^{-1} PH_M QGQH_M P|ES\hat{k}_1^-\rangle \tag{31}$$

where

$$QGQ = [E^- - Q\mathcal{H}Q]^{-1}, \tag{32}$$

$$Q\mathcal{H}Q = QH_MQ + QH_MP(E^- - PH_MP)^{-1}PH_MQ, \tag{33}$$

and H_M is the matter Hamiltonian. In Eq. (33), PH_MP is the scattering projection of the matter Hamiltonian and $P|ES\hat{k}_1^-\rangle$ are its solutions [29]. $Q\mathcal{H}Q$ in Eqs. (30)–(33) defines an effective Hamiltonian subject to which the resonance state evolves. In the limit of vanishing coupling of the bound state with the continuum, the second term in Eq. (33) vanishes, and the excited manifold reduces to a bound state, evolving subject to H_M. In the presence of coupling, $Q\mathcal{H}Q$ is non-Hermitian, with its imaginary part leading to decay of the bound state, giving rise to a finite width to the spectroscopic transition. Here the propagator corresponding to the second term of Eq. (33) describes (from right to left) transition from the bound to the continuum manifold, evolution subject to the continuum Hamiltonian, followed by transition back to the bound state.

Using Eqs. (29)–(31), the transition matrix elements in Eq. (5) are given as

$$\langle g|\bar{D}_\alpha|ES\hat{k}^-\rangle = \langle g|\bar{D}_\alpha|ES\hat{k}_1^-\rangle + \langle g|\bar{D}_\alpha\mathcal{F}(E)QGQH_M|ES\hat{k}_1^-\rangle, \tag{34}$$

where

$$\mathcal{F}(E) = I + (E^- - PH_MP)^{-1}PH_M. \tag{35}$$

The first term of Eq. (34) describes a direct transition from the bound state to the scattering projection of the structured continuum. The second term describes a resonance-mediated transition.

We consider experiments of the type discussed in Section III, where the signal is averaged over the scattering angles,

$$\delta_{ab} = \arg\int d\hat{k}\langle g|\bar{D}_a|ES\hat{k}^-\rangle\langle ES\hat{k}^-|\bar{D}_b|g\rangle \tag{36}$$

and hence it is convenient to expand the continuum state in partial waves,

$$|ES\hat{k}_1^-\rangle = \sum_{JMK}\sqrt{\frac{2J+1}{4\pi}}D^J_{KM}(\hat{k})|ESJMK\rangle \tag{37}$$

whereby the integration over $\hat{k}$ is analytical. In Eq. (37) J is the total angular momentum, M and K are its space-fixed and body-fixed z-projections, D^J_{KM} are rotation matrices [67], and the factor $\sqrt{(2J+1)/4\pi}$ normalizes the D^J_{KM} with respect to integration over $\hat{k} = (\phi_k, \theta_k)$. The notation in Eq. (37) corresponds to a symmetric top, but the formulation to follow applies equally to linear and asymmetric tops. In the latter cases the rotational state is specified by J, M, and the electronic angular momentum projection λ, or by J, M, and the asymmetric top quantum number τ, but the partial wave analysis holds. With Eq. (37), the direct transition amplitudes in Eq. (34) are given as

$$\langle g|\bar{D}_\alpha|ES\hat{k}_1^-\rangle = \sum_{JMK} \sqrt{\frac{2J+1}{4\pi}} D^J_{KM}(\hat{k}) \langle g|\bar{D}_\alpha|ESJMK\rangle \tag{38}$$

and the resonance-mediated components as

$$\begin{aligned}&\langle g|\bar{D}_\alpha \mathcal{F}(E) QGQH_M|ES\hat{k}_1^-\rangle \\ &\quad = \sum_{JMK} \sqrt{\frac{2J+1}{4\pi}} D^J_{KM}(\hat{k}) \langle g|\bar{D}_\alpha \mathcal{F}(E) QGQH_M|ESJMK\rangle\end{aligned} \tag{39}$$

Introducing a complete set of eigenstates of $QH_MQ, \{|i\rangle\}$, we express the resonance-mediated partial-wave amplitude in Eq. (39) as a sum of products of three physically distinct matrix elements,

$$\begin{aligned}&\langle g|\bar{D}_\alpha \mathcal{F}(E) QGQH_M|ESJMK\rangle \\ &\quad = \sum_{ii'} \langle g|\bar{D}_\alpha \mathcal{F}(E)|i\rangle\langle i|G|i'\rangle\langle i'|H_M|ESJMK\rangle\end{aligned} \tag{40}$$

The first element on the right-hand side of Eq. (40) describes excitation into a resonance manifold, which, given Eq. (35), is comprised of direct excitation of the $|i\rangle$ and excitation via the continuum with which the $|i\rangle$ are coupled. The second element describes the dynamics in the resonance manifold and allows for coupling between the resonances. The third element accounts for decay of the resonance into the continuum.

Substituting Eqs. (34), (38), and (39) into Eq. (36) and using the orthogonality of the rotation matrices, we have

$$\begin{aligned}\delta^S_{ab}(E) = \arg\Bigg\{ &\sum_{JMK} [\langle g|\bar{D}_a|ESJMK\rangle + \langle g|\bar{D}_a \mathcal{F}(E) QGQH_M|ESJMK\rangle] \\ &\times [\langle ESJMK|\bar{D}_b|g\rangle + \langle ESJMK|H_M QG^\dagger Q\mathcal{F}(E)^\dagger \bar{D}_b|g\rangle]\Bigg\}\end{aligned} \tag{41}$$

Integration over the scattering angles in Eq. (41) eliminates the interference among J-states that is present in the angle-resolved observable, leaving only interference within a given J-manifold. These latter interferences can be traced to structure and coupling mechanisms in the continuum, as discussed later.

B. The One- Versus Three-Photon Case

In order to relate the formal expression (41) to the experiment of Section III, we now specialize the discussion to the case of one- versus three-photon excitation, limiting attention for simplicity to fragmentation of diatomic systems, relevant to the experiments of Refs. 30, 32, 33, 35, and 39. We note later, however, that our conclusions are general and equally applicable to other excitation schemes and to ionization processes.

The reduced dipole operators are thus

$$\bar{D}_a = \bar{D}^{(1)} = \vec{\mu} \cdot \vec{\bar{\varepsilon}}$$
$$\bar{D}_b = \bar{D}^{(3)} = \sum_{r_1, r_2} \frac{\vec{\mu} \cdot \vec{\bar{\varepsilon}} | r_1 \rangle \langle r_1 | \vec{\mu} \cdot \vec{\bar{\varepsilon}} | r_2 \rangle \langle r_2 | \vec{\mu} \cdot \vec{\bar{\varepsilon}}}{(E_g + \omega_1 - E_1)(E_g + 2\omega_1 - E_2)} \tag{42}$$

where $\vec{\bar{\varepsilon}}$ is the phase-adjusted field (see Eq. (4)). We introduced in Eq. (42) two sets of bound intermediates $\{|r_k\rangle\}$ with eigenvalues E_k, where $k = 1, 2$ corresponds to the one- and two-photon levels, respectively. The sum over discrete indices is supplemented by integration over continuous variables in the case that scattering intermediates contribute.

Expressing the bound states in Eqs. (40) and (42) in terms of angular momentum components and expanding the dipole vector in spherical unit vectors, we obtain expressions for the dipole matrix elements in Eq. (42), in which the integration over the angular variables is analytical. This results in an expression for $\delta_{13}^S(E)$ that is somewhat more complex in form but significantly simpler to numerically evaluate [31]. Averaging over the initial angular momentum and magnetic levels, one finds after certain algebraic manipulations

$$\begin{aligned}
\delta_{13}^S(E) = \arg \sum_{J_g} W_{J_g} \Big\{ & \sum_J \tau^{(1)^*}(J_g|ESJ)\tau^{(3)}(J_g|ESJ) \\
& + \sum_{\nu J_i} [\tau^{(1)^*}(J_g|ESJ_i) X_E^{(3)}(J_g|\nu J_i)(E - E_{\nu J_i})^{-1} V(\nu J_i|ESJ_i) \\
& + X_E^{(1)^*}(J_g|\nu J_i)(E - E_{\nu J_i}^*)^{-1} V^*(\nu J_i|ESJ_i)\tau^{(3)}(J_g|ESJ_i) \\
& + \sum_{\nu'} X_E^{(1)^*}(J_g|\nu J_i)(E - E_{\nu J_i}^*)^{-1} V^*(\nu J_i|ESJ_i) X_E^{(3)}(J_g|\nu' J_i) \\
& \times (E - E_{\nu' J_i})^{-1} V(\nu' J_i|ESJ_i)] \Big\}
\end{aligned} \tag{43}$$

where J_g sums over the thermally populated initial angular momentum components, W_{J_g} are Boltzmann weight factors, and $\tau^{(1)}$ and $\tau^{(3)}$ are one- and three-photon matrix elements describing the direct one- and three-photon transitions. The $X_E^{(j)}$, $j = 1, 3$ are bound–quasibound amplitudes that signify, loosely speaking, j-photon matrix elements into a "modified" eigenstate, $|\Phi_{(E)}^{J_i}\rangle \sim \mathcal{F}(E)|J_i\rangle$, that is mixed with the continua [29]. In the limit of an isolated resonance and a single continuum, this state reduces to an incoming-wave analogue of the familiar state denoted by Φ in Fano's formalism [68]. From the structure of $|\Phi_{(E)}^{J_i}\rangle$ as a sum of a bound and a scattering component (see Eq. (35)), it follows that X_E consists of a bound–bound transition amplitude, denoted τ_b, and a self-energy-like component that imparts the transition with an energetic width,

$$X_E^{(j)}(J_g|J_i) = \tau_b^{(j)}(J_g|J_i) + \sum_S \left[P_v \int \frac{dE'}{E - E'} \tau^{(j)}(J_g|E'SJ_i)V(E'SJ_i|J_i) + i\pi\tau^{(j)}(J_g|EnSJ_i)V(ESJ_i|J_i) \right] \quad (44)$$

where P_v denotes the principal value of the integral. The $V(J_i|ESJ)$ in Eq. (43) are matrix elements of the interaction coupling the Q and P subspaces, and, in order to emphasize the dominant energy dependence of δ_{13}^S, we have diagonalized the effective Hamiltonian of Eq. (33) in the basis of eigenstates of QH_MQ, such that

$$QGQ = \sum_{vJ_i} \frac{|vJ_i\rangle\langle vJ_i|}{E - E_{vJ_i}} \quad (45)$$

where $|vJ_i\rangle$ is the (biorthogonal) set of eigenstates of $Q\mathcal{H}Q$ (Eq. (33)) and E_{vJ_i} are the corresponding eigenvalues. We refer the reader to Ref. 31 for derivation of Eq. (43) and a detailed discussion of its interpretation.

At large detunings only the first term of Eq. (43) survives. This term arises from interference among continuum states. Assuming that bound intermediates dominate the three-photon process (see Section IVC5 for discussion of quasibound intermediates), the structure of δ_{13}^S off-resonance derives only from the energy dependence of the phase shifts of the scattering partial waves composing the continuum. In the case of two coupled electronic states, for instance, an off-resonance δ_{13}^S measures the difference between the partial wave phase shifts of the coupled diabatic states (see Section IVC2 and Fig. 5b). The second and the third terms of Eq. (43) reflect, respectively, the interference of the direct one-photon process with the resonance-mediated three-photon process and the interference of the resonance-mediated one-photon process with the

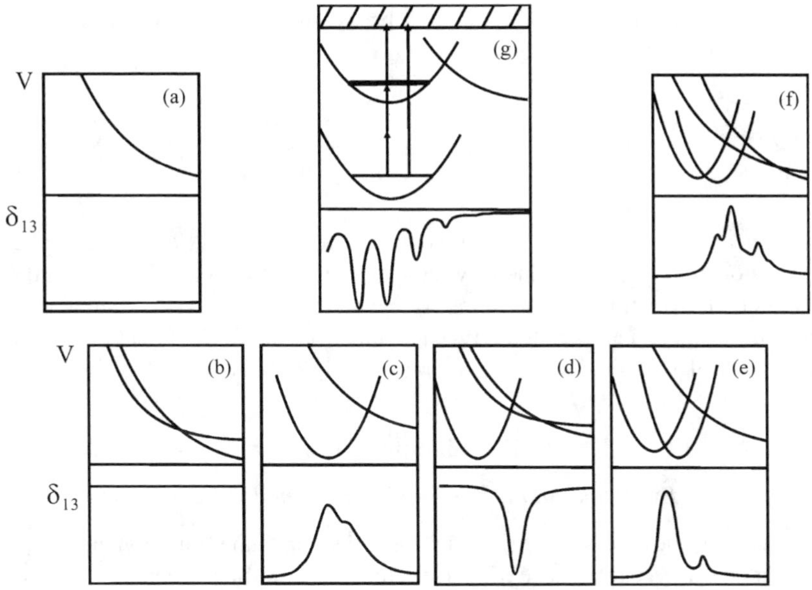

Figure 5. Schematic illustration of the energy dependence of the channel phase in different simple limits of Eq. (43), along with the corresponding potential energy curves. See Section IVC. (Reproduced with permission from Ref. 43, Copyright 2001 American Chemical Society.)

direct three-photon process. In the case of an isolated resonance interacting with an uncoupled continuum, the phase associated with these terms takes a particularly simple form that parallels the familiar Fano lineshape of the single-photon absorption probability (Section IVC4). The last term arises from resonant–resonant interference. Energy dependence of the direct–direct interference term, if present, is thus a measure of long-range interactions. The resonant–direct and resonant–resonant interferences are dominated by short-range, spectroscopic features. Their energy dependence, in particular, that of the latter, is significantly stronger than that of the direct-direct interference term, exhibiting extrema at the resonance positions.

The above discussion suggests separation of δ_{13}^S into a direct part,

$$\delta_{13,d}^S = \arg \sum_{J_g} W_{J_g} \sum_J \tau^{(1)^*}(J_g|ESJ)\tau^{(3)}(J_g|ESJ) \tag{46}$$

arising from the direct–direct interference term of Eq. (43), and a composite part, $\delta_{13,c}^S = \delta_{13}^S - \delta_{13,d}^S$, arising from the direct–resonant and resonant–resonant interferences. The former varies with energy on the scale of the deBroglie wavelength, hence providing a "smooth background" to the rapid variation of the latter. Using Eq. (46) one finds

$$\delta^S_{13,c} = \delta^S_{13} - \delta^S_{13,d} = \arg\left\{\frac{p^s_{13}e^{-i\phi}}{\sum_{J_g} W_{J_g} \sum_J \tau^{(1)^*}(J_g|ESJ)\tau^{(3)}(J_g|ESJ)}\right\} \quad (47)$$

where p^S_{13} is the one- versus three-photon interference term of Eq. (4). An explicit expression for $\delta^S_{13,c}$ is obtained by substituting Eq. (43) into Eq. (47) and is conveniently interpreted in terms of generalized asymmetry parameters, analogues of the Fano asymmetry parameter [68] in Ref. 31. Here we omit discussion of the general form and proceed to consider several of its simple limits.

C. Limiting Cases

1. *Excitation into a Structureless, Uncoupled Continuum*

Consider first the case where the scattering projection of the Hamiltonian, PH_MP, can induce only elastic scattering (referred to as an uncoupled or "elastic" continuum). In this situation the partial wave in Eq. (37) reduces to

$$\langle \mathbf{Q}R|\, ESJM\rangle = Y_{JM}(\hat{R})\langle \mathbf{Q}R|\, ESJM\rangle e^{i\delta^S_J} \quad (48)$$

where Y_{JM} are spherical harmonics, $\langle \mathbf{Q}R|\, ESJM\rangle$ are real, and δ^S_J are coordinate-independent phase factors (viz., the partial wave phase shifts discussed in Section II). Provided that the multiphoton process is not mediated by resonances, we have in this limit that the phases of $\tau^{(1)}$ and $\tau^{(3)}$ are equal. Hence the direct–direct interference term in Eq. (43) is real, $\delta^S_{13,d} = 0$, and the observable phase reduces to $\delta^S_{13,c}$.

For a structureless continuum (i.e., in the absence of resonances), assuming that the scattering projection of the potential can only induce elastic scattering, the channel phase vanishes. The simplest model of this scenario is depicted schematically in Fig. 5a. Here we consider direct dissociation of a diatomic molecule, assuming that there are no nonadiabatic couplings, hence no inelastic scattering. This limit was observed experimentally (e.g., in ionization of H_2S).

It is worth noting that the phase of $\langle g|\bar{D}_a|ES\hat{k}^-\rangle$ is not equal to that of $\langle g|\bar{D}_b|ES\hat{k}^-\rangle$ even in the above limit, a point that was the source of confusion in the previous literature [40]. It is only upon integration over scattering angles that interference among different partial waves is eliminated (see Section IVA), and an observed phase implies interference within a partial wave. Hence angle–resolved measurements may observe a nonzero phase regardless of the nature of the continuum.

2. *Excitation into a Structureless, Coupled Continuum*

Considering again the case of a structureless continuum, we have that δ_{13}^S arises from excitation of a superposition of continuum states, hence from coupling within PH_MP [69]. The simplest model of this class of problems, depicted schematically in Fig. 5b, is that of dissociation of a diatomic molecule subject to two coupled electronic dissociative potential energy curves. Here the channel phase can be expressed as

$$\delta_{13}^S = A + Be^{i(\delta_\xi - \delta_{\xi'})} + Ce^{-i(\delta_\xi - \delta_{\xi'})} \tag{49}$$

where δ_ξ are the partial wave phase shifts corresponding to the two diabatic states. In deriving Eq. (49) we assumed that the initial rotation is selected and that a single rotational branch is dipole-allowed. In this situation the channel phase arises from the interference between two deBroglie waves of different wavelengths—it probes the difference between the partial wave phase shifts comprising the continuum. Referring back to Eq. (10) and the discussion preceding it, we find that δ_{13}^S contains interesting information about the underlying potential energy curves. This information can be extracted from the observable by fitting the yield curve to a sum of three cosines, rather than to one cosine. It is interesting to note that similar information regarding the difference between two partial wave phase shifts has been determined using a very different experimental approach in Ref. 70.

3. *Excitation of an Isolated Resonance Embedded in an Uncoupled Continuum*

In the case that an isolated vibronic resonance interacts with the uncoupled (or "elastic") continuum of Eq. (48), Eq. (43) simplifies to

$$\delta_{13,c}^S = \delta_{13}^S = \tan^{-1}\left(\frac{2\sum_{J_gJ_i}(q_{J_g,J_i}^{(1)} - q_{J_g,J_i}^{(3)})(\epsilon_{J_i}^2+1)^{-1}\sigma_{J_g,J_i}}{1+\sum_{J_g,J_i}[3-\epsilon_{J_i}(q_{J_g,J_i}^{(1)} + q_{J_g,J_i}^{(3)}) + q_{J_g,J_i}^{(1)}q_{J_g,J_i}^{(3)}](\epsilon_{J_i}^2+1)^{-1}\sigma_{J_g,J_i}}\right) \tag{50}$$

This case is shown schematically in Fig. 5c. In Eq. (50), $q_{J_g,J_i}^{(j)}$ are generalized j-photon asymmetry parameters, defined, by analogy to the single-photon q parameter of Fano's formalism [68], in terms of the ratio of the resonance-mediated and direct transition matrix elements [31], ϵ_{J_i} is a reduced energy variable, and σ_{J_g,J_i} is proportional to the line strength of the spectroscopic transition. The structure predicted by Eq. (50) was observed in studies of HI and DI ionization in the vicinity of the $5d\delta$ resonance [30, 33]. In the case of a rotationally resolved experiment, where the parent state is prepared in a single level $|n_g J_g\rangle$ (e.g., by vibrational excitation) and the resonance vibration is resolved, Eq. (50) reduces to

$$\tan \delta_{13}^S(E \sim E_{n_i J_i}) = \frac{2(q^{(1)} - q^{(3)})}{[\epsilon - \frac{1}{2}(q^{(1)} + q^{(3)})]^2 + [4 - \frac{1}{4}(q^{(1)} - q^{(3)})^2]} \tag{51}$$

Equation (51) has a clear physical interpretation. Recalling the lineshape for a single excitation route, where fragmentation takes place both directly and via an isolated resonance [68], $p \propto (\epsilon + q)^2/(1 + \epsilon^2)$, we have that δ_{13}^S is maximized at the energy where interference of the direct and resonance-mediated routes is most constructive, $\epsilon = (q^{(1)} + q^{(3)})/2$. In the limit of a symmetric resonance, where $q^{(j)} \to \infty$, Eq. (51) vanishes, in accord with Eq. (53) and indeed with physical intuition. The numerator of Eq. (51) ensures that δ_{13}^S has the correct antisymmetry with respect to interchange of 1 and 3 and that it vanishes in the case that both direct and resonance-mediated amplitudes are equal for the one- and three-photon processes. At large detunings, $|\epsilon| \to \infty$, and δ_{13}^S of Eq. (51) approaches zero.

4. *Excitation of an Isolated Resonance Embedded in a Coupled Continuum*

Consider next the limit of weak dipole coupling of the initial state with the continuum and/or long-lived resonances, where the direct amplitudes are small as compared to the resonance-mediated ones in the vicinity of a resonance. Returning to Eq. (43) and neglecting the direct–direct and direct–resonant interferences as compared to the resonant–resonant interference at $E \approx E_{vJ_i}^R$, we have

$$\delta_{13}^S(E) \to \arg \sum_{J_g J_i v v'} W_{J_g \tau_b^{(1)*}} (J_g|vJ_i)(E - E_{vJ_i}^*)^{-1} V^*(vJ_i|ESJ_i) \times \tau_b^{(3)}(J_g|v'J_i)(E - E_{v'J_i})^{-1} V(v'J_i|ESJ_i) \tag{52}$$

where we used the fact that the dipole matrix elements for direct absorption to the continuum $(\tau^{(j)})$ are small as compared to those for excitation of the resonances $(\tau_b^{(j)})$. Clearly $\delta_{13,c}^S \neq 0$; that is, the interference term is complex, being the result of coupling among resonances.

In the limit of an isolated vibronic resonance $\tau_b^{(j)}$ are real, decay of the resonances arises only from interaction with the continuum, Eq. (52) reduces to

$$\delta_{13}^S(E \sim E_{n_i J_i}^R) \approx \arg \sum_{J_g J_i} W_{J_g} \frac{\tau_b^{(1)}(J_g|n_i J_i)|V(vJ_i|ESJ_i)|^2 \tau_b^{(3)}(J_g|n_i J_i)}{(E - E_{n_i J_i}^R)^2 + \Gamma_{n_i J_i}^2/4} = 0 \tag{53}$$

and on resonance δ_{13}^S vanishes.

The physical significance of Eq. (53) is clear. At an isolated resonance the excitation and dissociation processes decouple, all memory of the two excitation pathways is lost by the time the molecule falls apart, and the associated phase vanishes. The structure described by Eq. (53) was observed in the channel phase for the dissociation of HI in the vicinity of the (isolated) $5s\sigma$ resonance. The simplest model depicting this class of problems is shown schematically in Fig. 5d, corresponding to an isolated predissociation resonance. Figures 5e and 5f extend the sketches of Figs. 5c and 5d, respectively, to account qualitatively for overlapping resonances.

5. *Excitation of a Continuum Via a Low-Lying Resonance*

From Eq. (42) it is evident that an isolated resonance, $E_{n_kJ_k} = E^R_{n_kJ_k} - i\Gamma_{n_kJ_k}/2$ at the $k =$ one- or two-photon level, produces structure in the direct–direct interference term and hence in $\delta^S_{13,d}$ [40]. In the (uncommon) case where a single branch J_k contributes to the (one- or two-photon) transition, the phase of $\tau^{(1)^*}(J_g|ESJ)\tau^{(3)}(J_g|ESJ)$ undergoes a change of π at the resonance energy, $E = E_{J_g} + k\omega = E^R_{n_kJ_k}$:

$$\arg\{\tau^{(1)^*}(J_g|ESJ)\tau^{(3)}(J_g|ESJ)\} = \delta_\infty + \delta_{\text{res}}(E) \tag{54}$$

where

$$\delta_{\text{res}}(E) = -\arg(E - E_{n_kJ_k}) \tag{55}$$

is the resonant part of the partial wave phase shift, often referred to as the Breit–Wigner phase [71]. Near an isolated resonance (assuming that the resonance width is constant), the phase shift undergoes an abrupt change of π, reaching $\pi/2$ at the resonance energy, where the scattering cross section exhibits a sharp maximum. A useful, although qualitative, analogy is the phase of the classical forced harmonic oscillator. A harmonic oscillator that is driven by a periodic external force exhibits oscillations at the driving frequency with a phase that undergoes an abrupt change of π as the driving frequency goes through the natural oscillation frequency. Below resonance the displacement is in phase with the driving force. At resonance the displacement lags the applied force by $\pi/2$. Above resonance the displacement is π out of phase with the applied force. Another useful analogy is the Gouy phase of a focused light wave, which accumulates a π phase shift in the vicinity of a focal point (see Section III).

The first term on the right-hand side of Eq. (54) is a smooth "background" that vanishes in the case that PH_MP induces only elastic scattering. It follows that measurement of δ_{13}^S in the vicinity of a resonance at an intermediate level allows direct observation of the Breit–Wigner phase. This stands in contrast to phase measurements in the vicinity of an isolated resonance at the three-photon level, where, in the presence of a direct route, $\delta_{13,c}^S$ reflects direct–resonant as well as resonant–resonant interferences, while in its absence $\delta_{13,c}^S$ vanishes. In general, several branches contribute to the transition, of which a single one is resonant at a given photon energy. Hence $\arg\{\tau^{(1)^*}\tau^{(3)}\}$ drops by π to the red of each resonance energy, reaches $\delta_\infty - \pi/2$ on resonance, and returns to its asymptotic value to the blue of the resonance location.

The generalization to the case of a thermally averaged parent state describes an interesting modulation curve that reflects in position and width the rotational eigenvalue spectrum of the resonant intermediate [31]. This structure has been observed in studies of HI ionization in Ref. 33. A schematic cartoon depicting the excitation scheme and the form of the channel phase for the case of a thermally averaged initial state is shown in Fig. 5g.

V. EXAMPLES: WHAT DOES IT LOOK LIKE?

A. Atomic Systems

Photoionization of atoms provides an excellent arena for phase control studies because of the high symmetry of the target and the comparative ease of calculating the phase dependence of the observables. Different product channels may be detected by measuring the kinetic energies of the photoelectrons. By using a space-sensitive detector to record the angular distributions of the electrons, it is possible to measure the interference between transitions of even and odd parity. Also, as mentioned previously, it is possible to extract from the angular distributions the relative phases of the outgoing wave functions as well as those of the transition matrix elements.

This approach was used by Elliott and co-workers to control the ionization of alkali atoms by one- and two-photon excitation. Wang and Elliott [72] measured the interference between outgoing electrons in different angular momentum states. They showed, for example, that the angular flux of the p^2P and the d^2D continua of Rb is determined by the phase difference

$$\delta_{pd} = \phi_2 - \phi_1 + \delta_p - \delta_d \tag{56}$$

The fitted value of $\delta_p - \delta_d$ is in good agreement with the number calculated from the quantum defects of the atom and the phases of the Coulomb wavefunctions.

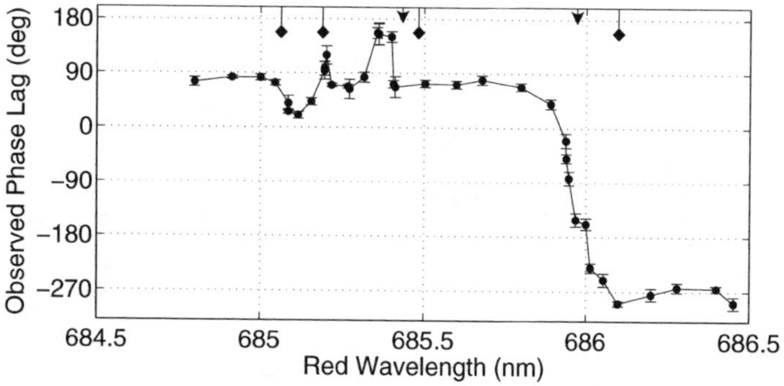

Figure 6. The measured phase lag of the photoelectron asymmetry parameter for the $6s^2S_{1/2}$ and the $5d^2D_{3/2}$ continua of Ba. (Reproduced with permission from Ref. 73, Copyright 2007 American Physical Society.)

In another study, Yamakazi and Elliott [73] measured the interference within a single angular momentum state arising from coupling to different continua. They defined an asymmetry parameter,

$$\alpha_{\text{asym}} = N_L/(N_L - N_R) \tag{57}$$

where $N_L(N_R)$ is the integrated electron count in the left (right) half of the angular distribution. This parameter is related to the matrix element describing the coupling of an autoionizing state to the continuum. The phase lag between α_{asym} for the $6s^2S_{1/2}$ and the $5d^2D_{3/2}$ continua of Ba is plotted in Fig. 6 as a function of the two-photon wavelength. A broad inelastic continuum between 684.8 and 685.9 nm and another starting at longer wavelengths are evident, with resonance peaks appearing in the first region.

B. Molecular Systems

Molecules provide a much richer landscape for studying coherence effects because of continua associated with bond breaking, rearrangements, isomerization, and so on. The simplest example is a diatomic molecule, where dissociation and ionization may compete with each other. An example that has been studied in some detail is HI. The potential energy diagram in Fig. 7 shows a manifold of dissociative continua near $2\omega_1$ and a series of autoionizing Rydberg states near $3\omega_1$. Measurements of the phase lag between the ionization and dissociation channels at various wavelengths provide examples of each of the cases described in Section IVC. The phase lag data in Fig. 8a were taken in the vicinity of the $5d(\pi, \delta)$ autoionizing resonances [45]. The circles show the phase lag between ionization and dissociation. Using the ionization of H_2S as a

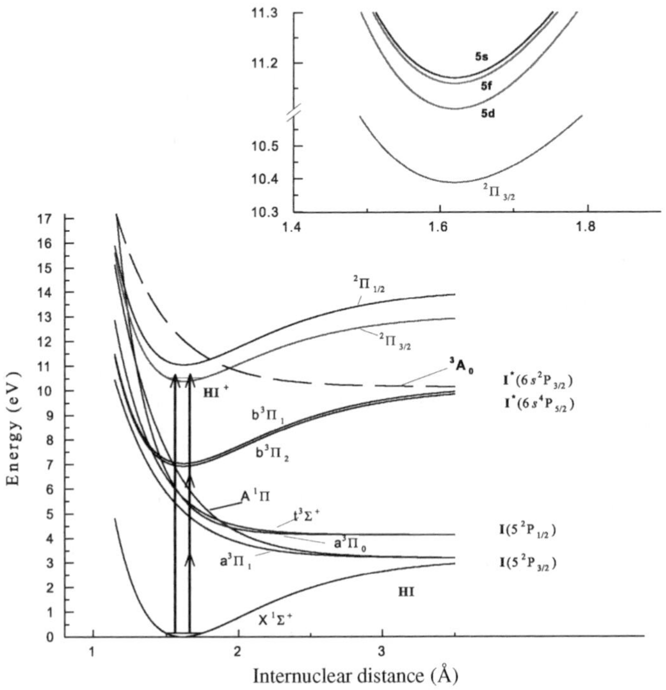

Figure 7. Potential energy diagram for HI, showing the two lowest ionization states ($^2\Pi_{3/2}$ and $^2\Pi_{1/2}$) coupled to a neutral dissociative continuum ($^3A_0$) at the three-photon ($3\omega_1$) level, as well as two low-lying Rydberg states ($b^3\Pi_1$ and $b^3\Pi_2$) predissociated by a manifold of repulsive states at the two-photon level. The inset shows a series of Rydberg states converging to the excited $^2\Pi_{1/2}$ ionic state.

reference, we are able to separate out the contributions from ionization (diamonds) and dissociation channel phases (triangles) of HI. The ionization phases lie above a baseline of zero, indicative of an uncoupled continuum (case 1 of Section IVC), where the scattering is elastic. The peaks lying near 356.2 and 355.2 nm are examples of resonances embedded in a continuum (case 3). The one-photon ionization spectrum in Fig. 8b shows that the 356.2-nm peak in the phase lag spectrum is associated with the $5d(\pi, \delta)$ resonance; the other peak has not been assigned. The featureless phase lag spectrum for dissociation varies slowly between $-90°$ and $-120°$, providing an example of dissociation in a coupled continuum (case 2 of Section IVC). The data in Fig. 9, taken in the vicinity of the $5s\sigma$ resonance, show a striking isotope effect in HI versus. DI [43]. The absence of a phase lag between the ionization of the two isotopes shows that the isotope effect occurs entirely in the dissociation continuum. The broad minimum in the phase lag spectrum of HI^+ versus H_2S^+ indicates,

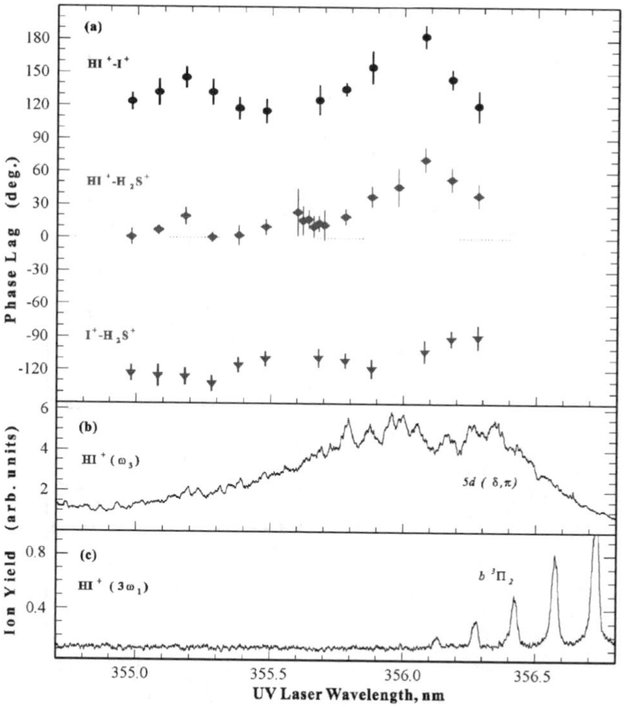

Figure 8. Phase lag spectrum of HI in the vicinity of the $5d(\pi, \delta)$ resonance. In panel (a), the circles show the phase lag between the ionization and dissociation channels. The diamonds and triangles separate the phase lag into contributions from each channel, using H_2S ionization as a reference. Panels (b) and (c) show the conventional one-photon (ω_3) and three-photon ($3\omega_1$) photoionization spectra. (Reproduced with permission from Ref. 45, Copyright 2002 American Institute of Physics.)

as before, that ionization occurs in an elastic continuum, whereas the very deep minima in the dissociation spectra show that these resonances are embedded in a coupled continuum (case 4 of Section IVC). Finally, a clear example of an intermediate resonance in the three-photon path (case 5) is shown in Fig. 10a [33]. The peaks in the phase lag spectrum for ionization occur precisely at the energies of the rotational levels of the $b^3\Pi_1$ state, resonant with $2\omega_1$ excitation, as shown in the conventional multiphoton ionization spectrum of Fig. 10b. (Similar rotational structure, originating from the $b^3\Pi_2$ state, is evident in Fig. 8c; however, the phase lag spectrum was not measured in that region.) Numerical studies of the phase lag between the ionization and dissociation of HI and H_2 have been reported in Refs. 42 and 44, respectively.

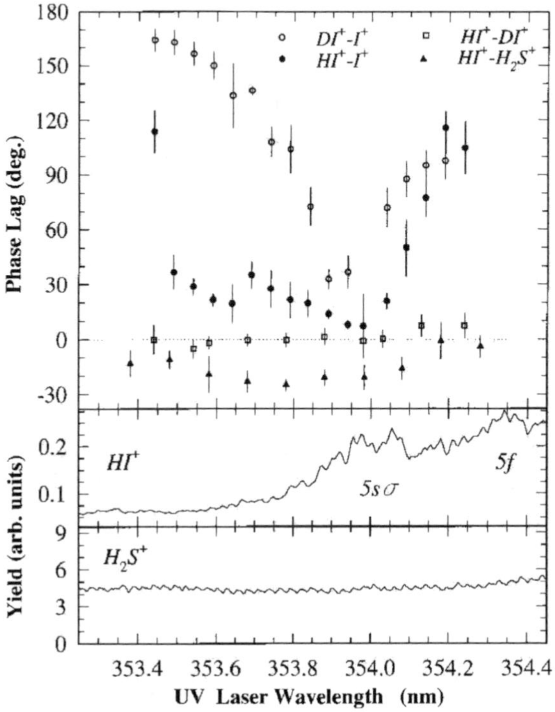

Figure 9. Phase lag spectrum of HI and DI in the vicinity of the $5s\sigma$ resonance. The top panel shows the phase lag between photoionization and photodissociation of HI (filled circles) and DI (open circles), the phase lag between the photoionization of HI and DI (squares), and the phase lag between the photoionization of HI and H_2S (triangles). The bottom two panels show the one-photon ionization spectra of HI and H_2S. (Reproduced with permission from Ref. 30, Copyright 1999 American Physical Society.)

Polyatomic molecules provide a still richer environment for studying phase control, where coupling between different dissociation channels can occur. Indeed, one of the original motivations for studying coherent control was to develop a means for bond-selective chemistry [25]. The first example of bond-selective two-pathway interference is the dissociation of dimethyl-sulfide to yield either H or CH_3 fragments [74]. The peak in Fig. 11 is indicative of a resonance embedded in an elastic continuum (case 4).

Another polyatomic molecule provided an opportunity to study the effect of the Gouy phase discussed in Section III [62]. Figure 12 depicts a slice of the potential energy surfaces of vinyl chloride, where the vertical arrows correspond to 532 nm photons. The two pathways for dissociation correspond to ω_3 versus $3\omega_1$, whereas those for ionization correspond to $\omega_3 + 2\omega_1$ versus $5\omega_1$ (i.e., $l = 2$, $m = 1$, $n = 3$). Figure 13 shows the phase lag for ionization versus dissociation as a function of the position of the laser beam waist along a line normal to the molecular beam axis. The dashed curve is the analytical model described previouslyin Eqs. (24)–(26), and the solid curve is a numerical calculation that takes into account the astigmatism of the laser focus and the Gaussian profile of the molecular beam. The good agreement with the data shows that the fragments are produced by dissociation of the neutral molecule at the $3\omega_1$ level rather than of the parent ion. The phase lag of $4.4 \pm 0.8°$ at $z_m = 0$ is due to a difference between the channel phases associated with the ionization and dissociation channels. Other polyatomic molecules for which branching has been controlled using the Gouy phase are acetone [75] and dimethyl sulfide [76].

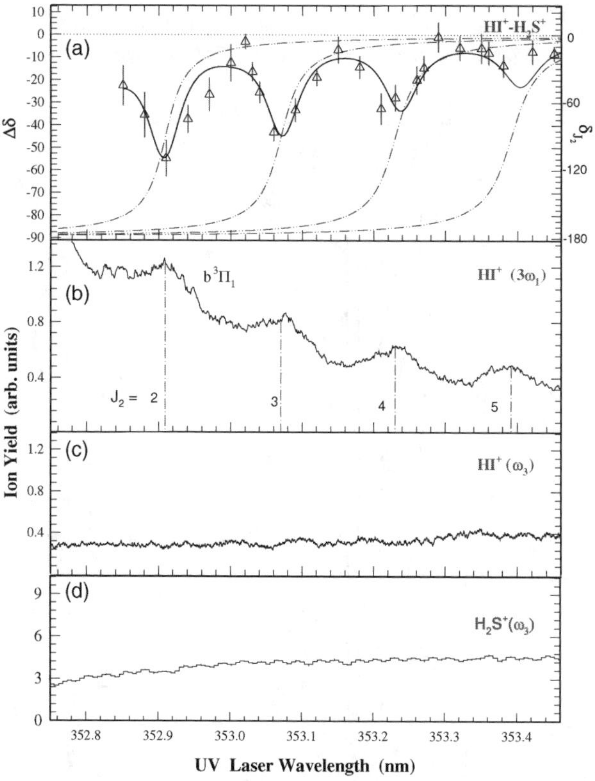

Figure 10. Phase lag spectrum of HI in the vicinity of the $b^3\Pi_1$ Breit–Wigner resonance. Panel (a) shows the phase lag between the photoionization of HI and H_2S. Panel (b) is the three-photon photoionization spectrum of HI, showing the rotational structure of the two-photon $b^3\Pi_1 - X^1\sum^+$ transition. The bottom two panels are the one-photon ionization spectra of HI and H_2S. (Reproduced with permission from Ref. 33, Copyright 2000 American Physical Society.)

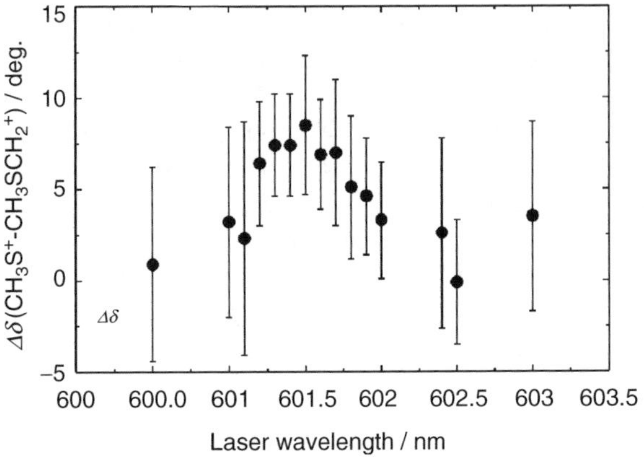

Figure 11. Phase lag between the CH_3S and CH_3SCH_2 photodissocation fragments of dimethylsulfide. (Reproduced with permission from Ref. 74, Copyright 2006 American Institute of Physics.)

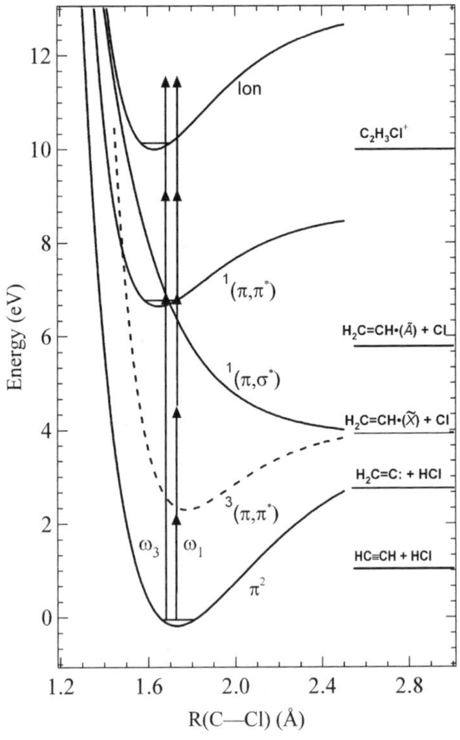

Figure 12. Slice of the potential energy surfaces of vinyl chloride, showing the excitation paths for photodissociation and photoionization. (Reproduced with permission from Ref. 62, Copyright 2006 American Physical Society.)

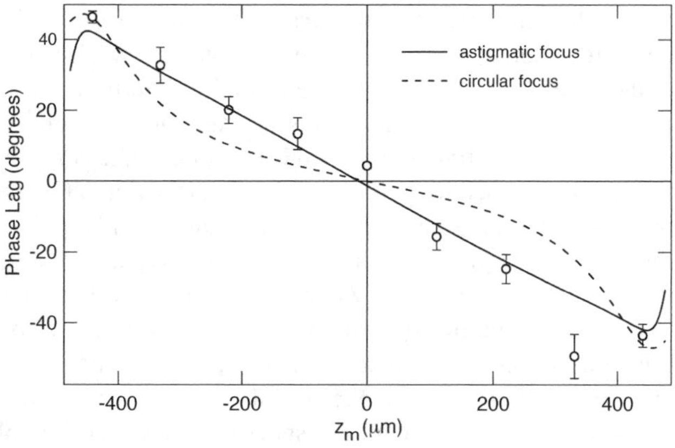

Figure 13. Phase lag between the photoionization and photodissociation of vinyl chloride resulting from the Gouy phase of the focused laser beam. The dashed curve shows the results of the analytical model discussed in the text, and the solid curve is a numerical calculation of the phase lag without adjustable parameters.

C. Extended Systems and Dissipative Environments

The previous sections focused on the case of isolated atoms or molecules, where coherence is fully maintained on relevant time scales, corresponding to molecular beam experiments. Here we proceed to extend the discussion to dense environments, where both population decay and pure dephasing [77] arise from interaction of a subsystem with a dissipative environment. Our interest is in the information content of the channel phase. It is relevant to note, however, that whereas the controllability of isolated molecules is both remarkable [24, 25, 27] and well understood [26], much less is known about the controllability of systems where dissipation is significant [78]. Although this question is not the thrust of the present chapter, this section bears implications to the problem of coherent control in the presence of dissipation, inasmuch as the channel phase serves as a sensitive measure of the extent of decoherence.

We consider a general dissipative environment, using a three-manifold model, consisting of an initial ($\{i\}$), a resonant ($\{r\}$), and a final ($\{f\}$) manifold to describe the system. One specific example of interest is an interface system, where the initial states are the occupied states of a metal or a semiconductor, the intermediate (resonance) states are unoccupied surface states, and the final (product) states are free electron states above the photoemission threshold. Another example is gas cell atomic or molecular problems, where the initial, resonant, and final manifolds represent vibronic manifolds of the ground, an excited, and an ionic electronic state, respectively.

We focus on the simplest case scenario where δ^S arises solely from an intermediate state resonance, see Fig. 5g, and comment elsewhere on the extension to more complex scenario, where δ^S probes interfering resonances at either or both intermediate and final levels of excitation as well as coupling mechanisms in the final continuum manifold. To simplify the connection of the present with the previous sections, we consider first (Section VC1) the energy-domain case of continuous wave (CW) fields (or, equivalently, pulsed fields of duration long with respect to all system time scales). To make connection with other interferometric experiments, we next (Section VC2) generalize the discussion to the time domain. Attention is focused on the case of one- versus two-photon excitation (assuming that the final state is at least partially angle-resolved). This case allows us to make useful connection with the literature of two-photon photoemission (2PPE) spectroscopies and simplifies the interpretation of the results. As discussed later, however, our conclusions are general.

Our analysis is based on solution of the quantum Liouville equation in occupation space. We use a combination of time-dependent and time-independent analytical approaches to gain qualitative insight into the effect of a dissipative environment on the information content of $\delta(E)$, complemented by numerical solution to go beyond the range of validity of the analytical theory. Most of the results of Section VC1 are based on a perturbative analytical approach formulated in the energy domain. Section VC2 utilizes a combination of analytical perturbative and numerical nonperturbative time-domain methods, based on propagation of the system density matrix. Details of our formalism are provided in Refs. 47 and 48 and are not reproduced here.

1. Energy Domain

The signal consists, as in Eq. (2), of three components, describing the one-photon route, the two-photon route, and the interference term between the one- and two-photon routes from the bound state to the continuum. Analytical expressions for the one- and two-photon terms, along with Feynman diagrams that assist in the interpretation of their physical content, are provided elsewhere and are interesting in other contexts. Here we limit attention to the interference term, which is the subject of the earlier sections. In order to account for this term within perturbation theory, one needs to expand the density matrix to at least the third order in powers of the electric fields, since it involves one interaction with the one-photon field, $\epsilon^{(1)}$ and two interactions with the two-photon field, $\epsilon^{(2)}$. By doing so in all possible permutations, we obtain six pathways in Liouville space, consisting of three complex conjugate pairs that provide each one (real arithmetic) pathway [47]. A schematic illustration of the different Liouville space couplings and a diagrammatic representation of the associated processes are displayed in Fig. 14.

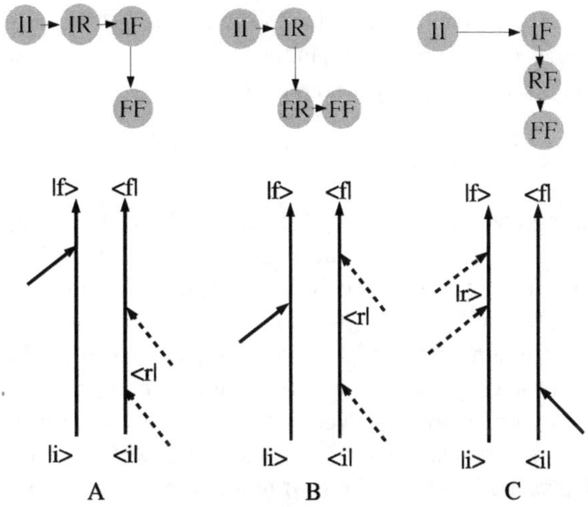

Figure 14. Liouville space coupling schemes and their respective double-sided Feynman diagrams for three of the six pathways in Liouville space which contribute to p_{12}^S. The complex conjugates are not shown. All pathways proceed only via coherences, created by the interactions with the two fields shown as incoming arrows. Solid curves pertain to $\epsilon^{(1)}$ and dashed curves to $\epsilon^{(2)}$. (Reproduced with permission from Ref. 47, Copyright 2005 American Institute of Physics.)

The three distinct physical processes pictured in Fig. 14 translate into three closed form equations for the interference route from the initial to the final state [47]. For the purpose of this chapter, these are more conveniently recast into a single equation as

$$p_{12}^S = \sum_{i,r} \frac{2\mu_{ir}\mu_{rf}\mu_{if}\epsilon_0^{(1)}|\epsilon_0^{(2)}|^2\sqrt{X^2+Y^2}\rho_{ii}^0}{\hbar^3\Gamma_{ff}(\Gamma_{ir}^2+\Omega_{ir}^2)(\Gamma_{if}^2+\Omega_{if}^2)(\Gamma_{rf}^2+\Omega_{rf}^2)} \cos(\phi+\delta^S) \tag{58}$$

where

$$X = (\Gamma_{ir}\Gamma_{rf} + \Omega_{rf}\Omega_{ir})(\Gamma_{if}^2+\Omega_{if}^2) + (\Gamma_{ir}\Gamma_{if} - \Omega_{if}\Omega_{ir})(\Gamma_{rf}^2+\Omega_{rf}^2) + (\Gamma_{if}\Gamma_{rf} - \Omega_{if}\Omega_{rf})(\Gamma_{ir}^2+\Omega_{ir}^2) \tag{59}$$

$$Y = (\Gamma_{rf}\Omega_{ir} - \Gamma_{ir}\Omega_{rf})(\Gamma_{if}^2+\Omega_{if}^2) + (\Gamma_{ir}\Omega_{if} + \Gamma_{if}\Omega_{ir})(\Gamma_{rf}^2+\Omega_{rf}^2) - (\Gamma_{if}\Omega_{rf} + \Gamma_{rf}\Omega_{if})(\Gamma_{ir}^2+\Omega_{ir}^2) \tag{60}$$

and the channel phase δ^S is

$$\delta^S = \tan^{-1}(X/Y) \tag{61}$$

In Eqs. (58)–(60), μ_{ab} denote electronic matrix elements of the dipole operator, $\mu_{ab} = \langle b|\mathbf{r}|a\rangle$, $a, b = i, r, f$; ρ^0_{ii} is the initial probability of occupancy of a state $|i\rangle$; and the Γ_{ab} govern the dissipative processes,

$$\Gamma_{ab} = \Gamma_{\text{pd}} + \tfrac{1}{2}(\Gamma_{aa} + \Gamma_{bb}), \quad a, b = i, r, f \tag{62}$$

with Γ_{pd} denoting the pure dephasing [75] rate and Γ_{aa} the lifetime of level a. Finally, $\Omega_{ir} = ((\epsilon_r - \epsilon_i)/\hbar) - \omega_1$ and $\Omega_{rf} = ((\epsilon_f - \epsilon_r)/\hbar) - \omega_1$ denote the detunings of the intermediate state from resonance with the initial and final states, respectively, and $\Omega_{if} = ((\epsilon_f - \epsilon_i)/\hbar) - 2\omega_1$.

The interference term is thus of the form of Eq. (4), where in the present case δ^S reflects the molecular as well as the environment properties. Within our occupation space formalism, it is given as a simple analytical function of system parameters, and its physical content is explicit. In the isolated molecule limit, where pure dephasing vanishes, Eq. (58) reduces to

$$p^S_{12}(\Gamma_{\text{pd}} \to 0) = \sum_{i,r} \frac{2\mu_{ir}\mu_{rf}\mu_{if}\epsilon_0^{(1)}|\epsilon_0^{(2)}|^2\sqrt{\Gamma_f^2 + \Omega_{ir}^2}\,\rho^0_{ii}}{\hbar^3(\Gamma_r^2 + \Omega_{ir}^2)(\Gamma_f^2 + \Omega_{if}^2)} \cos(\phi + \delta^S) \tag{63}$$

where $2\Gamma_r = \Gamma_{rr}$ and the channel phase δ^S depends only on the intermediate state lifetime and its detuning from resonance with the initial state,

$$\delta(\Gamma_{\text{pd}} \to 0) = \tan^{-1}\left(\frac{\Gamma_{rr}}{2\Omega_{ir}}\right) \tag{64}$$

Thus in the zero dephasing case, δ^S reduces to the Breit–Wigner phase of the intermediate state resonance, elaborated on in the previous sections. In the dissipative environment, it is sensitive also to decay and decoherence mechanisms, as illustrated later.

As seen in Eqs. (59)–(61), dephasing processes introduce two new time scales into the dynamics, in addition to the intermediate state lifetime that determines the structure of δ^S in the isolated molecule case. One is the time scale of pure dephasing, and the other is the lifetime of the final state. Equation (64) illustrates that the Γ_{ff} dependence of δ^S is a condensed phase effect that vanishes in the limit of no dephasing. The more careful analysis later shows that the qualitative behavior of the channel phase is dominated by the $\Gamma_{\text{pd}}/\Gamma_{rr}$ and $\Gamma_{\text{pd}}/\Gamma_{ff}$ ratios, that is, by the rate of dephasing as compared to the system time scales.

Figure 15 illustrates the interplay between the time scales of intermediate state decay, final state decay, and dephasing in determining the photon energy dependence of δ^S. The final state in Fig. 15 is chosen to satisfy the resonance condition when the intermediate state is resonantly excited, $\epsilon_f - \epsilon_i = 2(\epsilon_r - \epsilon_i)$.

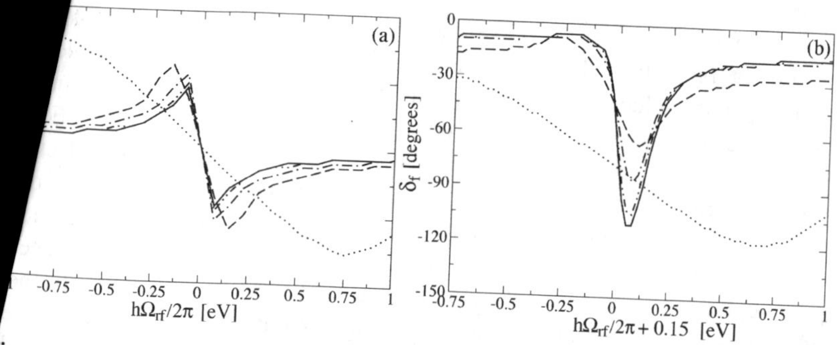

igure 18. The channel phase versus the final state energy for $\Gamma_{rr}^{-1} = 40\,\text{fs}$ and $\Gamma_{\text{pd}}^{-1} = 40\,\text{fs}$. pulse durations are as in Fig. 17. Panel (a) corresponds to resonant excitation of the intermediate e, $\hbar\omega_1 = 1.425\,\text{eV}$, and panel (b) to red-detuned excitation, $\hbar\omega_1 = 1.275\,\text{eV}$. (Reproduced with mission from Ref. 48, Copyright 2006 American Institute of Physics.)

ime-shifted pulses has been used to probe decoherence processes in systems such as surface states [80], holes [81], and chemisorbed states [82]. The similarity is evident: the variable time delay between the two pulses translates into a controllable phase difference between the two coherent excitation routes that plays an analogous role to ϕ, although it is not independent of the system properties.

As applied so far, the interferometric method uses pulses of identical photon energies, but we envision a modification where the two equal energy excitation routes correspond to different frequencies, for example, $\omega_2 = 2\omega_1$. Modifying the scheme of the previous sections by introducing a variable time delay between the two pulses, we find that pathway B of Fig. 14 and its complex conjugate disappear, since simultaneous interactions with both electric fields are needed at any given time to complete these pathways. Also, depending on which pulse field arrives first, only one of the pathways A or C of Fig. 14 and its complex conjugate will be available and contribute to the interference process. The first pulse creates a coherence $|i\rangle\langle f|$, and the second, time-delayed pulse probes how much of this coherence has survived the pure dephasing process. Pathway C of Fig. 14 corresponds to a process where the one-photon pulse arrives first, at a time T_1, and is followed by the two-photon pulse at a time T_2. In the delta pulse limit, the associated interference term takes the form

$$P_{12}^{S}(t) = -\sum_{i,r} \frac{2}{\hbar^3} \mu_{ir}\mu_{kf}\mu_{if}\epsilon^{(1)}|\epsilon^{(2)}|^2 \rho_{ii}^{o} \exp[-\Gamma_{ff}(t - T_2)] \times \exp[-\Gamma_{if}(T_2 - T_1)] \sin(\phi + \Omega_{if}[T_2 - T_1]) \quad (66)$$

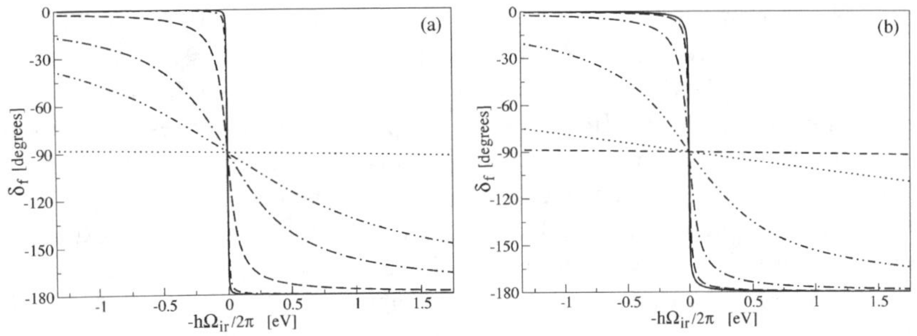

Figure 15. Channel phase versus photon energy ($\hbar\omega_1$) centered at the resonance energy ($\epsilon_r - \epsilon_i = 1.425\,\text{eV}$) for lifetime $\Gamma_{rr}^{-1} = 400\,\text{fs}$. (a) $\Gamma_{ff}^{-1} = 1800\,\text{fs}$ and the pure dephasing time scale Γ_{pd}^{-1} varies as: infinite (solid), 3000 fs (dash-dash-dotted), 300 fs (dashed), 100 fs (dot-dashed), 60 fs (dot-dot-dashed), and 10 fs (dotted). (b) $\Gamma_{\text{pd}}^{-1} = 300\,\text{fs}$ and Γ_{ff}^{-1} varies as: 18 fs (solid), 180 fs (dashed), 1800 fs (dot-dashed), 1.8×10^4 fs (dot-dot-dashed), 1.8×10^5 fs (dotted), and 1.8×10^6 fs (dash-dash-dotted). (Reproduced with permission from Ref. 47, Copyright 2005 American Institute of Physics.)

In the limit where $\Gamma_{\text{pd}}^{-1} \gg \Gamma_{ff}^{-1}$, δ^S takes the Breit–Wigner shape, observed in the isolated molecule limit, whereas in the limit of fast dephasing $\Gamma_{ff}^{-1} \gg \Gamma_{\text{pd}}^{-1}$, δ^S becomes independent of photon energy. In the latter limit the information regarding the phase properties of the material is lost, as is also the controllability of the process with coherent light.

It is interesting to note (see Eqs. (59)–(61)) that pure decoherence introduces dependence of the channel phase on the final state energies. This dependence can be utilized to obtain new insights into the resonance properties, as illustrated in Fig. 16. Here we explore the photon energy dependence of δ_f for final state

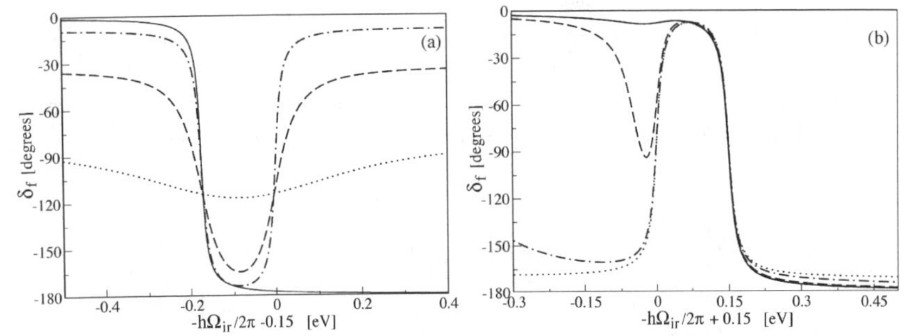

Figure 16. Channel phase versus photon energy ($\hbar\omega_1$) centered at intermediate state peak for $\Gamma_{rr}^{-1} = 40\,\text{fs}$ (a) $\epsilon_f - \epsilon_i = 1.6\,\text{eV}$ (above resonance) and $\Gamma_{ff}^{-1} = 10^{10}\,\text{fs}$. The dephasing time Γ_{pd}^{-1} varies as: 2 fs (dotted), 20 fs (dashed), 200 fs (dot-dashed), and no dephasing (solid). (b) $\epsilon_f - \epsilon_i = 1.275\,\text{eV}$ (below resonance) and $\Gamma_{\text{pd}}^{-1} = 200\,\text{fs}$. The final state lifetime Γ_{ff}^{-1} varies as: 6 fs (solid), 60 fs (dashed), 600 fs (dot-dashed), and 6000 fs (dotted). (Reproduced with permission from Ref. 47, Copyright 2005 American Institute of Physics.)

energies detuned from resonance with the initial state $\epsilon_f - \epsilon_i \neq 2\omega_1$. In Fig. 16a the final state decay rate is fixed and the pure dephasing rate is varied from well below to well above the resonance decay rate. In the limit of very slow dephasing, δ_f takes the Breit–Wigner shape that characterizes the isolated resonant case. Pure dephasing processes force the sigmoidal structure into a hairpin bend, where the bending edge is related to the resonance energy. As Γ_{pd} grows, the hairpin structure gradually flattens out, and in the limit of extremely fast decoherence the phase information is again lost, δ_f becoming a constant function of the photon energy. Figure 16b complements this picture by displaying the dependence of the off-resonance channel phase on the final state lifetime. The sigmoidal shape obtained in the limit of short lifetime, $\Gamma_{ff}^{-1} \ll \Gamma_{\text{pd}}^{-1}$, is gradually distorted as Γ_{ff}^{-1} grows, developing the π change of phase at the resonance energy observed also in Fig. 16a. In the large Γ_{ff}^{-1} limit, δ_f stabilizes on the symmetric hairpin structure—in essence the combination of two sigmoidal curves marking the intermediate and initial state energies. Mathematically, this feature can be shown to arise from interference between sequential and coherent pathways in Liouville space (Fig. 14). From a physical perspective, it can be understood by analogy to the emergence of spontaneous light emission from the interplay between Raman and fluorescence processes in optics.

2. *Time Domain*

We proceed by generalizing the stationary state approach of the previous subsection to the time domain. To that end we first consider the effects of a finite laser pulsewidth and next explore the effect of a finite time delay between the pulses. The latter signal reduces in the case of identical excitation pathways to the method of two-photon interferometry that has been studied extensively in the literature of solids and surfaces [79].

Using a perturbative analysis of the time-dependent signal, and focusing on the interference term between the one- and two-photon processes in Fig. 14, we consider first the limit of ultrashort pulses (in practice, short with respect to all time scales of the system). Approximating the laser pulse as a delta function of time, we have

$$p_{12}^{S}(t) \to -\sum_{i,r} \frac{6}{\hbar^3} \mu_{ir}\mu_{rf}\mu_{if}\epsilon_0^{(1)}|\epsilon_0^{(2)}|^2 \rho_{ii}^{o} \cos(\phi + \pi/2) \exp[-\Gamma_{ff}(t - t_0)] \qquad (65)$$

Equation (65) illustrates that in the limit of ultrashort pulses the two-pathway method loses its value as a coherence spectroscopy; δ^S is fixed at $\pi/2$ irrespective of the system parameters. From the physical perspective, when the excitation is much shorter than the system time scales, the channel phase carries no imprint of the system dynamics since the interaction time does not suffice to observe dynamical processes.

Figure 17 shows the signal versus the re
ϕ, as obtained through time propagation o
density matrix elements. As the pulse durati
converges, as expected, to the delta functi
durations comparable to the system time scales
short with respect to the pure dephasing time, dep
the signal carries the same information as in
(compare solid and dashed curves in Fig. 17a). As
pure dephasing time, Fig. 17b, pure dephasing becon
of off-resonant excitation, considered in Fig. 17b, pur
signal since it broadens the levels involved in the tra
intensity to nonresonant processes.

As pointed out earlier, pure dephasing processes introdu
channel phase on the final state energies. It is thus interes
dependence of δ^S on the detuning of the two-photon field fr
the $|r\rangle$ to $|f\rangle$ transition, $\Omega_{rf} = ((\epsilon_f - \epsilon_r)/\hbar - \omega_1$. This func
Fig. 18 for $\hbar\omega_1$ on-resonance (panel a) and red detuned from res
with the $|i\rangle \to |r\rangle$ transition ($\epsilon_r - \epsilon_i = 1.425$ eV). In both cases
temporal shaping of the channel phase, as the increasing pulsew
the resonance lifetime and pure dephasing time scales to eme
interference of distinct Liouville space pathways. When the pu
exceeds the system time scales, δ^S converges to the CW limit (c
dot-dashed and solid curves).

Before concluding the discussion of the time-domain two-pathway a
it is appropriate to expand on the relation of this experiment to the two
interferometry of Refs. 78–80, where excitation with two phase-l

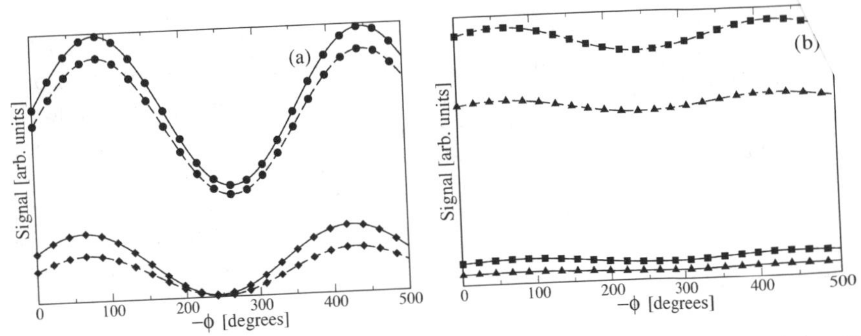

Figure 17. The total signal versus the relative phase of the laser fields for the case of no pure dephasing (solid curves) and for $\Gamma_{\text{pd}}^{-1} = 30$ fs (dashed curves). The pulsewidth for the two cases is varied as follows: (a) 4 fs (circles) and 30 fs (diamonds); and (b) 150 fs (squares) and 200 fs (triangles). (Reproduced with permission from Ref. 48, Copyright 2006 American Institute of Physics.)

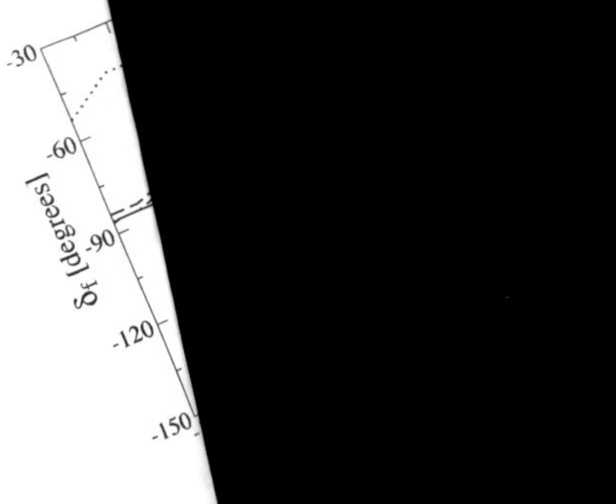

The
stat
pe

where $t > T_2 > T_1$ and $\Omega_{if} = ((\epsilon_f - \epsilon_i)/\hbar - 2\omega_1$. Thus the interference term decays as a function of the time delay, and its rate of decay is determined by the decoherence rate between the initial and final states. (The same result, differing only in the sign of Ω_{if}, holds if the order of the pulses is reversed, as depicted by path A in Fig. 14.) The analytical result of Eq. (66) is derived in the δ-pulse approximation, but its conclusions hold equally in the finite-pulse case. Numerical illustrations are provided elsewhere. We mention here only that, while the approach is similar to the familiar method of two-photon interferometry [79], already successfully utilized to probe a variety of dephasing mechanisms, the distinguishability of the two optical pathways provides potentially new information. In particular, we find [48] that the combination of the two-pathway excitation method with the time-domain interferometry method could be used to disentangle relaxation from pure dephasing and determine the rates of both processes.

VI. OUTLOOK: WHERE FROM HERE?

In the previous sections we reviewed the problem of the channel phase, an outgrowth of the energy-domain, two-pathway approach to phase control, which we regard as a "coherence spectroscopy." Having placed this topic in context with other coherent means of inquiring into molecular properties in Section I, we provided a qualitative introduction to the structure and information content of the channel phase in Section II, and proceeded in Sections III and IV with a detailed description of an experimental method of measuring it and an account of the underlying theory. Finally, we provided examples in Section V ranging from isolated atoms and diatomic molecules to polyatomic chemistry and extended dissipative systems, including both theoretical and experimental research.

Although coherent control is now a mature field, much remains to be accomplished in the study of the channel phase. There is no doubt that coherence plays an important role in large polyatomic molecules as well as in dissipative systems. To date, however, most of the published research on the channel phase has focused on isolated atoms and diatomic molecules, with very few studies addressing the problems of polyatomic and solvated molecules. The work to date on polyatomic molecules has been entirely experimental, whereas the research on solvated molecules has been entirely theoretical. It is important to extend the experimental methods from the gas to the condensed phase and hence explore the theoretical predictions of Section VC. Likewise interesting would be theoretical and numerical investigations of isolated large polyatomics. A challenge to future research would be to make quantitative comparison of experimental and numerical results for the channel phase. This would require that we address a sufficiently simple system, where both the experiment and the numerical calculation could be carried out accurately.

We began our analysis in Section II and ended it in Section VC2 by making the connection of the time- and energy-domain approaches to both coherence spectroscopy and coherent control. It is appropriate to remark in closing that new experimental approaches that combine time- and energy-domain techniques are currently being developed to provide new insights into the channel phase problem. We expect that these will open further avenues for future research.

Acknowledgments

We thank the National Science Foundation for its generous support of much of the research reviewed here, and several students and postdoctoral fellows for their contribution to this work. Particularly appreciated is the assistance of Vishal Barge and Prof. Zhan Hu with reproduction of several of the figures.

References

1. D. Meshulach and Y. Silberberg, *Nature* **396**, 239 (1998).
2. T. C. Weinacht, J. Ahn, and P. H. Bucksbaum, *Nature* **397**, 233 (1999).
3. A. Baltuska, T. Udem, M. Uiberacker, M. Hentschel, E. Goulielmakis, C. Gohle, R. Holzwarth, V. S. Yakoviev, A. Scrinzi, T. W. Hansch, and F. Krausz, *Nature* **421**, 611 (2003).
4. M. D. Eisaman, A. Andre, F. Massou, M. Fleischhauer, A. S. Zibrov, and M. D. Lukin, *Nature* **438**, 837 (2005).
5. M. Bajcsy, A. S. Zibrov, and M. D. Lukin, *Nature* **426**, 638 (2003).
6. J. R. Petta, A. C. Johnson, J. M. Taylor, E. A. Laird, A. Yacoby, M. D. Lukin, C. M. Marcus, M. P. Hanson, and A. C. Gossard, *Science* **309**, 2180 (2005).
7. C. Daniel, J. Full, L. Gonzalez, C. Lupulescu, J. Manz, A. Merli, S. Vajda, and L. Woste, *Science* **299**, 536 (2003).
8. R. J. Levis, G. M. Menkir, and H. Rabitz, *Science* **292**, 709 (2001).
9. V. I. Prokhorenko, A. M. Nagy, S. A. Waschuk, L. S. Brown, R. R. Birge, and R. J. D. Miller, *Science* **313**, 1257 (2006).
10. J. M. Dela Cruz, I. Pastirk, M. Comstock, V. V. Lozovoy, and M. Dantus, *Proc. Natl. Acad. Sci. USA* **101**, 16996 (2004).
11. T. Brixner, N. H. Damrauer, P. Niklaus, and G. Gerber, *Nature* **414**, 57 (2001).
12. J. L. Herek, W. Wohlleben, R. J. Cogdell, D. Zeidler, and M. Motzkus, *Nature* **417**, 533 (2002).
13. D. Leibfried, E. Knill, S. Seidelin, J. Britton, R. B. Blakestad, J. Chiaverini, D. B. Hume, W. M. Itano, J. D. Jost, C. Langer, R. Ozeri, R. Reichle, and D. J. Wineland, *Nature* **438**, 639 (2005).
14. J. P. Reithmaier, G. Sek, A. Loffler, C. Hofmann, S. Kuhn, S. Reitzenstein, L. V. Keldysh, V. D. Kulakovskii, T. L. Reinecke, and A. Forchel, *Nature* **432**, 197 (2004).
15. T. Yamamoto, Y. A. Pashkin, O. Astafiev, Y. Nakamura, and J. S. Tsai, *Nature* **425**, 941 (2003).
16. J. L. Cirac and P. Zoller, *Nature* **404**, 579 (2000).
17. J. M. Geremia, J. K. Stockton, and H. Mabuchi, *Science* **304**, 270 (2004).
18. N. Dudovich, D. Oron, and Y. Silberberg, *Nature* **418**, 512 (2002).
19. T. Feurer, J. C. Vaughan, and K. A. Nelson, *Science* **299**, 374 (2003).
20. M. F. DeCamp, D. A. Reis, P. H. Bucksbaum, B. Adams, J. M. Caraher, R. Clarke, C. W. S. Conover, E. M. Dufresne, R. Merlin, V. Stoica, and J. K. Wahlstrand, *Nature* **413**, 825 (2001).

21. R. Bartels, S. Backus, E. Zeek, L. Misoguti, G. Vdovin, I. P. Christov, M. M. Murnane, and H. C. Kapteyn, *Nature* **406**, 164 (2000).

22. N. H. Bonadeo, J. Erland, D. Gammon, D. Park, D. S. Katzer, and D. G. Steel, *Science* **282**, 1473 (1998).

23. The list of Refs. 1–22 is clearly far from complete. We limited attention rather arbitrarily to articles that appeared in *Science*, *Nature* or the *Proc. Natl. Acad. Sci. USA* during the past few years.

24. S. A. Rice and M. Zhao, *Optical Control of Molecular Dynamics*, Wiley, Hoboken, NJ, 2000; D. J. Tannor and S. A. Rice, *Adv. Chem. Phys.* **70**, 441 (1988); R. J. Gordon and S. A. Rice, *Annu. Rev. Phys. Chem.* **48**, 501 (1997).

25. M. Shapiro and P. Brumer, *Principles of the Quantum Control of Molecular Processes*, Wiley-Interscience, Hoboken, NJ, 2003; P. Brumer and M. Shapiro, *Annu. Rev. Phys. Chem.* **43**, 257 (1992); M. Shapiro and P. Brumer, *Acc. Chem. Res.* **22**, 407(1989).

26. H. A. Rabitz, M. M. Hsieh, and C. M. Rosenthal, *Science* **303**, 1998 (2004).

27. A. D. Bandrauk, Y. Fujimura, and R. J. Gordon (eds.), *Laser Control and Manipulation of Molecules*, American Chemical Society Oxford University Press, New York 2002; W. Potz and A. Schroeder (eds.), *Coherent Control in Atoms, Molecules, and Semiconductors*, Kluwer Academic Publishers, Dordrecht, 1999.

28. F. Langhojer D. Cardoza, M. Baertschy, and T. Weinacht, *J. Chem. Phys.* **122**, 014102 (2005).

29. T. Seideman, *J. Chem. Phys.* **108**, 1915 (1998).

30. J. A. Fiss, L. Zhu, R. J. Gordon, and T. Seideman, *Phys. Rev. Lett.* **82**, 65 (1999).

31. T. Seideman, *J. Chem. Phys.* **111**, 9168 (1999).

32. J. A. Fiss, A. Khachatrian, L. Zhu, R. Gordon, and T. Seideman, *Faraday Discuss. Chem. Soc.* **113**, 61 (1999).

33. J. A. Fiss, A. Khachatrian, K. Truhins, L. Zhu, R. Gordon, and T. Seideman, *Phys. Rev. Lett.* **85**, 2096 (2000).

34. M. Shapiro, J. W. Hepburn, and P. Brumer, *Chem. Phys. Lett.* **149**, 451 (1988).

35. L. Zhu, V. D. Kleiman, X. Li, L. Lu, K. Trentelman, and R. J. Gordon, *Science* **270**, 77 (1995).

36. Z. Wang and D. S. Elliott, *Phys. Rev. Lett.* **87**, 173001 (2001).

37. We follow the convention adopted by the literature on coherent control via the one- versus three-photon excitation method, where the frequency of the three-photon field is denoted ω_1 and that of the one-photon field is denoted ω_3.

38. H. Lefebvre-Brion, *J. Chem. Phys.* **106**, 2544 (1997).

39. L. Zhu, K. Suto, J. A. Fiss, R. Wada, T. Seideman, and R. J. Gordon, *Phys. Rev. Lett.* **79**, 4108 (1997).

40. S. Lee, *J. Chem. Phys.* **107**, 2734 (1997).

41. R. J. Gordon, L. Zhu, and T. Seideman, *Acc. Chem. Rese.* **32**, 1007 (1999).

42. H. Lefebvre-Brion, T. Seideman, and R. J. Gordon, *J. Chem. Phys.* **114**, 9402 (2001).

43. R. J. Gordon, L. Zhu, and T. Seideman, *J. Phys. Chem. A* **105**, 4387 (2001).

44. A. Apalategui, A. Saenz, and P. Lambropoulos, *Phys. Rev. Lett.* **86**, 5454 (2001).

45. A. Khachatrian, R. Billotto, L. Zhu, R. J. Gordon, and T. Seideman, *J. Chem. Phys.* **116**, 9326 (2002).

46. R. J. Gordon, L. Zhu, and T. Seideman, *Comm. Atom. Mol. Phys.* **2**, D262 (2002).

47. S. Ramakrishna and T. Seideman, Coherence spectroscopy in dissipative media: a Liouville space approach, *J. Chem. Phys.* **122**, 084502 (2005).

48. S. Ramakrishna and T. Seideman, Coherent spectroscopy in dissipative media: time domain studies of channel phase and signal interferometry, *J. Chem. Phys.* **124**, 244503 (2006).

49. M. Sukharev and T. Seideman, Phase and polarization control as a route to plasmonic nanodevices, *Nanoletters* **6**, 715 (2006).

50. M. Sukharev and T. Seideman, Coherent control approaches to light guidance in the nanoscale, *J. Chem. Phys.* **124**, 144707 (2006).

51. A. Pelzer, S. Ramakrishna, and T. Seideman, Optimal control of molecular alignment in dissipative media, *J. Chem. Phys.* in press.

52. N. Dudovich, B. Dayan, S. M. G. Faeder, and Y. Silberberg, *Phys. Rev. Lett.* **86**, 47 (2001).

53. K. A. Walowicz, I. Pastkirk, V. L. Lozovoy, and M. Dantus, *J. Phys. Chem. A* **106**, 9369 (2002).

54. A. H. Zewail, *Femtochemistry: Ultrafast Dynamics of the Chemical Bond*, World Scientific, Singapore, 1994.

55. E. Merzbacher, *Quantum Mechanics*, 3rd ed., Wiley, Hoboken, NJ, 1988, Chap. 7.

56. See, for instance, R. D. Levine, *Quantum Mechanics of Molecular Rate Processes*, Clarendon Press, Oxford, 1969.

57. I. Fischer, M. J. J. Vrakking, D. M. Villeneuve, and A. Stolow, *Chem. Phys.* **207**, 331 (1996).

58. S.-P. Lu, S. M. Park, Y. Xie, and R. J. Gordon, *J. Chem. Phys.* **96**, 6613 (1992).

59. A. Yariv, *Quantum Electronics*, 2nd ed., Wiley, Hoboken, NJ, 1989.

60. R. W. Boyd, *J. Opt. Soc. Am.* **70**, 877 (1980).

61. R. J. Gordon, S.-P. Lu, S. M. Park, K. Trentelman, Y. Xie, L. Zhu, A. Kumar, and W. J. Meath, *J. Chem. Phys.* **98**, 9481 (1993).

62. V. J. Barge, Z. Hu, J. Willig, and R. J. Gordon, *Phys. Rev. Lett.* **97**, 263001 (2006).

63. R. J. Gordon and V. J. Barge, *J. Chem. Phys.* **127**, 204302 (2007).

64. E. Hecht, *Optics*, 4th ed., Addison Wesley, San Francisco, 2002, p. 261.

65. W. C. Wiley and I. H. McLaren, *Rev. Sci. Intrum.* **26**, 1150 (1955).

66. H. Nagai, H. Ohmura, T. Nakanaga, and M. Tachiya, *J. Photochem. Photobiol. A: Chemistry*, **182**, 209 (2006).

67. A. R. Edmonds, *Angular Momentum in Quantum Mechanics*, 2nd ed., Princeton University Press, Princeton, NJ, 1960.

68. U. Fano, *Phys. Rev.* **124**, 1866 (1961).

69. It is not essential that the electronic curves cross (at some level of approximation). In case they do not, however, they should become energetically degenerate in the atomic limit to be simultaneously observed. In the latter case kinetic (radial derivative) coupling is generally nonnegligible, because it is likely to become comparable to the energy separation before the potential curves become a constant function of the internuclear distance. See, for instance, S. J. Singer, K. F. Freed, and Y. B. Band, *J. Chem. Phys.* **79**, 6060 (1983).

70. T. P. Rakitzis, S. A. Kandel, A. J. Alexander, Z. H. Kim, and R. N. Zare, *Science* **281**, 1346 (1998).

71. R. G. Newton, *Scattering Theory of Waves and Particles*, McGraw-Hill, New York, 1966.

72. Z.-M. Wang and D. S. Elliott, *Phys. Rev. Lett.* **87**, 173001 (2001).

73. R. Yamakazi and D. S. Elliott, *Phys. Rev. Lett.* **98**, 053001 (2007).

74. H. Nagai, H. Ohmura, F. Ito, T. Nakanaga, and M. Tachiya, *J. Chem. Phys.* **124**, 034304 (2006).

75. V. Barge, Z. Hu, and R. J. Gordon (manuscript in preparation).

76. V. Barge and R. J. Gordon (manuscript in preparation).

77. We conform here to the convention of the condensed phase literature, where "dephasing" is used for lose of phase coherence (e.g., through collisions). It should be noted that this convention differs from that used in the wavepacket literature, where "dephasing" is used for a coherence-preserving process where anharmonicity of the underlying eigenvalue spectrum causes the wavepacket components to step out of phase but revivals occur provided that the motion is regular.

78. J. Gong, A. Ma, and S. A. Rice, *J. Chem. Phys.* **122**, 204505 (2005).

79. H. Petek and S. Ogawa, *Prog. Surf. Sci.* **56**, 239 (1998).

80. H. Petek, A. P. Heberle, W. Nessler, H. Nagano, S. Kubota, S. Matsunami, N. Moriya, and S. Ogawa, *Phys. Rev. Lett.* **79**, 4649 (1997).

81. H. Petek, H. Nagano, and S. Ogawa, *Phys. Rev. Lett.* **83**, 832 (1999).

82. H. Petek, H. Nagano, M. J. Weida, and S. Ogawa, *J. Phys. Chem. B* **105**, 6767 (2001).

TIME-DEPENDENT TRANSITION STATE THEORY

THOMAS BARTSCH

Department of Mathematical Sciences, Loughborough University, Loughborough LE11 3TU, United Kingdom

JEREMY M. MOIX AND RIGOBERTO HERNANDEZ

Center for Computational Molecular Science and Technology, Georgia Institute of Technology, Atlanta, Georgia 30332 USA

SHINNOSUKE KAWAI

Molecule and Life Nonlinear Sciences Laboratory, Research Institute for Electronic Science (RIES), Hokkaido University, Sapporo 060-0812 Japan

T. UZER

Center for Nonlinear Science, Georgia Institute of Technology, Atlanta, Georgia 30332 USA

CONTENTS

Advances in Chemical Physics, Volume 140, edited by Stuart A. Rice
Copyright © 2008 John Wiley & Sons, Inc.

I. INTRODUCTION

Reaction rate theory, broadly defined, characterizes the long-time dynamics of systems with several metastable states. It is therefore relevant to the understanding of a wide range of systems in physics, chemistry, and biology. The centerpiece of classical rate theory is transition state theory (TST), which has provided a guiding framework to the theory of chemical reactions for decades (see Refs. 1–4 for early work in the field and Refs. 5–9 for reviews). TST serves a dual purpose in reaction rate theory: it provides an intuitive picture of the reaction dynamics and, through this picture, it guides the development of efficient computational schemes for the numerical calculation of reaction rates. The fundamental concept of TST is the transition state (TS) or activated complex that marks the transition from "reactants" and "products." It has been realized only recently that this fundamental idea is widely applicable not only in theoretical chemistry, where it was developed, but also to problems as diverse as the ionization of atoms [10, 11], the rearrangement of clusters [12], conductance through microjunctions [13], asteroid capture [14], and even phase transitions in cosmology [15]. Accordingly, the term "reaction" should in the following be understood in this broad sense.

TST is traditionally based on the assumption that the reaction rate is determined by the dynamics in a small area of phase space and that in this "bottleneck" area a dividing surface can be identified that all reactive trajectories must cross and that no trajectory can cross more than once. If such a surface can be found, TST will yield the exact reaction rate (apart from quantum effects) [16, 17]. Approximate dividing surfaces that are located near the bottleneck, but are not strictly free of recrossings, still provide a portrait of the activated complex and an upper bound for the reaction rate. For this reason, they have been used routinely in the chemistry community (e.g., see Refs. 7,18, and 19). Nevertheless, a no-recrossing dividing surface remains the gold standard, and a prescription for its construction has been sought for a long time. It was found in the 1970s for the special case of systems with two degrees of freedom [20–22]. A general method that yields a no-recrossing TS dividing surface for reactions in arbitrarily many degrees of freedom has been described only recently [10–12, 14, 23–35]. It is based on a phase-space rather than a configuration-space picture of the dynamics and employs a detailed description of invariant geometric objects in phase space. Apart from allowing one to construct a no-recrossing dividing surface and thereby solving the fundamental problem of TST, this geometric approach provides impenetrable barriers that separate reactive trajectories from those that are nonreactive. Knowing these separatrices, one can therefore predict whether or not a given initial condition will lead to a reaction, without having to carry out any detailed propagation of trajectories.

The powerful techniques of geometric TST are well adapted to the description of autonomous gas-phase reactions, which form also the prototypical setting for traditional TST. Most reactions of practical relevance, however, do not fall into this class. Instead, the reactive species are typically subject to the influence of a time-dependent environment. If the physical insight and the computational efficiency offered by TST are to be retained for those systems, its conceptual framework must be extended to incorporate external driving forces. Many reactions take place in simple liquids in which the external driving forces can be treated using continuum methods, and hence *approximate* transition state structures have been known for some time [7, 18, 19, 36]. A detailed dynamical picture of *exact* time-dependent transition states in phase space has recently begun to emerge [37–40]. Here we survey these developments to emphasize the generality of the underlying ideas, which might not be readily apparent in the original papers that focus on special cases.

The paradigmatic example of an externally driven reactive system is that of a chemical reaction in a liquid. This case is particularly challenging because the driving force arises from the interaction with a macroscopic number of solvent modes. The force exerted by the solvent onto the solute fluctuates seemingly randomly around an average value that depends on the instantaneous configuration of the solute. The quasideterministic average part gives rise to an effective renormalization of the free energies. The computation of these shifts poses a complicated many-body problem that has been approached using, for example, linear-response theory and/or projection operator methods [41–48]. Once they are known, however, their impact on the reaction dynamics can be discussed within the established framework of TST. The time-dependent fluctuations around the mean force, by contrast, require an extension of that framework. Many advances in multidimensional and variational transition state theory were motivated by the search for such an extension [7, 18, 19, 36], and indeed, such problems were also the original motivation for the recent development of time-dependent TST. Its fundamental concepts apply nevertheless to deterministic as well as stochastic driving, although the details of the dynamics are quite different. The main difference is due to the quasideterministic damping that results from a stochastic force according to the fluctuation-dissipation theorem, but is absent under deterministic driving. Deterministic time-dependent TST can thus be firmly rooted in the familiar Hamiltonian setting.

The study of deterministically driven reactive systems has become prominent as the development of laser technology over the past decades has led to laser pulses whose duration is comparable to the time scale of molecular motion (i.e., pico- or femtosecond) [49–52] and is now heading toward the attosecond time scale of electronic dynamics [53, 54]. It is now becoming feasible to control chemical reactions on a microscopic level through the application of judiciously shaped laser pulses [55]. To understand the influence of a laser pulse onto the reaction dynamics, and to determine a pulse shape that would steer the reaction toward a desired outcome, one needs a conceptual framework capable of describing the dynamics in microscopic detail. The time-dependent TST offers the prospect of providing such a framework, and this scenario has motivated its development for deterministic driving so far, without being the only possible area of application.

Reaction rates in the presence of a random force were derived by Kramers [56] for the limits of weak and strong damping by the environment. His derivations did not use TST ideas. Indeed, it might initially appear that TST is ill equipped to handle the influence of the environment because the fluctuating force will cause the reactive system to move randomly back and forth, so that any dividing surface in phase space will be crossed many times by a typical reactive trajectory [57]. It therefore seems that any calculation of a reaction rate that is based on a TST approach is bound to overestimate the rate, and indeed powerful computational schemes such as transition path sampling [58, 59] are available that dispense with the notion of the transition state altogether. In fact, however, the Kramers rate formula can be obtained from TST approaches if an explicit infinite-dimensional model of the heat bath is introduced and the dividing surface is located in the infinite-dimensional phase space of the extended system [44, 60, 61]. This observation led to a solution of the Kramers turnover problem, that is, a rate formula that interpolates between the limits of low and high friction [61–65]. Remarkably, this result can be derived without recourse to a microscopic model of the heat bath. To this end, Graham [64] introduced a macroscopic reaction coordinate for the dynamics in the vicinity of the barrier top that reappears here as part of a complete dynamically adapted coordinate system. That coordinate system was subsequently used by Martens [66] to describe the noise-averaged macroscopic dynamics in the geometric language of dynamical-systems theory. The time-dependent framework presented here allows one to bridge the gap between macroscopic and microscopic descriptions of the dynamics in that it gives easy access to microscopic information that characterizes the dynamics for a specific realization of the fluctuating force, but at the same time does not require one to introduce an explicit model of the heat bath.

Following most earlier work on either traditional or geometric TST, we assume that reactants and products are separated in phase space by a bottleneck (or barrier) on the potential energy landscape. The location of this barrier in configuration space is generally marked by a col or saddle point of the potential energy. In an autonomous system, the saddle corresponds to a fixed point of the dynamics in phase space that is of crucial importance to geometric TST. In the presence of external driving, that fixed point ceases to exist, and one of the central challenges in the development of time-dependent TST is to find a geometric object in phase space that can replace it. We define a transition state (TS) trajectory that satisfies this requirement under a wide variety of conditions. Several authors have previously discussed how fixed points in autonomous Hamiltonian systems can be generalized to driven systems (e.g., see Refs. 67–69). However, these studies were largely focused on stable fixed points and restricted to periodic or quasiperiodic driving. In these cases, the fixed point is replaced by a periodic orbit or an invariant torus. TST requires the treatment of unstable fixed points that correspond to rank-1 saddle points of the potential energy. In addition, we aim to present a framework that is applicable to as broad a class of external driving forces as possible. Throughout the following development, we assume that, although the fixed point on the barrier top is lost, the bottleneck itself remains intact; that is, the external driving is not strong enough to obliterate the wells and barriers in the deterministic potential altogether.

This chapter is organized as follows. In Section II, we briefly summarize the findings of the geometric TST for autonomous Hamiltonian systems to the extent that it is needed for the present discussion. Readers interested in a more detailed exposition are referred to Ref. 35, where the field has recently been reviewed in depth. We restrict our discussion to classical mechanics. Semiclassical extensions of geometric TST have been developed in Refs. 70–75. Section III discusses the notion of the TS trajectory in general and its incarnation in different specific settings. Section IV demonstrates how the TS trajectory allows one to carry over the central concepts of geometric TST into the time-dependent realm.

II. GEOMETRIC TST

The geometric approach to multidimensional TST has been developed independently along two different lines that were labeled the "top–down" and "bottom–up" approaches in Ref. 35. Both approaches are equivalent. They lead to similar computational procedures using Lie–Deprit normal form transformations [76–78]. The bottom–up approach [12, 24–30] starts by considering individual trajectories. It constructs a coordinate system in which the complexity of the trajectories is reduced through a constant of motion and, in particular, a recrossing-free dividing surface can be specified. The top–down approach [10, 11, 14, 23, 33–35], which we mainly follow here, proceeds in several steps. First, the dynamics is linearized around the saddle point that marks the barrier between reactants and products. The linearized equations of motion can be solved explicitly, and the dynamics can be described in detail. Second, invariant-manifold theory is invoked to conclude that certain qualitative features that are represented by invariant manifolds remain intact in the full nonlinear system. These manifolds are calculated explicitly using normal form theory. In this manner, one can construct a normally hyperbolic invariant manifold (NHIM) of the reactive dynamical system that can be identified as the geometric representation of the classical TS. This notion provides a precise statement of the earlier idea to identify the TS with a classical bound state above the barrier [20]. It strictly satisfies the fundamental requirement of TST, that, at least locally, no trajectory can recross the TS after crossing it once. In two degrees of freedom, the NHIM coincides with the well-known periodic-orbit dividing surface (PODS) that has been widely applied in reaction-rate studies [20–22]. Unlike earlier approaches, however, the NHIM formulation of TST can be applied in phase spaces of arbitrary dimensionality. The stable and unstable manifolds of the NHIM serve as separatrices between reactive and nonreactive trajectories. They generalize the cylindrical manifolds that play the same role in two-dimensional systems [79–81]. They are typically calculated in the vicinity of the barrier through a normal form expansion and can then be continued into the well regions numerically. They thus allow one to distinguish reactive from nonreactive regions of the reactant phase space even far away from the barrier and to discuss the nonlocal dynamics that ultimately leads to recrossings [31, 82–87].

We assume that the dynamics of the reactive system is described by a Hamiltonian of the form

$$H(\boldsymbol{p},\boldsymbol{q}) = \tfrac{1}{2}\boldsymbol{p}^2 + U(\boldsymbol{q}) \tag{1}$$

where $\boldsymbol{q} = (q_1,\ldots,q_N)$ is a set of N mass-weighted coordinates and $\boldsymbol{p} = (p_1,\ldots,p_N)$ is the set of conjugate momenta. Equation (1) covers most applications in chemistry, but it is not completely general. The phase-space approach to TST allows one to handle more general Hamiltonians easily [11, 35]. It will become clear later that the same holds for its time-dependent generalization. We nevertheless restrict the discussion to Eq. (1) for simplicity.

One of the fundamental assumptions of TST states that the reaction rate is determined by the dynamics in a small neighborhood of the saddle point, which we place at $\boldsymbol{q} = 0$. It is then a reasonable approximation to expand the potential $U(\boldsymbol{q})$ in a Taylor series around the saddle and retain only the lowest-order terms. Because $\nabla_{\boldsymbol{q}} U(\boldsymbol{q} = 0) = 0$ at the saddle point itself, these are of second order and lead to the Hamiltonian

$$H(\boldsymbol{p},\boldsymbol{q}) = \tfrac{1}{2}\boldsymbol{p}^2 - \tfrac{1}{2}\sum_{i,j} q_i \Omega_{ij} q_j \tag{2}$$

where the potential energy $U(\boldsymbol{q} = 0)$ at the saddle point has been set to zero and the matrix $\boldsymbol{\Omega}$ of force constants is given by

$$\Omega_{ij} = -\left.\frac{\partial^2 U}{\partial q_i\, \partial q_j}\right|_{\boldsymbol{q}=0} \tag{3}$$

Although any coordinate system can in principle be chosen, it is convenient to choose one for which

$$\boldsymbol{\Omega} = \begin{pmatrix} \omega_b^2 & & & \\ & -\omega_2^2 & & \\ & & \ddots & \\ & & & -\omega_N^2 \end{pmatrix} \tag{4}$$

is diagonal. In Eq. (2), q_1 is the reaction coordinate that corresponds to the single unstable degree of freedom of the dynamics around the saddle point. The instability is characterized by the barrier frequency ω_b. The frequencies $\omega_2, \ldots, \omega_N$ describe the frequencies of oscillations in the stable transverse normal modes $q_2, \ldots, q_N$.

In the normal-mode coordinates, the Hamiltonian in Eq. (2) reads

$$H(\boldsymbol{p}, \boldsymbol{q}) = \tfrac{1}{2}(p_1^2 - \omega_b^2 q_1^2) + \tfrac{1}{2}(p_2^2 + \omega_2^2 q_2^2) + \cdots + \tfrac{1}{2}(p_N^2 + \omega_2^2 q_N^2) \tag{5}$$

Therefore, the normal modes $(p_1, q_1), \ldots, (p_N, q_N)$ decouple, which makes the dynamics in the barrier region extremely simple [70]. It can be put into an even simpler form by introducing the coordinates

$$Q_1 = \frac{1}{\sqrt{2}}\left(\sqrt{\omega_b}\, q_1 + \frac{p_1}{\sqrt{\omega_b}}\right), \quad P_1 = \frac{1}{\sqrt{2}}\left(\frac{p_1}{\sqrt{\omega_b}} - \sqrt{\omega_b}\, q_1\right) \tag{6a}$$

and

$$Q_j = \frac{1}{\sqrt{2}}\left(\sqrt{\omega_j}\, q_j + \mathrm{i}\frac{p_j}{\sqrt{\omega_j}}\right), \quad P_j = \frac{1}{\sqrt{2}}\left(\frac{p_j}{\sqrt{\omega_j}} - \mathrm{i}\sqrt{\omega_j}\, q_j\right) \tag{6b}$$

for $j = 2, \ldots, N$. In these coordinates, the Hamiltonian reads

$$H(\boldsymbol{P}, \boldsymbol{Q}) = \omega_b P_1 Q_1 + \sum_{j=2}^{N} \mathrm{i}\omega_j P_j Q_j \tag{7}$$

and yields the equations of motion

$$\dot{P}_1 = -\omega_b P_1, \quad \dot{Q}_1 = \omega_b Q_1 \tag{8a}$$

$$\dot{P}_j = -\mathrm{i}\omega_j P_j, \quad \dot{Q}_j = \mathrm{i}\omega_j Q_j \quad \text{for} \quad j = 2, \ldots, N \tag{8b}$$

The phase space is formally decomposed into one-dimensional invariant subspaces. This is only a formal decomposition because the coordinates Q_j

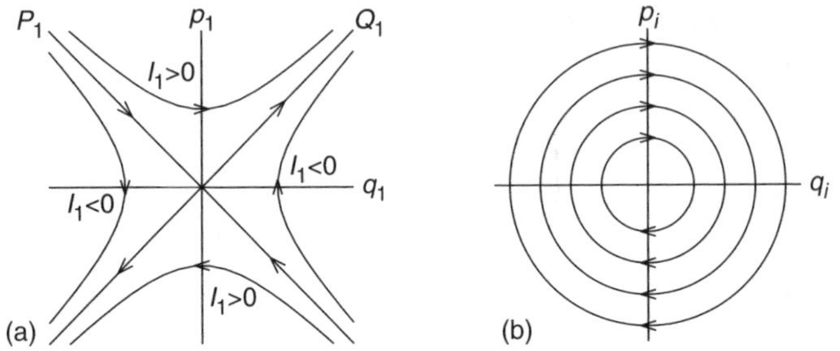

Figure 1. Phase-space portrait of the dynamics described by the linearized Hamiltonian in Eq. (5), projected onto (a) the reactive mode and (b) a bath mode.

and momenta P_j of the bath modes are complex. It is nevertheless useful to describe the linear and nonlinear dynamics in the autonomous case and is indispensable for the discussion of driven systems.

Because the degrees of freedom decouple in the linear approximation, it is easy to describe the dynamics in detail. There is the motion across a harmonic barrier in one degree of freedom and $N-1$ harmonic oscillators. Phase-space plots of the dynamics are shown in Fig. 1. The transition from the "reactant" region at $q_1 < 0$ to the "product" region at $q_1 > 0$ is determined solely by the dynamics in (p_1, q_1), which in the traditional language of reaction dynamics is called the reactive mode.

The point $q_1 = p_1 = 0$ (or $P_1 = Q_1 = 0$) is a fixed point of the dynamics in the reactive mode. In the full-dimensional dynamics, it corresponds to all trajectories in which only the motion in the bath modes is excited. These trajectories are characterized by the property that they remain confined to the neighborhood of the saddle point for all time. They correspond to a bound state in the continuum, and thus to the transition state in the sense of Ref. 20. Because it is described by the two independent conditions $q_1 = 0$ and $p_1 = 0$, the set of all initial conditions that give rise to trajectories in the transition state forms a manifold of dimension $2N-2$ in the full $2N$-dimensional phase space. It is called the central manifold of the saddle point. The central manifold is subdivided into level sets of the Hamiltonian in Eq. (5), each of which has dimension $2N-1$. These energy shells are normally hyperbolic invariant manifolds (NHIM) of the dynamical system [88]. Following Ref. 34, we use the term NHIM to refer to these objects. In the special case of the two-dimensional system, every NHIM has dimension one. It reduces to a periodic orbit and reproduces the well-known PODS [20–22].

The trajectories of the NHIM are characterized by zero values of both P_1 and Q_1. A nontrivial dynamics in the reactive mode is obtained if either of these coordinates takes a nonzero value. According to the equation of motion (8),

P_1 exponentially decreases to zero as $t \to \infty$, whereas Q_1 increases. Trajectories with $Q_1 = 0$ therefore asymptotically approach the central manifold $P_1 = Q_1 = 0$ as $t \to \infty$, or, more precisely, they approach the NHIM of the appropriate energy within the central manifold. Due to this behavior, the set of all initial conditions with $Q_1 = 0$ is called the stable manifold of the NHIM. Similarly, trajectories with $P_1 = 0$ asymptotically approach the NHIM as $t \to -\infty$. They are said to form the unstable manifold of the NHIM.

A typical trajectory has nonzero values of both P_1 and Q_1. It is part of neither the NHIM itself nor the NHIM's stable or unstable manifolds. As illustrated in Fig. 1a, these typical trajectories fall into four distinct classes. Some trajectories cross the barrier from the reactant side $q_1 < 0$ to the product side $q_1 > 0$ (reactive) or from the product side to the reactant side (backward reactive). Other trajectories approach the barrier from either the reactant or the product side but do not cross it. They return on the side from which they approached (nonreactive trajectories). The boundaries or separatrices between regions of reactive and nonreactive trajectories in phase space are formed by the stable and unstable manifolds of the NHIM. Thus once these manifolds are known, one can predict the fate of a trajectory that approaches the barrier with certainty, without having to follow the trajectory until it leaves the barrier region again. This predictive value of the invariant manifolds constitutes the power of the geometric approach to TST, and when we are discussing driven systems, we mainly strive to construct time-dependent analogues of these manifolds.

A convenient quantitative characterization of the stable and unstable manifolds themselves as well as of reactive and nonreactive trajectories can be obtained by noting that the special form of the Hamiltonian in Eq. (5) allows one to separate the total energy into a sum of the energy of the reactive mode and the energies of the bath modes. All these partial energies are conserved. The value of the energy

$$E_1 = \tfrac{1}{2}(p_1^2 - \omega_b^2 q_1^2) = \omega_b P_1 Q_1 \tag{9}$$

in the reactive mode distinguishes reactive from nonreactive trajectories: a trajectory for which E_1 is positive (i.e., larger than the potential energy at the saddle point) will cross the barrier, a trajectory with negative E_1 will be reflected by it.

Instead of the reactive-mode energy E_1, it is convenient to study the associated action variable

$$I_1 = E_1/\omega_b = P_1 Q_1 \tag{10}$$

The stable and unstable manifolds are then described by $I_1 = 0$, reactive trajectories by $I_1 > 0$, and nonreactive trajectories by $I_1 < 0$.

In addition to describing the TS and the separatrices between reactive and nonreactive trajectories that are the central discovery of geometric TST, one can also easily specify a recrossing-free dividing surface, the centerpiece of traditional TST. As Fig. 1 shows, the surface defined by $q_1 = 0$ is crossed only by reactive trajectories, and no trajectory crosses it more than once. If one restricts the surface to one of the half-planes $q_1 = 0, p_1 > 0$ or $q_1 = 0, p_1 < 0$, one can study forward or backward reactive trajectories, respectively. In contrast to the NHIM and its stable and unstable manifolds, the dividing surface is not uniquely defined by the dynamics. Any sufficiently small deformation of the surface just specified would also be free of recrossings. This nonuniqueness is due to the fact that the dividing surface can obviously not be invariant under the dynamics because trajectories have to cross it. For a rate calculation, however, it suffices to know any recrossing-free dividing surface, and one is free to choose it conveniently.

So far, the discussion of the dynamics and the associated phase-space geometry has been restricted to the linearized Hamiltonian in eq. (5). However, in practice the linearization will rarely be sufficiently accurate to describe the reaction dynamics. We must then generalize the discussion to arbitrary nonlinear Hamiltonians in the vicinity of the saddle point. Fortunately, general theorems of invariant manifold theory [88] ensure that the qualitative features of the dynamics are the same as in the linear approximation: for every energy not too high above the energy of the saddle point, there will be a NHIM with its associated stable and unstable manifolds that act as separatrices between reactive and nonreactive trajectories in precisely the manner that was described for the harmonic approximation.

Although general results guarantee the existence of the invariant objects that are used in geometric TST, calculating them for a specific nonlinear Hamiltonian system can be a difficult task. It was shown in Ref. 34 that normal form theory, a particular form of classical perturbation theory, provides an effective means to carry out these calculations. Normal form theory allows one to construct a coordinate system $(\bar{P}_1, \bar{Q}_1, \ldots, \bar{P}_N, \bar{Q}_N)$ in which, to an arbitrary degree of approximation, the reactive mode $(\bar{P}_1, \bar{Q}_1)$ decouples from the bath modes. The action variable $\bar{I}_1 = \bar{P}_1 \bar{Q}_1$ is conserved and distinguishes reactive from nonreactive trajectories. The separatrices are then given by $\bar{I}_1 = 0$, as in the linear approximation. In these coordinates, the central manifold (i.e., the collection of NHIMs) can be characterized by

$$\mathcal{M} = \{(\bar{\boldsymbol{P}}, \bar{\boldsymbol{Q}}) : \bar{P}_1 = \bar{Q}_1 = 0\} \tag{11a}$$

and the associated stable and unstable manifolds by

$$\mathcal{W}^{\mathrm{s}} = \{(\bar{\boldsymbol{P}}, \bar{\boldsymbol{Q}}) : \bar{Q}_1 = 0\} \tag{11b}$$

$$\mathcal{W}^{\mathrm{u}} = \{(\bar{\boldsymbol{P}}, \bar{\boldsymbol{Q}}) : \bar{P}_1 = 0\} \tag{11c}$$

The normal form coordinate systems also allow one to define a TS dividing surface

$$\mathcal{T} = \{(\bar{\boldsymbol{P}}, \bar{\boldsymbol{Q}}) : \bar{P}_1 = \bar{Q}_1\} \tag{11d}$$

as the surface on which

$$\bar{q}_1 = \frac{1}{\sqrt{2\omega_b}}(\bar{Q}_1 - \bar{P}_1) = 0 \tag{12}$$

This surface is locally free of recrossings. It remains possible, however, that a trajectory leaves the vicinity of the barrier, descends into the product well, and later approaches the barrier again and recrosses the dividing surface. This sequence of events would phenomenologically be described as a reaction that is followed by a back-reaction. If enough time elapses between subsequent crossings, this possibility does not violate the no-recrossing assumption of TST because the initial reaction that is indicated by the crossing of the dividing surface has actually taken place. The dynamics that leads to nonlocal recrossings can be investigated by calculating the separatrices far away from the barrier [31, 82–87]. As its counterpart in the linearized dynamics, the dividing surface in Eq. (12) is not uniquely defined by the dynamics. Nevertheless, geometric TST allows one to construct a dividing surface free of recrossings, which has formerly been impossible in general.

The algorithm used in the normal form transformation is not discussed here in detail. It is described in Refs. 34 and 35. A time-dependent generalization of this normal form procedure is presented in Section IVB.

III. THE TRANSITION STATE TRAJECTORY

The geometric version of TST laid out in Section II is centered around the NHIM that defines the dividing surface and its stable and unstable manifolds that act as separatrices. The NHIMs at different energies are in turn organized by the saddle point. It forms a fixed point of the dynamics—that is it is itself an invariant object—and it provides the Archimedean point in which the geometric phase-space structure is anchored.

In the presence of external driving that is not strong enough to wipe out the troughs and barriers of the autonomous potential altogether, the location of the reaction bottleneck is still roughly characterized by the saddle point. However, the saddle point is no longer a dynamical fixed point. It does not provide an invariant object in phase space that could serve as the carrier of a TS dividing surface or of the separatrices. In order to generalize the geometric picture of autonomous TST to time-dependent settings, it is essential to find an invariant object in the phase space of the time-dependent dynamics that can play the role of the saddle point and serve as the anchor of more complicated structures. This need is satisfied by the transition state trajectory that we first introduced for stochastically driven systems [38].

In an autonomous system, because the saddle point is unstable in the reactive degree of freedom, most trajectories will either descend into one of the wells in the future or have ascended from a well in the past. Typically, they do both. The fixed point (at the threshold energy) and the trajectories within the NHIM (at higher energies) are distinguished from all other trajectories by the property that they are trapped in the barrier region. It is for this reason that they are of special importance for the dynamics in the barrier region, and it is this characteristic property that has to be preserved in the transition to driven systems. We are thus led to the following general definition:

> A trajectory that remains in the vicinity of the barrier for all times in the future and in the past, without ever descending on either side, is called a *transition state trajectory*.

The exact meaning of this condition depends on the details of the external driving and has to be specified on a case-by-case basis. The notion of the TS trajectory is thereby given enough flexibility to offer a guiding concept for the study of different types of driven systems.

In an autonomous system, any trajectory within the NHIM satisfies the definition of the TS trajectory. Thus the TS trajectory cannot in general be unique. In the particular case of two-dimensional systems, the periodic orbits giving rise to the PODS [20–22] at any energy above the saddle point are TS trajectories. At high energies, the large-amplitude oscillation of the PODS will explore positions far from the saddle point along the stable direction, although never entering the reactant or product exit channels. Moreover, if the nonlinearities are strong, the PODS can be markedly displaced from the saddle point [89], so that it might appear to be placed on the slope of the potential, although in fact it represents the classical bound state that TST requires [20]. This familiar example illustrates that the above definition should not be understood to require the TS trajectory to remain in the immediate neighborhood of the saddle point for all time. Instead, what is postulated is that the TS trajectory dances around the unstable ridge of the potential energy surface and never descends into one of the wells so as to settle there.

Because energy is not conserved in a time-dependent system, it is not meaningful to ascribe a certain energy to a TS trajectory in the way that the fixed point and the NHIM in an autonomous system exist at different energies. Instead, there is typically a single TS trajectory that is uniquely defined by the driving force. It can be regarded as the generalization of the saddle point, whereas an analogue of the NHIM is absent in the time-dependent setting. In special circumstances, if there exist undamped oscillatory modes, there are several TS trajectories that span a NHIM, as described later.

A. The TS Trajectory in White Noise

The concept of the TS trajectory was first introduced [37] in the context of stochastically driven dynamics described by the Langevin equation of motion

$$\ddot{\boldsymbol{q}}_\alpha(t) = -\nabla_q U(\boldsymbol{q}_\alpha(t)) - \boldsymbol{\Gamma}\dot{\boldsymbol{q}}_\alpha(t) + \xi_\alpha(t) \tag{13}$$

In Eq. (13), the vector $\boldsymbol{q}$ denotes a set of mass-weighted coordinates in a configuration space of arbitrary dimension N, $U(\boldsymbol{q})$ is the potential of mean force governing the reaction, $\boldsymbol{\Gamma}$ is a symmetric positive-definite friction matrix, and $\xi_\alpha(t)$ is a stochastic force that is assumed to represent white noise that is Gaussian distributed with zero mean. The subscript α in Eq. (13) is used to label a particular noise sequence $\xi_\alpha(t)$. For any given α, there are infinitely many different trajectories $\boldsymbol{q}_\alpha(t)$ that are distinguished by their initial conditions at some time t_0. If the stochastic force is assumed to arise from a heat bath at a given temperature T, its statistical properties determine the friction matrix $\boldsymbol{\Gamma}$ through the fluctuation-dissipation theorem [41, 42, 48, 90, 91]

$$\left\langle \xi_\alpha(t)\xi_\alpha^{\mathrm{T}}(t') \right\rangle_\alpha = 2k_{\mathrm{B}}T\boldsymbol{\Gamma}\delta(t-t') \tag{14}$$

where the angular brackets denote the average over the instances α of the fluctuating force, and the Dirac δ function appears because white noise is local in time—that is, the strengths of the stochastic force at different times are statistically uncorrelated. The fluctuation-dissipation theorem implies the presence of damping in any system that is driven by an equilibrium heat bath. Stochastically driven systems are therefore fundamentally different from deterministically driven Hamiltonian systems, where no dissipation occurs.

The deterministic force in Eq. (13) is formally assumed to be derived from a potential $U(\boldsymbol{q})$. However, it will become clear in the following that this assumption is not necessary: it would be straightforward to include velocity-dependent forces that arise from, for example, a magnetic field.

As a first step toward a TST treatment of the stochastically driven dynamics, it is crucial to assume, just as in the autonomous case, that the deterministic dynamics has a fixed point that marks the location of an energetic barrier between reactants and products. In the case of Eq. (13), the fixed point is given by a saddle point $\boldsymbol{q}_0$ of the potential $U(\boldsymbol{q})$. The reaction rate is determined by the dynamics in a small neighborhood of the saddle point, so that the deterministic force can be linearized around the saddle point to yield

$$\ddot{\boldsymbol{q}}_\alpha(t) = \boldsymbol{\Omega}\boldsymbol{q}_\alpha(t) - \boldsymbol{\Gamma}\dot{\boldsymbol{q}}_\alpha(t) + \xi_\alpha(t) \tag{15}$$

where the first term on the right-hand side involves the symmetric force constant matrix in Eq. (3).

The linearization of the deterministic dynamics allows one to solve the equations of motion explicitly. Equation (15) can be rewritten as a first-order equation of motion in the $2N$-dimensional phase space with the coordinates

$$z = \begin{pmatrix} \boldsymbol{q} \\ \boldsymbol{v} \end{pmatrix} \tag{16}$$

where $\boldsymbol{v} = \dot{\boldsymbol{q}}$, as

$$\dot{z}_\alpha(t) = \boldsymbol{A}z_\alpha(t) + \begin{pmatrix} 0 \\ \xi_\alpha(t) \end{pmatrix} \tag{17}$$

with the $2N$-dimensional constant matrix

$$\boldsymbol{A} = \begin{pmatrix} \boldsymbol{0} & \boldsymbol{I} \\ \boldsymbol{\Omega} & -\boldsymbol{\Gamma} \end{pmatrix} \tag{18}$$

where $\boldsymbol{I}$ is the $N \times N$ identity matrix. If the matrix $\boldsymbol{A}$ is diagonalized, the Langevin equation (17) decomposes into a set of $2N$ independent scalar equations of motion

$$\dot{z}_{\alpha j}(t) = \lambda_j z_{\alpha j}(t) + \xi_{\alpha j}(t) \tag{19}$$

where λ_j are the eigenvalues of $\boldsymbol{A}$, the coordinates $z_{\alpha j}$ are the components of z in the basis $\boldsymbol{V}_j$ of eigenvectors, and $\xi_{\alpha j}$ are the corresponding components of $(0, \xi_\alpha(t))^{\mathrm{T}}$.

In the absence of damping ($\boldsymbol{\Gamma} = \boldsymbol{0}$), there will be the real eigenvalues $\pm\omega_{\mathrm{b}}$ corresponding to the reactive mode and the purely imaginary eigenvalues $\pm i\omega_i$ describing the oscillations in the bath modes, as discussed in Section II. Nonzero damping will shift the real eigenvalues toward smaller values. The bath mode oscillations will be damped, and the corresponding eigenvalues will acquire negative real parts if the damping is weak. They will become negative real if the damping is so strong that some modes are overdamped. In summary, there will always be one positive real eigenvalue, and all other eigenvalues will have negative real parts. No purely imaginary eigenvalues can exist in the presence of damping.

The set (19) of equations of motion can be solved explicitly [37] by

$$z_{\alpha j}(t) = c_{\alpha j}e^{\lambda_j t} + z^{\ddagger}_{\alpha j}(t) \tag{20}$$

where $c_{\alpha j}$ are arbitrary constants and $z^{\ddagger}_{\alpha j}(t)$ is a particular solution given by

$$z^{\ddagger}_{\alpha j}(t) = \begin{cases} \int_{-\infty}^{t} e^{\lambda_j (t-\tau)} \xi_{\alpha j}(\tau)\, d\tau, & \text{if} \quad \mathrm{Re}\,\lambda_j < 0 \\ -\int_{t}^{\infty} e^{\lambda_j (t-\tau)} \xi_{\alpha j}(\tau)\, d\tau, & \text{if} \quad \mathrm{Re}\,\lambda_j > 0 \end{cases} \tag{21}$$

Note that because of the decaying exponential factors the integrals in Eq. (21) exist unless the stochastic force $\xi_{\alpha j}(t)$ increases exponentially in strength for large positive or negative times, which occurs only with probability zero. To ensure convergence, one must choose the particular solution of Eq. (21) as an integral over the past driving force if $\mathrm{Re}\,\lambda_j < 0$ and as an integral over future driving if $\mathrm{Re}\,\lambda_j > 0$.

We identify the trajectory $z^{\ddagger}_{\alpha}(t)$, which is obtained by setting $c_{\alpha j} = 0$ in Eq. (20) for all components, as the TS trajectory. To justify this identification, we must demonstrate that this trajectory remains in the vicinity of the barrier for all times, without ever descending on either side, and we must make precise in what sense this statement is true. In the case of stochastic driving, we adopt a statistical interpretation of this condition.

The definition (21) of the TS trajectory in white noise can be rewritten as follows [40]. For a function $f(t)$ with the Fourier transform

$$\hat{f}(\omega) = \frac{1}{\sqrt{2\pi}} \int_{-\infty}^{\infty} f(t) \exp(-\mathrm{i}\omega t)\, dt \tag{22}$$

and for a complex number μ, let

$$S[\mu, f](t) = \frac{1}{\sqrt{2\pi}} \int_{-\infty}^{\infty} \frac{\hat{f}(\omega)}{-\mu + \mathrm{i}\omega} \exp(\mathrm{i}\omega t)\, d\omega \tag{23}$$

In terms of this functional, the definition (21) of the TS trajectory can be written

$$z^{\ddagger}_{\alpha j}(t) = S[\lambda_j, \xi_{\alpha j}](t) \tag{24}$$

This form, while less explicit than Eq. (21), allows one to treat stable and unstable modes on an equal footing and provides an efficient way to evaluate the TS trajectory numerically for a given instance of the noise. It also proves convenient to use this notation in the calculation of a TS trajectory for systems under the influence of nonwhite noise or deterministic driving.

The TS trajectory $z^{\ddagger}_{\alpha}(t)$ given by Eq. (21) depends on time t and on the instance of the noise that is represented by the subscript α. Properly speaking, therefore, it defines a statistical ensemble of trajectories. For a fixed time t_0, it specifies a random variable $z^{\ddagger}_{\alpha}(t_0)$ with certain statistical properties. Its distribution can be shown [38] to be a multidimensional Gaussian with zero mean and a covariance matrix that depends on the noise strength and on the eigenvalues λ_j. The characteristic property that singles out the ensemble (21) is that its distribution is independent of the time t_0 at which it is evaluated. This is remarkable because the ensemble is localized in the barrier region, which is dynamically unstable. If an ensemble of trajectories is chosen arbitrarily, the trajectories will typically leave the barrier region after a while, and the ensemble will evolve into a bimodal distribution that is peaked in the reactant and product wells. Only the ensemble (21) avoids this decay and remains stationary with its peak at the position of the barrier. It is therefore a suitable analogue of the fixed point in an autonomous deterministic setting. It represents what is known in the mathematical literature (e.g., see Ref. 92) as an invariant measure of the random dynamical system given by the linearized Langevin equation (15).

Instead of studying an ensemble of trajectories at a given time, we can consider the time dependence of the TS trajectory $z^{\ddagger}_{\alpha_0}(t)$ for a given instance α_0 of the noise. Because the white noise that drives the dynamics is ergodic, the distribution of this trajectory, if it is sampled over a sufficiently long time, will be the same Gaussian distribution that was found for the ensemble at any given time. In this sense, even a single trajectory in the ensemble can be said to remain in the barrier region for all time. Although the Gaussian distribution implies that the TS trajectory will eventually move arbitrarily far from the saddle point, it will return quickly and spend most of the time in the barrier region. To achieve this, the TS trajectory may only leave the barrier region when it is about to experience a restoring force from the external noise that will drive it back. Indeed, the explicit expression (21) for the TS trajectory shows that its stable components depend on the *past* driving force, but the unstable component depends on the *future*. This ability to "look ahead" enables the TS trajectory to adjust its motion to the noise in such a delicate way that it will never leave the barrier region permanently. Other trajectories, by contrast, will typically slide down the potential slope and, because the linearized model does not have wells, are very unlikely ever to return.

This dependence of the TS trajectory on the past and future driving is illustrated in Fig. 2, which shows the time dependence of the S-functional $S[\mu, \mathcal{E}](t)$ for a fixed driving field $\mathcal{E}(t)$ and different values of the eigenvalue μ. To demonstrate the properties of the S-functional clearly, we have chosen a smooth driving field $\mathcal{E}(t)$ that is given by Eq. (81) (with $A_0 = 1$, $\omega = 1$, and $N = 2$). It is zero for $|t| > 2\pi$.

Note that the S-functionals are nonzero even in the time range when the field vanishes. For $\operatorname{Re}\mu < 0$, as in the example of Fig. 2b, the S-functional depends on the past of the driving fields. Consequently, it is zero before the onset of the pulse. After the end of the pulse, $S[-1, \mathcal{E}]$ tends to zero

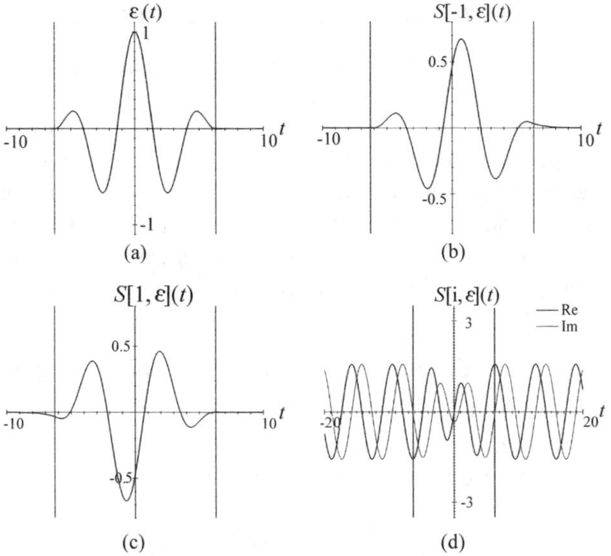

Figure 2. Illustration of the S-functional (23) for (a) the external driving field and the eigenvalues (b) $\mu = -1$, (c) $\mu = 1$, and (d) $\mu = \mathrm{i}$. (The latter case is discussed in detail in Section IIIC.)

exponentially, but it remains nonzero for arbitrarily large times. It is also clear from Fig. 2b that the S-functional lags the driving field: it takes its maximum only after the maximum of the field has been attained. By contrast, if $\mathrm{Re}\,\mu > 0$, as shown in Fig. 2c, $S[\mu, \mathcal{E}]$ depends on the future of the noise. It is zero after the end of the pulse, but nonzero before the pulse sets in. The S-functional $S[1, \mathcal{E}]$ leads the driving field: it takes its maximum earlier than the field. Figure 2d shows an example for an S-functional with purely imaginary eigenvalue μ. The behavior of the functional in this case will be discussed in detail in Section IIIC.

The time-dependent view of the TS trajectory also allows one to show that in this case it is unique: all other trajectories can be obtained by choosing nonzero values of the constants $c_{\alpha j}$ in Eq. (20). The exponential contributions then cause $z_{\alpha j}(t)$ to diverge without return either for large positive or for large negative times because all eigenvalues λ_j have nonzero real parts. Of course, the uniqueness of the TS trajectory has to be understood in the sense of statistical ensembles, as described earlier: for every instance α of the noise, there is a unique trajectory that remains close to the barrier for all time, again in a statistical sense. This uniqueness is a consequence of the presence of damping in the Langevin equation (13), which ensures that there are no purely imaginary eigenvalues. The damping is due to the fluctuation-dissipation theorem, Eq. (14), and is a specific property of stochastic driving.

B. The TS Trajectory in Colored Noise

So far, we have assumed that the noise source $\xi_\alpha(t)$ represents white noise; that is, its values at different instances in time are statistically independent. At first sight, it might seem that this assumption has never actually been used and that the results of the previous section remain valid if the noise is not white. This is wrong, however, because the fluctuation-dissipation theorem demands that for colored noise the damping term in the Langevin equation must be modified to incorporate memory friction effects. The Langevin equation (13) must then be replaced by the generalized Langevin equation [6, 42, 43, 48, 91, 93–95]

$$\ddot{\boldsymbol{q}}_\alpha(t) = -\nabla_{\boldsymbol{q}} U(\boldsymbol{q}_\alpha(t)) - \int_{-\infty}^{t} d\tau\, \boldsymbol{\Gamma}(t-\tau)\dot{\boldsymbol{q}}_\alpha(\tau) + \xi_\alpha(t) \tag{25}$$

where the symmetric-matrix valued function $\boldsymbol{\Gamma}(t)$ is related to the fluctuating force by the generalized fluctuation-dissipation theorem

$$\left\langle \xi_\alpha(t)\xi_\alpha^{\mathrm{T}}(s)\right\rangle_\alpha = k_{\mathrm{B}}T\,\boldsymbol{\Gamma}(|t-s|) \tag{26}$$

Because the friction force in Eq. (25) depends on the entire prehistory of the trajectory rather than only the present velocity, Eq. (25) cannot be solved as an initial value problem. In particular, the phase space of the generalized Langevin equation cannot easily be specified, as it was done in Eq. (16) for the white noise case. Looking at the equation of motion (25), one might expect that dimension to be infinite because the entire trajectory up to the present time must be specified before its further evolution can be calculated. However, only those initial trajectories can be prescribed that themselves satisfy Eq. (25), and this restriction makes the phase-space dimension finite or at least countably infinite in many cases, to which we restrict our discussion. For a certain class of friction kernels $\boldsymbol{\Gamma}(t)$, an explicit construction of the phase space was given by Martens [66]. A convenient implicit description can be obtained through a solution of the linearized generalized Langevin equation

$$\ddot{\boldsymbol{q}}_\alpha(t) = \boldsymbol{\Omega}\boldsymbol{q}_\alpha(t) - \int_{-\infty}^{t} d\tau\, \boldsymbol{\Gamma}(t-\tau)\dot{\boldsymbol{q}}_\alpha(\tau) + \xi_\alpha(t) \tag{27}$$

with $\boldsymbol{\Omega}$ as in Eq. (3).

The explicit solution of Eq. (27), which uses a Fourier transform or a bilateral Laplace transform, is described in detail in Ref. 38. Its eigenvalues and eigenvectors are determined by the nonlinear eigenvalue equation

$$(\lambda_j^2 + \lambda_j \hat{\boldsymbol{\Gamma}}(\lambda_j) - \boldsymbol{\Omega})\boldsymbol{v}_j = 0 \tag{28}$$

where

$$\hat{\boldsymbol{\Gamma}}(\mu) = \int_0^\infty dt\, e^{-\mu t} \boldsymbol{\Gamma}(t) \tag{29}$$

is the Laplace transform of the friction kernel $\boldsymbol{\Gamma}(t)$. The dimension of the phase space is given by the number of solutions of Eq. (28). (Note that the dimension of the eigenvectors $\boldsymbol{v}_j$ in Eq. (28) is the dimension N of configuration space, not the dimension of phase space.) In general, we expect to find one positive real eigenvalue (which coincides with the Grote–Hynes reaction frequency [61, 95]) and several eigenvalues with negative real parts.

In terms of a complete set of solutions $(\lambda_j, \boldsymbol{v}_j)$ of the eigenvalue equation (28), the general solution to the linearized equation of motion (27) can be written

$$\boldsymbol{q}_\alpha(t) = \sum_j c_{\alpha j} \boldsymbol{v}_j\, e^{\lambda_j t} + \boldsymbol{q}_\alpha^\ddagger(t) \tag{30}$$

where

$$\boldsymbol{q}_\alpha^\ddagger(t) = \sum_j S[\lambda_j, \boldsymbol{\mu}_j \xi_\alpha](t) \tag{31}$$

The constant matrices $\boldsymbol{\mu}_j$ act as projection operators onto the different eigenspaces. They are given in Ref. 38. The solution Eq. (30) is entirely analogous to Eq. (20) in the white noise case. To obtain a trajectory that remains in the vicinity of the barrier for all times, we again have to set $c_{\alpha j} = 0$ and identify Eq. (31) as the TS trajectory. It satisfies the condition of the general definition in that it provides, at fixed time, a random ensemble of trajectories that is stationary in time, and at fixed noise sequence α a trajectory that spends most of its time close to the barrier.

C. The TS Trajectory Under Deterministic Driving

We have so far focused on reactive systems that are subject to stochastic driving by a heat bath. For these systems, the fluctuation-dissipation theorem requires that the dynamics must be dissipative. Conversely, the presence of damping implies the influence of a stochastic force, so that systems that suffer only deterministic driving will in general be conservative. Of course, in view of the wide applicability of TST, it is conceivable that the system of interest is macroscopic and that damping is important while noise is negligible. In such a setting, the theory developed in Section III A can be applied if the stochastic driving is replaced by a deterministic external force. We can therefore discount this possibility here and assume, following Ref. 40, that the dynamics of a deterministically driven system is described by the time-dependent Hamiltonian

$$H(\boldsymbol{p},\boldsymbol{q},t) = H^{\text{sys}}(\boldsymbol{p},\boldsymbol{q}) + H^{\text{ext}}(\boldsymbol{p},\boldsymbol{q},t) \tag{32}$$

where H^{sys} is the autonomous Hamiltonian of the isolated system and H^{ext} describes the interaction with the time-dependent external field. Again, for the sake of simplicity, we assume that the system Hamiltonian has the form in Eq. (1), although the results remain valid for more general Hamiltonians.

As before, we make the fundamental assumption of TST that the reaction is determined by the dynamics in a small neighborhood of the saddle, and we accordingly expand the Hamiltonian around the saddle point to lowest order. For the system Hamiltonian, we obtain the second-order Hamiltonian of Eq. (2), which takes the form of Eq. (7) in the complexified normal-mode coordinates, Eq. (6). In the external Hamiltonian, we can disregard terms that are independent of $\boldsymbol{p}$ and $\boldsymbol{q}$ because they have no influence on the dynamics. The leading time-dependent terms will then be of the first order. Using complexified coordinates, we obtain the approximate Hamiltonian

$$H_0(\boldsymbol{P},\boldsymbol{Q},t) = \omega_{\text{b}} P_1 Q_1 + \sum_{j=2}^{N} \mathrm{i}\omega_j P_j Q_j - \sum_{j=1}^{N} \xi_j(t) Q_j - \sum_{j=1}^{n} \eta_j(t) P_j \tag{33}$$

with certain functions $\xi_j(t)$ and $\eta_j(t)$ that depend on time, but not on coordinates or momenta. These functions can be determined explicitly once a driving Hamiltonian H^{ext} has been chosen to describe a specific system. The Hamiltonian in Eq. (33) yields the equations of motion

$$\dot{P}_1 = -\omega_{\text{b}} P_1 + \xi_1(t), \quad \dot{Q}_1 = \omega_{\text{b}} Q_1 - \eta_1(t) \tag{34a}$$

$$\dot{P}_j = -\mathrm{i}\omega_j P_j + \xi_j(t), \quad \dot{Q}_j = \mathrm{i}\omega_j Q_j - \eta_j(t) \quad \text{for } j = 2,\ldots,N \tag{34b}$$

These equations are formally identical to the equations (19) for the diagonal coordinates of the Langevin equation. The diagonal coordinates are in both cases determined only by the deterministic part of the dynamics. In the Hamiltonian setting they can naturally be identified as coordinates or momenta, which is impossible for the dissipative dynamics.

The equations of motion (34) can be solved explicitly to yield

$$P_1(t) = c_1 e^{-\omega_{\text{b}} t} + P_1^{\ddagger}(t), \quad Q_1(t) = \tilde{c}_1 e^{\omega_{\text{b}} t} + Q_1^{\ddagger}(t) \tag{35a}$$

$$P_j(t) = c_j e^{-\mathrm{i}\omega_j t} + P_j^{\ddagger}(t), \quad Q_j(t) = \tilde{c}_j e^{\mathrm{i}\omega_j t} + Q_j^{\ddagger}(t) \quad \text{for } j = 2,\ldots,N \tag{35b}$$

with arbitrary constants c_j and $\tilde{c}_j$ and

$$P_1^\ddagger(t) = S[-\omega_b, \xi_1](t), \quad Q_1^\ddagger(t) = -S[\omega_b, \eta_1](t) \tag{36a}$$

$$P_j^\ddagger(t) = S[-i\omega_j, \xi_j](t), \quad Q_j^\ddagger(t) = -S[i\omega_j, \eta_j](t) \quad \text{for } j = 2, \ldots, N \tag{36b}$$

Assuming that the $P_j^\ddagger(t)$ and $Q_j^\ddagger(t)$ can be interpreted as a TS trajectory, which is discussed later, we can conclude as before that $c_1 = \tilde{c}_1 = 0$ if the exponential instability of the reactive mode is to be suppressed. Coordinate and momentum of the TS trajectory in the reactive mode, if they exist, are therefore unique. For the bath modes, however, difficulties arise. The exponentials in Eq. (35b) remain bounded for all times, so that their coefficients c_j and $\tilde{c}_j$ cannot be determined from the condition that we impose on the TS trajectory. Consequently, the TS trajectory cannot be unique. The physical cause of the nonuniqueness is the presence of undamped oscillations, which cannot be avoided in a Hamiltonian setting. In a dissipative system, by contrast, all oscillations are typically damped, and the TS trajectory will be unique.

Even worse than the nonuniqueness is the observation that Eq. (36b) is only a formal solution of the equation of motion, but does not actually define a TS trajectory: for a purely imaginary eigenvalue $i\omega$, the integrand in the definition of the $S[i\omega,f]$ in Eq. (23) has a pole on the integration path, so that the integral diverges unless the Fourier transform $\hat{f}(\omega) = 0$. A true solution of the equations of motion can only be obtained if this singularity is suitably regularized. This can be done in several ways. An obvious choice is to add an infinitesimal positive or negative real part to the eigenvalues of the bath modes, that is, to replace the $S[i\omega,f]$ by one of

$$S_\pm[i\omega,f] = S[i\omega \pm \varepsilon,f], \qquad \varepsilon > 0 \tag{37}$$

or by an arbitrary linear combination $\lambda S_+[i\omega,f] + (1-\lambda)S_-[i\omega,f]$. These regularizations differ by

$$S_-[i\omega,f](t) - S_+[i\omega,f](t) = \sqrt{2\pi}\hat{f}(\omega)e^{i\omega t} \tag{38}$$

which is a multiple of the exponential in the general solution (36b). It thus turns out that the nonuniqueness of the TS trajectory and the singularity in its definition are related: different ways to regularize the singularity correspond to different choices of the arbitrary constants in Eq. (36b). All solutions of the equations of motion in the bath modes will serve equally well as components of the TS trajectory, and any regularization of the S-functionals in Eq. (36b) can be used to compute such a solution.

At this point in the argument, it remains to be checked if the solutions (36) actually define a TS trajectory, that is, whether this trajectory remains in the vicinity of the saddle point for all times. Whether or not this is the case, and in what sense, will depend on the properties of the driving forces $\xi_j(t)$ and $\eta_j(t)$ that have so far been left unspecified.

In Ref. 40 it was assumed that the driving forces vanish asymptotically for $t \to \pm\infty$. Under this condition it can easily be derived from the explicit representation (21) of the S-functional that in the reactive mode $P_1(t)$ and $Q_1(t)$ tend to zero for $t \to \pm\infty$ and that the bath modes approach the autonomous dynamics

$$P_j(t) \to c_j^{\pm} e^{-\mathrm{i}\omega_j t}, \quad Q_j(t) \to \tilde{c}_j^{\pm} e^{\mathrm{i}\omega_j t} \quad \text{as } t \to \pm\infty \tag{39}$$

In particular, the TS trajectory remains bounded for all times, which satisfies the general definition. The constants $c_j^{\pm}$ and $\tilde{c}_j^{\pm}$ in Eq. (39) depend on the specific choice of the TS trajectory. Because the saddle point of the autonomous system becomes a fixed point for large positive and negative times, one might envision an ideal choice to be one that allows the TS trajectory to come to rest at the saddle point both in the distant future and in the remote past. However, this is impossible in general because the driving force will transfer energy into or out of the bath modes in such a way that

$$c_j^+ - c_j^- = \sqrt{2\pi}\hat{\xi}_j(-\omega_j) \quad \text{and} \quad \tilde{c}_j^+ - \tilde{c}_j^- = -\sqrt{2\pi}\hat{\eta}_j(\omega_j) \tag{40}$$

depend only on the driving, and not on the trajectory. A "symmetric" TS trajectory was chosen in Ref. 40 such that the amplitudes of the bath-mode oscillations are the same for large positive and negative times, as illustrated in Fig. 2d. Strictly speaking, this is an arbitrary convention because all possible TS trajectories remain bounded. In practice, however, it is useful to require that the excursions of the TS trajectory from the saddle point remain as small as possible. This additional consideration favors the symmetric choice.

Similarly, if the external driving force is periodic or quasiperiodic, the trajectory defined by Eq. (36) will have the same property and obviously satisfy the requirements for a TS trajectory. It is therefore possible in many important cases to define a TS trajectory. Nevertheless, it cannot be done for all driving fields. If, for example, the strength of the driving grows without bounds as $t \to \infty$, the trajectory of Eq. (36) will do the same and will not qualify as a TS trajectory. Although this counterexample might on physical grounds appear contrived, it illustrates that a suitable regularity condition on the driving at temporal infinity is required to ensure the existence of a TS trajectory. The same can be said for the case of stochastic driving that was discussed earlier. Although individual instances of the fluctuating forces in the ordinary and generalized Langevin equations are extremely irregular, the noise sources are uniform in time in the sense that their statistical distributions are time independent. This statistical regularity allows us to define a TS trajectory that remains confined to the barrier region in the two different statistical interpretations described earlier.

IV. TIME-DEPENDENT INVARIANT MANIFOLDS

With the identification of the TS trajectory, we have taken the crucial step that enables us to carry over the constructions of the geometric TST into time-dependent settings. We now have at our disposal an invariant object that is analogous to the fixed point in an autonomous system in that it never leaves the barrier region. However, although this dynamical boundedness is characteristic of the saddle point and the NHIMs, what makes them important for TST are the invariant manifolds that are attached to them. It remains to be shown that the TS trajectory can take over their role in this respect. In doing so, we follow the two main steps of time-independent TST: first describe the dynamics in the linear approximation, then verify that important features remain qualitatively intact in the full nonlinear system.

From a geometric point of view, the autonomous fixed point is the organizing center for the hierarchy of invariant manifolds. From a technical point of view, it is also the expansion center around which all Taylor series expansions are carried out. If the TS trajectory is to take over the role of the fixed point, this observation suggests that it be used as a time-dependent coordinate origin. We therefore introduce the relative coordinates

$$\Delta \boldsymbol{q}(t) = \boldsymbol{q}_\alpha(t) - \boldsymbol{q}_\alpha^\ddagger(t) \tag{41}$$

in the dissipative stochastic case and

$$\begin{aligned} \Delta P_j(t) &= P_j(t) - P_j^\ddagger(t) \\ \Delta Q_j(t) &= Q_j(t) - Q_j^\ddagger(t) \quad \text{for } j = 1, \ldots, N \end{aligned} \tag{42}$$

for the Hamiltonian case of deterministic driving. We henceforth study the dynamics in the *moving* relative coordinate system instead of the original space-fixed system. At least in the linear approximation, the relative dynamics will turn out to be very simple and allow us to specify the invariant manifolds and the dividing surface that are known from autonomous TST. Once these objects are known in relative coordinates, they can be referred back to the space-fixed coordinate system by Eq. (41) or (42). From a geometric point of view, the objects in the relative phase space are attached to the TS trajectory, and the latter carries these TS structures around through phase space. The location of the invariant manifolds is determined by the instantaneous position of the TS trajectory. The construction therefore yields time-dependent *moving invariant manifolds* and, in the same way, a *moving dividing surface*. As illustrated later, these moving geometric objects can serve the same purposes as their autonomous counterparts.

A. Stochastically Moving Manifolds

Following the general procedure of geometric TST, we start by discussing the linearized dynamics in relative coordinates. If the definition (41) is substituted into the linearized Langevin equation (13), it yields an equation of motion for the relative coordinate:

$$\Delta\ddot{\boldsymbol{q}}(t) = \boldsymbol{\Omega}\,\Delta\boldsymbol{q}(t) - \boldsymbol{\Gamma}\Delta\dot{\boldsymbol{q}}(t) \tag{43}$$

This is the same equation of motion that is satisfied by the original coordinate $\boldsymbol{q}_\alpha(t)$, except that *the stochastic driving term is absent*. The relative dynamics is therefore deterministic. We have chosen the notation accordingly and left out the index α in the definition (41) of $\Delta\boldsymbol{q}$ (although, of course, we cannot expect the relative dynamics to remain noiseless in the full nonlinear system). Although noiseless, the relative dynamics is still dissipative because Eq. (43) retains the damping term.

As before, the dynamics can most easily be described in phase space. The relative phase space is spanned by the $2N$ coordinates

$$\Delta z(t) = z_\alpha(t) - z_\alpha^\ddagger(t) = \begin{pmatrix} \Delta\boldsymbol{q}(t) \\ \Delta\boldsymbol{v}(t) \end{pmatrix} \tag{44}$$

that satisfy the equation of motion

$$\Delta\dot{z}(t) = \boldsymbol{A}\,\Delta z(t) \tag{45}$$

with the matrix $\boldsymbol{A}$ as in Eq. (18). In diagonal coordinates, Eq. (45) decomposes into the set of scalar equations

$$\Delta\dot{z}_j(t) = \lambda_j\,\Delta z_j(t) \tag{46}$$

with the solution

$$\Delta z_j(t) = c_j e^{\lambda_j t} \quad (c_j \text{ arbitrary constants}) \tag{47}$$

The eigenvalues λ_j are the same as those used in the construction of the TS trajectory (24).

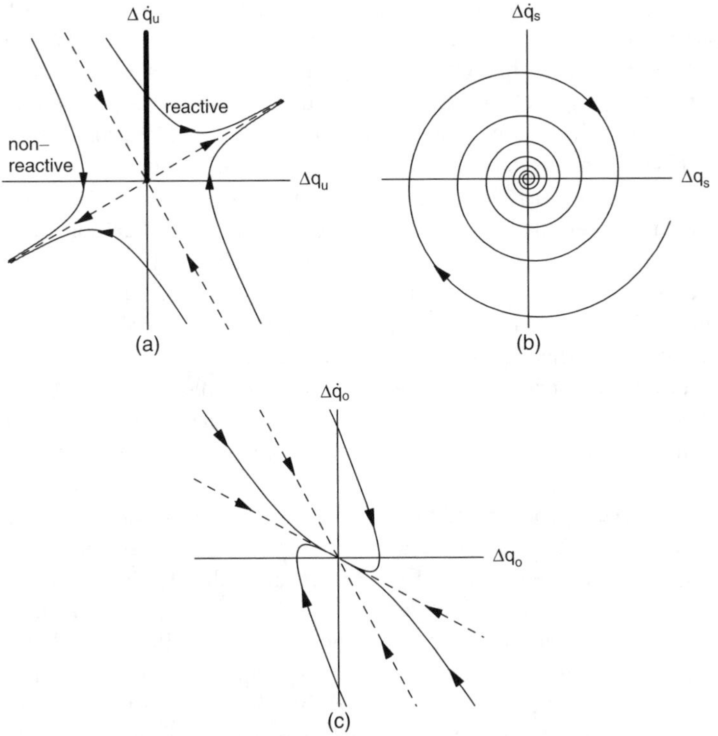

Figure 3. Phase portrait of the noiseless dynamics (43) corresponding to the linear Langevin equation (15) (a) in the unstable reactive degree of freedom, (b) in a stable oscillating bath mode, and (c) in an overdamped bath mode. (From Ref. 37.)

The simplest case occurs in one degree of freedom, where there is one positive real eigenvalue λ_u and one negative eigenvalue λ_s. The phase portrait of the dynamics is shown in Fig. 3a. The eigenvectors corresponding to the positive and negative eigenvalues, respectively, span one-dimensional unstable and stable manifolds of the saddle point. They act as separatrices between reactive and nonreactive trajectories. The knowledge of the invariant manifolds allows one to determine the ultimate fate of a specific trajectory from its initial conditions. In addition, the line $\Delta q_u = 0$ in the quadrant of reactive trajectories acts as a surface of no recrossing.

In the absence of damping (and in units where $\omega_b = 1$), the invariant manifolds bisect the angles between the coordinate axes. The presence of damping destroys this symmetry. As the damping constant increases, the unstable manifold rotates toward the Δq_u-axis, the stable manifold toward the $\Delta \dot{q}_u$-axis. In the limit of infinite damping the invariant manifolds coincide with the axes. Thus the fraction of phase space on the reactant side that corresponds to reactive trajectories decreases with increasing friction. This is intuitively clear because if the dissipation is strong, a trajectory must start at a given initial position with a high velocity to overcome the friction and cross the barrier.

In multiple dimensions, the dynamics of the single unstable degree of freedom is the same as in the one-dimensional case shown in Fig. 3a. Transverse damped oscillations, as shown in Fig. 3b, must be added to this picture. For small damping their presence manifests itself through $N - 1$ complex conjugate pairs of eigenvalues λ_j. For stronger damping, some of the transverse modes can become overdamped, as illustrated in Fig. 3c, so that further eigenvalues become negative real. In any case, there is exactly one unstable eigenvalue with positive real part. This eigenvalue is actually real and corresponds to the system sliding down the barrier. In all other directions, the dynamics is stable, and at least one eigenvalue is negative real. The dynamics in the distant past is determined by the eigenvalue or pair of eigenvalues with the largest negative real part.

The eigenvector corresponding to the most stable eigenvalue together with the unstable eigenvector span a plane in phase space in which the dynamics is given by the phase portrait of Fig. 3a. Because the dynamics in all other, "transverse" directions is stable, the separatrices between reactive and nonreactive trajectories that we identified for the one-dimensional dynamics together with the transverse subspace form separatrices in the high-dimensional phase space. In a similar manner, a no-recrossing curve in the plane together with the transverse directions form a no-recrossing surface in the full phase space.

The invariant manifolds have thus been specified in the relative phase space. According to the general prescription, moving invariant manifolds in the original, space-fixed phase space are obtained by attaching the manifolds in relative phase space to the TS trajectory. Just as a trajectory can be classified as reactive or nonreactive through the location of its initial conditions with respect to the invariant manifolds in the relative dynamics, it can be classified in the original coordinates by comparing its initial conditions to the *instantaneous* location of the moving manifolds. The ultimate fate of a trajectory depends on the future behavior of the noise, and the TS trajectory, which also depends on the future noise, positions the manifolds in such a way that they predict the reaction correctly. In this sense the TS trajectory, or more specifically the component that corresponds to the single positive eigenvalue, encodes the information on the noise that determines reactivity or nonreactivity in the most economical way.

An example for the predictive power of the TS trajectory is shown in Fig. 4. This figure shows a random instance of the TS trajectory (black) and a reactive trajectory (red) for the linearized Langevin equation (15) with $N = 2$ degrees of

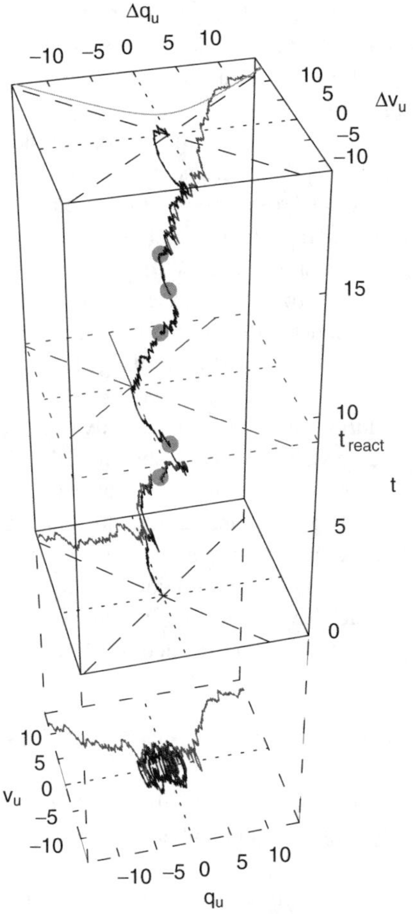

Figure 4. A random instance of the TS trajectory (black) and a reactive trajectory (red) under the influence of the same noise in a system with $N = 2$ degrees of freedom, projected onto the reactive degree of freedom. See text for a detailed description. (From Ref. 38.) (*See color insert.*)

freedom. Units were chosen so that $\omega_b = 1$ and $k_B T = 1$. The transverse frequency is $\omega_2 = 1.5$, and the friction is isotropic, $\mathbf{\Gamma} = \gamma \boldsymbol{I}$, with $\gamma = 0.2$. Only the projection onto the unstable degree of freedom is displayed as it alone determines reactivity. The bottom of the figure shows the two trajectories in phase space. Above this, their time evolution is illustrated using the same axes. The TS trajectory moves back and forth around the saddle point without ever leaving its vicinity as it must by construction. The transition path (red) approaches the TS trajectory from the reactant side, remains in its neighborhood for a while, and then wanders off to the product side. Because the moving invariant manifolds (dashed lines) are known, it can be predicted at the initial time $t = 0$ that the trajectory will actually lead to a reaction instead of returning to the reactant side of the saddle.

From both the time-dependent plot and the time-independent projection, it is clear that the transition path crosses the space-fixed "dividing surface" $q_u = 0$ several times. These crossings are indicated by thick green dots. As expected, therefore, the fixed surface is not free of recrossings and thus does not satisfy the fundamental requirement for an exact TST dividing surface. The moving surface, by contrast, is crossed only once, at the reaction time $t_{react} = 8.936$ that is marked by the blue cut. The solid blue line in this cut shows the instantaneous position of the dividing surface; dotted lines indicate coordinate axes.

That the moving dividing surface is crossed only once is further confirmed by the trajectory in relative coordinates that is shown as the light green curve in the top face of the figure (for graphical reasons not to scale). It exhibits the behavior that is shown schematically for a reactive trajectory in Fig. 3a. In these coordinates it is obvious that the moving dividing surface is recrossing-free.

To further illustrate the conceptual and computational advantages offered by the moving dividing surface, extensive simulations of several quantities relevant to rate theory calculations were performed [39] on the anharmonic model potential

$$U = -\tfrac{1}{2}\omega_x^2 x^2 + \tfrac{1}{2}\omega_y^2 y^2 + kx^2 y^2 \tag{48}$$

If the anharmonicity is absent (i.e., for $k = 0$), the TS trajectory and the invariant manifolds carried along with it can be calculated exactly from the prescriptions given above.

As is well known in calculations of rare events [6, 16, 96], it is notoriously inefficient to start the simulation of trajectories with initial conditions sampled in one of the metastable states. Instead, initial conditions should be sampled at the transition state, which ensures that all trajectories cross the transition state at least once and thereby drastically improves the sampling of reactive events. In the model system (48), such an ensemble is given by

$$f(x, y, v_x, v_y) = (2\pi k_B T)^{3/2}\, \omega_y \exp\{-(\omega_y^2 y^2 + v_x^2 + v_y^2)/2k_B T\}\delta(x) \tag{49}$$

The crucial ingredient in a reaction rate calculation is the identification of reactive trajectories. To this end, initial conditions sampled from Eq. (49) are propagated forward and backward to a time $\pm T_{int}$. Those trajectories that begin on the reactant side of the barrier at $t = -T_{int}$ and end on the product side at $t = +T_{int}$ are then regarded as (forward) reactive. The identification of reactive trajectories is only approximate if $T_{\rm int}$ is small because a trajectory might cross and recross the dividing surface many times before or after the observed interval. A reliable identification can be obtained for large $T_{\rm int}$, but the reliability comes at the cost of an increased numerical effort for a longer simulation.

Traditionally, the reactant and product sides of the barrier would be defined by the space-fixed dividing surface $x = 0$. Time-dependent TST allows one to use the reactant and product sides of the moving dividing surface instead. Because that surface cannot be recrossed, this criterion brings the immediate advantage that a trajectory can reliably be identified as reactive once it has been observed to cross the moving surface. The time at which this identification can be made depends on the trajectory, and therefore a sufficiently long (maximum) integration time $T_{\rm int}$ is still needed to identify the nonreactive trajectories, which never cross the moving surface at all. Nevertheless, it will be shown later that the numerical effort that is required with the moving surface is considerably smaller than with the traditional fixed surface.

Reactive trajectories can be identified a priori, without any numerical simulation, if the moving separatrices are used instead of the standard dividing surfaces. In relative coordinates, the portion of the barrier ensemble that is forward reactive can immediately be identified from Fig. 3: those trajectories are reactive whose initial velocity is so large that it lies above both the stable and unstable manifolds. Because the initial conditions in the ensemble (49) lie at $\Delta x(0) = -x^{\ddagger}_{\alpha}$, this reactivity criterion reads explicitly

$$\Delta v_x > \Delta v_{x\alpha}^{\min} := \begin{cases} -x^{\ddagger}_{\alpha}\lambda_{\rm s} & : \quad x^{\ddagger}_{\alpha} > 0 \\ -x^{\ddagger}_{\alpha}\lambda_{\rm u} & : \quad x^{\ddagger}_{\alpha} < 0 \end{cases} \tag{50}$$

or

$$v_x > v_{x\alpha}^{\min} := v^{\ddagger}_{x\alpha} + \Delta v_{x\alpha}^{\min} \tag{51}$$

Since the fate of a trajectory is determined not only by its initial condition but also by the fluctuating force acting on the system, the critical velocity $v_{x\alpha}^{\min}$ depends on the instance α of the noise. For a fixed instance α, the criterion (51) allows one to calculate the probability for a trajectory randomly sampled from the ensemble (49) to be reactive:

$$\begin{aligned} P_{\rm f} &= \int dx \int_{v_x > v_{x\alpha}^{\min}} dv_x \int dy \int dv_y\, f(x, y, v_x, v_y) \\ &= \frac{1}{2}\operatorname{erfc}\left(\frac{v_{x\alpha}^{\min}}{\sqrt{2\,k_{\rm B}T}}\right) \end{aligned} \tag{52}$$

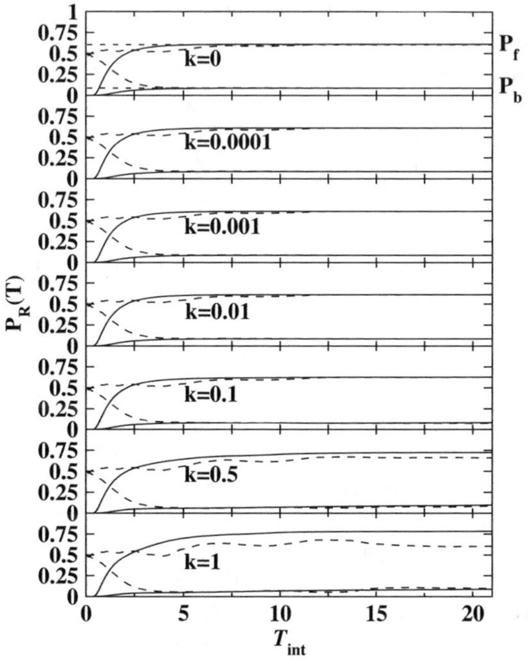

Figure 5. Reaction probabilities for a given instance of the noise as a function of the total integration time $T_{\rm int}$ for different values of the anharmonic coupling constant k. The solid lines represent the forward and backward reaction probabilities calculated using the moving dividing surface and the dashed lines correspond to the results obtained from the standard fixed dividing surface. In the top panel the dotted lines display the analytic estimates provided by Eq. (52). The results were obtained from 15,000 barrier ensemble trajectories subject to the same noise sequence evolved on the reactive potential (48) with barrier frequency $\omega_x = 0.75$, transverse frequency $\omega_y = 1.5$, a damping constant $\gamma = 0.2$, and temperature $k_B T = 1$. (From Ref. 39.)

with the complementary error function [97]

$$\mathrm{erfc}(x) = \frac{2}{\sqrt{\pi}} \int_x^{\infty} \exp(-t^2)\, dt \tag{53}$$

In a similar manner one can obtain the probability P_b for a trajectory to be backward reactive or the probability for it to be nonreactive [39].

Figure 5 displays the reaction probabilities for a fixed instance of the noise. The results were obtained with the algorithm described earlier, using either the moving or the traditional fixed dividing surface to identify reactive trajectories. In the harmonic limit (top panel), one obtains the analytic predictions (52) using either method if the integration time $T_{\rm int}$ is sufficiently large. However, the details of the convergence toward that limit are markedly different in the two cases. If the fixed surface is used, all trajectories start on the dividing surface and are classified as either forward or backward reactive for short integration times depending on the direction in which they are crossing the surface. By contrast, the moving surface starts at some distance from the initial positions of the trajectories, so that no crossings of the moving surface take place for short integration times, and the computed reaction probabilities are zero. They rise monotonically to their asymptotic values as T_{int} increases and thus always provide a strict lower bound to the true probabilities. This is in contrast to the probabilities calculated from the fixed surface, which fluctuate because of recrossings. In addition, it can be seen that the probabilities obtained from the moving surface converge considerably faster, which allows one to significantly reduce the computational effort.

If the barrier potential is anharmonic, for example, the potential (48) with $k \neq 0$, the prescriptions for computing the moving manifolds that have been given so far cannot be applied immediately. A computational scheme that allows one to construct the exact separatrices and a moving dividing surface strictly free of recrossings in the presence of noise and anharmonicities has not been devised so far. (However, see Section IVB for an analogous calculation in a deterministically driven system.) If the anharmonicities are not too strong, one can use the TS trajectory and associated manifolds that are calculated for a harmonic approximation. The dividing surface thus obtained will no longer be strictly free of recrossings. Nevertheless, it can be expected to provide a useful approximation and to retain the computational advantages that it exhibits in the harmonic limit. The lower panels of Fig. 5 display the results of numerical simulations for the reaction probabilities evaluated using the fixed and moving dividing surfaces for increasing values of the anharmonic coupling. It is clear that even with substantial anharmonicity, the moving dividing surface remains nearly free of recrossings and is still capable of providing a reliable criterion to identify reactive trajectories. In addition, the reaction probabilities converge quickly and monotonically, as in the harmonic limit. If the anharmonicities become too large, however, the harmonic approximation becomes inadequate and deviations from the true result are evident. A more rigorous treatment of anharmonic barriers therefore remains an important goal of further development.

The moving invariant manifolds determine the reactivity or nonreactivity of an individual trajectory under the influence of a specific noise sequence. They thus provide the most detailed microscopic information on the reaction dynamics that one can possibly possess. In practice, though, one is more often interested in macroscopic quantities that are obtained by averaging over the noise. To illustrate that such quantities can easily be derived from the microscopic information encoded in the TS trajectory, we calculate the probability for a trajectory starting at a point $(\boldsymbol{q}, \boldsymbol{v})$ in the space-fixed phase space to end up on the product side of the barrier for large times $t \to \infty$. Referring to the relative phase space in Fig. 3a, we find that this will be the case if the relative coordinate in the unstable degree of freedom $\Delta z_u = z_u - z_{\alpha u}^{\ddagger} > 0$. With the initial condition $(\boldsymbol{q}, \boldsymbol{v})$, the value of the diagonal coordinate z_u is known: the transformation from position–velocity coordinates to diagonal coordinates is found by diagonalizing the matrix $\boldsymbol{A}$ in Eq. (18). For simple cases, it is given explicitly in Ref. 37. Therefore the probability that a trajectory with given initial condition will reach the product region in the future is determined only by the unstable coordinate z_u of the initial condition and is the same as the probability that the corresponding component of the TS trajectory $z_{\alpha u}^{\ddagger} < z_u$. As reported in Section IIIA, the statistical distribution of $z_{\alpha u}^{\ddagger}$ is Gaussian with zero mean and a variance σ_u^2 that depends on the strength of the noise, on the temperature, and on the force constant matrix $\boldsymbol{\Omega}$ and that is given in Ref. 37. The required probability is therefore

$$\begin{aligned} P_+ &= \text{Prob}(\text{product at } t \to \infty) \\ &= \text{Prob}(z_{\alpha u}^{\ddagger} < z_u) \\ &= \frac{1}{2}\text{erfc}\left(-\frac{z_u}{\sqrt{2}\,\sigma_u}\right) \end{aligned} \tag{54}$$

This result is exact in the harmonic limit; otherwise it can be used to obtain approximate values, as shown in Fig. 5, without requiring the propagation of an ensemble of initial points for each noise sequence.

Although the discussion of this section formally concerned the Langevin equation (15) and therefore white noise, it carries over almost unchanged to correlated noise, if one keeps in mind that in this generalized case, the phase space is defined implicitly by the nonlinear eigenvalue equation (28). Again, one can combine the unique mode with positive real eigenvalue and a mode with negative real eigenvalue into a reactive degree of freedom, resulting in the dynamics shown in Fig. 3a. The remaining eigenvalues will either occur in complex conjugate pairs, corresponding to a damped oscillatory mode as in Fig. 3b, or be negative real, corresponding to overdamped modes as in Fig. 3c.

B. Deterministically Moving Manifolds

1. *The Harmonic Approximation*

Let us begin by studying the relative dynamics in the Hamiltonian case, that is, for deterministic driving. The shift (42) to the moving origin is a time-dependent canonical transformation [98]. It transforms the linearized Hamiltonian (33) into

$$H_0(\Delta\boldsymbol{P}, \Delta\boldsymbol{Q}) = \omega_b\, \Delta P_1\, \Delta Q_1 + \sum_{j=2}^{N} i\omega_j\, \Delta P_j\, \Delta Q_j \tag{55}$$

This is exactly the autonomous linearized Hamiltonian (7), the dynamics of which was discussed in detail in Section II. One therefore finds the TS dividing surface and the full set of invariant manifolds described earlier: one-dimensional stable and unstable manifolds corresponding to the dynamics of the variables ΔQ_1 and ΔP_1, respectively, and a central manifold of dimension $2N-2$ that itself decomposes into two-dimensional invariant subspaces spanned by ΔP_j and ΔQ_j. However, all these manifolds are now moving manifolds that are attached to the TS trajectory. Their actual location in phase space at any given time is obtained from their description in terms of relative coordinates by the time-dependent shift of origin, Eq. (42).

The most important of these manifolds for the purposes of TST are, as before, the surface given by $\Delta Q_1 = \Delta P_1$ that serves as a recrossing-free dividing surface and the stable and unstable manifolds that act as separatrices between reactive and nonreactive trajectories. The latter are given by $\Delta Q_1 = 0$ (stable manifolds) or $\Delta P_1 = 0$ (unstable manifolds), respectively. Together, they can be characterized as the zeros of the reactive-mode action

$$I_1 = \Delta P_1 \, \Delta Q_1 \tag{56}$$

which is a constant of motion. These invariant manifolds have the same diagnostic power that was illustrated in Fig. 4 for stochastic dynamics: knowing their position at any given time allows one to distinguish reactive from nonreactive trajectories with certainty. We postpone an illustration of these separatrices until Section IVB3, where a fully anharmonic system is discussed.

2. *Nonlinear Corrections*

Because in an autonomous system many of the invariant manifolds that are found in the linear approximation do not remain intact in the presence of nonlinearities, one should expect the same in the time-dependent case. In particular, the separation of the bath modes will not persist but will give way to irregular dynamics within the center manifold. At the same time, one can hope to separate the reactive mode from the bath modes and in this way to find the recrossing-free dividing surfaces and the separatrices that are of importance to TST. As was shown in Ref. 40, this separation can indeed be achieved through a generalization of the normal form procedure that was used earlier to treat autonomous systems [34].

To this end, we expand the time-dependent Hamiltonian (32), written in diagonal coordinates $\boldsymbol{P}$ and $\boldsymbol{Q}$, around the saddle point of the autonomous potential to obtain

$$H(\boldsymbol{P}, \boldsymbol{Q}, t) = \sum_{|\boldsymbol{j}|+|\boldsymbol{k}|=2}^{\infty} \alpha_{\boldsymbol{jk}} \boldsymbol{P}^{\boldsymbol{j}} \boldsymbol{Q}^{\boldsymbol{k}} + \sum_{|\boldsymbol{j}|+|\boldsymbol{k}|=1}^{\infty} \beta_{\boldsymbol{jk}}(t) \boldsymbol{P}^{\boldsymbol{j}} \boldsymbol{Q}^{\boldsymbol{k}} \tag{57}$$

(We are using standard multi-index notation: for $\boldsymbol{j} = (j_1, \dots, j_N)$, we set $|\boldsymbol{j}| = j_1 + \dots + j_N$ and $\boldsymbol{P}^{\boldsymbol{j}} = P_1^{j_1} \dots P_N^{j_N}$.) As discussed in Section IIIC, the expansion of the autonomous Hamiltonian H^{sys} (with coefficients $\alpha_{\boldsymbol{jk}}$) starts with terms of degree two, while that of the driving Hamiltonian H^{ext} (with coefficients $\beta_{\boldsymbol{jk}}(t)$) starts with terms of degree one. The terms of lowest degree in both parts are given by Eq. (33).

To carry out the normal form calculation, we introduce a formal expansion parameter ε to identify the orders of perturbation theory via

$$\boldsymbol{P} \mapsto \varepsilon \boldsymbol{P}, \quad \boldsymbol{Q} \mapsto \varepsilon \boldsymbol{Q}, \quad H \mapsto H/\varepsilon^2 \tag{58}$$

The scaling (58) assigns order ε^0 to the leading term of the autonomous Hamiltonian H^{sys}. The leading term of the time-dependent part is assigned order ε^{-1}. This term is treated exactly in the linear approximation (33). It should therefore have order ε^0. To achieve this, we also have to scale

$$\beta_{\boldsymbol{jk}}(t) \mapsto \varepsilon \beta_{\boldsymbol{jk}}(t) \tag{59}$$

The scaling prescription (59) embodies the assumption that the external force is so weak that it does not drive the TS trajectory out of the phase-space region in which the normal form expansion is valid. In the autonomous version of geometric TST, one generally assumes that this region is sufficiently large to make the normal form expansion a useful tool for the computation of the geometric objects. Once this assumption has been made, the additional condition imposed by Eq. (59) is only a weak constraint.

The Taylor expansion (57) of the Hamiltonian can now be rewritten as an expansion in orders of ε:

$$H(\boldsymbol{P}, \boldsymbol{Q}, t) = H_0(\boldsymbol{P}, \boldsymbol{Q}, t) + \sum_{n=1}^{\infty} \varepsilon^n H_n(\boldsymbol{P}, \boldsymbol{Q}, t) \tag{60}$$

with expansion coefficients

$$H_n(\boldsymbol{P}, \boldsymbol{Q}, t) = \sum_{|\boldsymbol{j}|+|\boldsymbol{k}|=n+2} \alpha_{\boldsymbol{jk}} \boldsymbol{P}^{\boldsymbol{j}} \boldsymbol{Q}^{\boldsymbol{k}} + \sum_{|\boldsymbol{j}|+|\boldsymbol{k}|=n+1} \beta_{\boldsymbol{jk}}(t) \boldsymbol{P}^{\boldsymbol{j}} \boldsymbol{Q}^{\boldsymbol{k}} \tag{61}$$

Under the shift (42) to the TS trajectory as a moving origin, the leading term H_0 takes the form of Eq. (55). The higher terms transform into *inhomogeneous* polynomials

$$H_n(\Delta \boldsymbol{P}, \Delta \boldsymbol{Q}, t) = \sum_{l=0}^{n} \sum_{|\boldsymbol{j}|+|\boldsymbol{k}|=l} h_{\boldsymbol{jk}n}(t) \Delta \boldsymbol{P}^{\boldsymbol{j}} \Delta \boldsymbol{Q}^{\boldsymbol{k}} \tag{62}$$

in $\Delta \boldsymbol{P}$ and $\Delta \boldsymbol{Q}$, where the expansion coefficients $h_{\boldsymbol{jk}n}(t)$ are given in terms of the coefficients in Eq. (61) and the TS trajectory.

To achieve the desired separation of the reactive degree of freedom from the bath modes, we use time-dependent normal form theory [40, 99]. As a first step, the phase space is extended through the addition of two auxiliary variables: a canonical coordinate τ, which takes the same value as time t, and its conjugate momentum P_τ. The dynamics on the extended phase space is described by the Hamiltonian

$$\begin{aligned} K(\Delta \boldsymbol{P}, P_\tau, \Delta \boldsymbol{Q}, \tau) &= H(\Delta \boldsymbol{P}, \Delta \boldsymbol{Q}, \tau) + P_\tau \\ &= K_0^{(0)}(\Delta \boldsymbol{P}, P_\tau, \Delta \boldsymbol{Q}, \tau) + \sum_{n=1}^{\infty} \varepsilon^n K_n^{(0)}(\Delta \boldsymbol{P}, \Delta \boldsymbol{Q}, \tau) \end{aligned} \tag{63}$$

where the leading term is

$$K_0^{(0)}(\Delta \boldsymbol{P}, P_\tau, \Delta \boldsymbol{Q}, \tau) = H_0(\Delta \boldsymbol{P}, \Delta \boldsymbol{Q}) + P_\tau \tag{64}$$

and the higher-order terms

$$K_n^{(0)}(\Delta \boldsymbol{P}, \Delta \boldsymbol{Q}, \tau) = H_n(\Delta \boldsymbol{P}, \Delta \boldsymbol{Q}, \tau) \tag{65}$$

are the same as in the original Hamiltonian. The extended dynamics is formally autonomous.

We recursively define a sequence of partially normalized Hamiltonians

$$K^{(\nu)}(\Delta \boldsymbol{P}, P_\tau, \Delta \boldsymbol{Q}, \tau) = K_0^{(\nu)}(\Delta \boldsymbol{P}, P_\tau, \Delta \boldsymbol{Q}, \tau) + \sum_{n=1}^{\infty} \varepsilon^n K_n^{(\nu)}(\Delta \boldsymbol{P}, \Delta \boldsymbol{Q}, \tau) \tag{66}$$

as follows. For $\nu > 0$, if

$$K_\nu^{(\nu-1)}(\Delta \boldsymbol{P}, \Delta \boldsymbol{Q}, t) = \sum_{l=0}^{n} \sum_{|\boldsymbol{j}|+|\boldsymbol{k}|=l} h_{\boldsymbol{jk}\nu}^{(\nu-1)}(t) \Delta \boldsymbol{P}^{\boldsymbol{j}} \Delta \boldsymbol{Q}^{\boldsymbol{k}} \tag{67}$$

is the term of order ν in $K^{(\nu-1)}$, set

$$f_\nu(\Delta \boldsymbol{P}, \Delta \boldsymbol{Q}) = \sum_{\boldsymbol{j},\boldsymbol{k}} w_{\boldsymbol{jk}\nu}(\tau) \Delta \boldsymbol{P}^{\boldsymbol{j}} \Delta \boldsymbol{Q}^{\boldsymbol{k}} \tag{68}$$

where the sum runs over a subset of the monomials in Eq. (67) that remains to be determined,

$$w_{\boldsymbol{jk}\nu}(\tau) = S[\gamma_{\boldsymbol{jk}}, h_{\boldsymbol{jk}\nu}^{(\nu-1)}](\tau) \tag{69}$$

and

$$\gamma_{jk} = \omega_b(j_1 - k_1) + i\sum_{l=2}^{n} \omega_l(j_l - k_l) \tag{70}$$

Let

$$F_\nu = \{\cdot, f_\nu\} \tag{71}$$

be the operator of Poisson bracket with the function f_ν and define

$$K^{(\nu)} = \exp(\varepsilon^\nu F_\nu) K^{(\nu-1)} \tag{72}$$

Thus $K^{(\nu)}$ is obtained from $K^{(\nu-1)}$ through a canonical transformation with the generating function $\varepsilon^\nu f_\nu$. It is easy to verify [40] that this transformation eliminates all those monomials $\boldsymbol{j}, \boldsymbol{k}$ from $K^{(\nu)}$ that are included in the generating function (68) and leaves all terms of order smaller than ε^ν unchanged. The generating function $\varepsilon^\nu f_\nu$ used in the νth step will therefore be chosen to transform the terms of order ν into "normal form," and that normal form will remain unchanged throughout all subsequent steps. We are free to define the precise meaning of the term "normal form" by choosing which monomials should be eliminated from the Hamiltonian and which monomials are to be retained.

Only those monomials $\boldsymbol{j}, \boldsymbol{k}$ for which the coefficient $w_{jk\nu}$ in Eq. (69) is well defined can be included in the generating function, and thus eliminated from the normal form. It is clear from the definition (23) of the S-functional that this is always the case if γ_{jk} has a nonzero real part (i.e., if $j_1 \neq k_1$). For $j_1 = k_1$ (i.e., purely imaginary γ_{jk}), the S-functional in Eq. (69) is not well defined by the general definition (23). It was shown in Section IIIC how different regularizations of the S-functional can be used to define the bath-mode components of the TS trajectory, which correspond to imaginary eigenvalues. That discussion assumed implicitly that the function f in $S[i\omega, f]$ has a Fourier transform that is regular at ω. The functions $h_{jk\nu}^{(\nu-1)}$ in Eq. (69) do not satisfy this condition: they are given in terms of the components of the TS trajectory, the Fourier transforms of which, by their definition (36) in terms of S-functionals, have poles at the bath-mode frequencies ω_j. Products of these components will therefore have Fourier transforms with poles at integer linear combinations of the ω_j. If $i\gamma_{jk\nu}$ coincides with such a pole, it is impossible to regularize the S-functional in Eq. (69). The corresponding "resonant" monomials therefore cannot be eliminated from the normal form. This impossibility can be interpreted as a resonance between the external driving and the internal bath modes of the system. By analogy with autonomous normal forms, it is to be expected that resonances are dense and that their density will mar the convergence of the normal form transformation if one attempts to eliminate all nonresonant terms. For simplicity, we do not discuss these convergence properties further and instead retain all terms with imaginary γ_{jk}.

This choice still allows us to achieve the separation of the reactive degree of freedom from the bath mode that is required for TST because we can eliminate all monomials with $j_1 \neq k_1$. Thus if we carry out the normal form transformation up to an arbitrarily high order M, we find normal form coordinates

$$\Delta\bar{P}_j = \exp(-\varepsilon F_1)\exp(-\varepsilon^2 F_2)\cdots\exp(-\varepsilon^M F_M)\Delta P_j \tag{73a}$$

$$\Delta\bar{Q}_j = \exp(-\varepsilon F_1)\exp(-\varepsilon^2 F_2)\cdots\exp(-\varepsilon^M F_M)\Delta Q_j \tag{73b}$$

and a normal form Hamiltonian

$$\bar{K} = \exp(\varepsilon^M F_M)\cdots\exp(\varepsilon^2 F_2)\exp(\varepsilon F_1)K \tag{74}$$

that depends on $\Delta\bar{P}_1$ and $\Delta\bar{Q}_1$ only through the action variable

$$\bar{I}_1 = \Delta\bar{P}_1\Delta\bar{Q}_1 \tag{75}$$

where, as usual, we neglect terms of order higher than M in the Hamiltonian. It can be shown [40] that the normal form transformation (73) leaves the time coordinate τ and its conjugate momentum P_τ invariant and that the normal form Hamiltonian has the form

$$\bar{K}(\Delta\bar{\boldsymbol{P}}, P_\tau, \Delta\bar{\boldsymbol{Q}}, \tau) = \bar{H}(\Delta\bar{\boldsymbol{P}}, \Delta\bar{\boldsymbol{Q}}, \tau) + P_\tau \tag{76}$$

We can therefore revert from the formally autonomous description in the extended phase space to an explicitly time-dependent dynamics in the original phase space with a time-dependent normal form Hamiltonian

$$\bar{H}(\Delta\bar{\boldsymbol{P}}, \Delta\bar{\boldsymbol{Q}}, \tau) = \bar{K}(\Delta\bar{\boldsymbol{P}}, P_\tau, \Delta\bar{\boldsymbol{Q}}, \tau) - P_\tau \tag{77}$$

Just as $\bar{K}$, the Hamiltonian $\bar{H}$ depends on $\Delta\bar{P}_1$ and $\Delta\bar{Q}_1$ only through the action variable $\bar{I}_1$, which is a constant of the motion.

The normal form coordinates allow us to define the time-dependent manifolds

$$\mathcal{M} = \{(\Delta\bar{\boldsymbol{P}}, \Delta\bar{\boldsymbol{Q}}, t) : \Delta\bar{P}_1 = \Delta\bar{Q}_1 = 0\} \tag{78a}$$

$$\mathcal{W}^{\mathrm{s}} = \{(\Delta\bar{\boldsymbol{P}}, \Delta\bar{\boldsymbol{Q}}, t) : \Delta\bar{Q}_1 = 0\} \tag{78b}$$

$$\mathcal{W}^{\mathrm{u}} = \{(\Delta\bar{\boldsymbol{P}}, \Delta\bar{\boldsymbol{Q}}, t) : \Delta\bar{P}_1 = 0\} \tag{78c}$$

$$\mathcal{T} = \{(\Delta\bar{\boldsymbol{P}}, \Delta\bar{\boldsymbol{Q}}, t) : \Delta\bar{P}_1 = \Delta\bar{Q}_1\} \tag{78d}$$

These definitions closely mimic their deterministic counterparts (11). They are applicable only within the domain of validity of the normal form transformation and can be extended beyond that range through numerical integration of the equations of motion. Because the action $\bar{I}_1$ is conserved, the phase-space portrait of the dynamics, when projected onto the reactive degree of freedom $\Delta\bar{Q}_1 - \Delta\bar{P}_1$, is the same as that of the autonomous dynamics in Fig. 1a. The manifolds (78) thus play the familiar roles. The manifold $\mathcal{M}$ is the center manifold of the TS trajectory. It is still normally hyperbolic, but it differs from its time-independent analogue in that it is not foliated into energy shells because energy is not conserved in a driven system. The manifolds $\mathcal{W}^{\mathrm{s}}$ and $\mathcal{W}^{\mathrm{u}}$ are the stable and unstable manifolds of the NHIM $\mathcal{M}$. They can collectively be described as the zeros of the reactive-mode action $\bar{I}_1$ and act as separatrices between reactive and nonreactive trajectories. Reactive trajectories are characterized by $\bar{I}_1 > 0$, nonreactive trajectories by $\bar{I}_1 < 0$. Finally, $\mathcal{T}$ is a TS dividing surface that is strictly free of recrossings.

Through the normal form transformation (73) and the time-dependent shift of origin (42), the relation between the original phase-space coordinates $\boldsymbol{P}, \boldsymbol{Q}$ and the normal form coordinates $\Delta\bar{\boldsymbol{P}}, \Delta\bar{\boldsymbol{Q}}$, in which the manifolds (78) are defined, is known explicitly. At any given time, one can therefore easily calculate the instantaneous location of the invariant manifolds and decide immediately whether a given initial condition lies on the reactive or nonreactive side of the separatrices. The nonlinear normal form transformation (73) is time dependent. This complicates our picture of the moving manifolds somewhat: in the presence of nonlinearities, the invariant manifolds not only change their positions as they are carried around by the TS trajectory, but they also change their shape over time. Nevertheless, the basic functions of all manifolds and the explicit constructive nature of the algorithm that computes them are preserved.

3. The Driven Hénon–Heiles System

The diagnostic power of the time-dependent invariant manifolds $\mathcal{W}^{\mathrm{u}}$ and $\mathcal{W}^{\mathrm{s}}$ as separatrices between reactive and nonreactive trajectories was illustrated in Ref. 40 for the example of a driven Hénon–Heiles system described by the Hamiltonian

$$H = \tfrac{1}{2}({p_x}^2 + {p_y}^2) + V_{\mathrm{HH}}(x, y) + \mathcal{E}(t)\exp(-\alpha x^2 - \beta y^2) \tag{79}$$

with the deterministic Hénon–Heiles potential [100]

$$V_{\mathrm{HH}}(x, y) = \tfrac{1}{2}(x^2 + y^2) + x^2 y - \tfrac{1}{3} y^3 \tag{80}$$

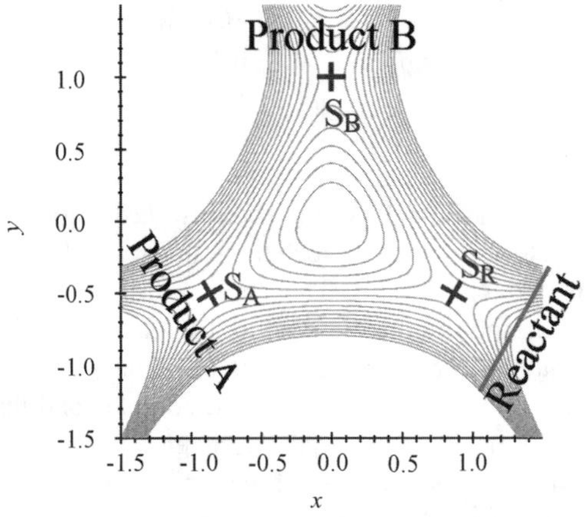

Figure 6. The deterministic potential (80) of the driven Hénon–Heiles system. A central well is separated from three asymptotic channels by three saddle points S_A, S_B, and S_R. Also indicated is the surface, Eq. (82), on which initial conditions are sampled. (From Ref. 40.)

and the external driving field

$$\mathcal{E}(t) = -\frac{\partial}{\partial t}A(t) \tag{81a}$$

$$A(t) = \begin{cases} A_0 \cos^2\left(\frac{\omega t}{2N}\right) \sin(\omega t) & \text{if} |t| < N\pi/\omega \\ 0 & \text{otherwise} \end{cases} \tag{81b}$$

The numerical parameters in Eqs. (79) and (81) were set to $\alpha = 2$, $\beta = 4$, $N = 4$, $\omega = 3$, and $A_0 = 0.1$.

The deterministic potential (80) is illustrated in Fig. 6. It has a minimum at the origin and three saddle points that separate a central well from three asymptotic regions. The latter can be interpreted as a reactant channel and two different product channels and are labeled accordingly in the figure. The central region then corresponds to an intermediate activated complex, and the system can serve as a simple model for multichannel chemical reactions. In the absence of external driving, the dynamics within the central well is regular for sufficiently low energy. As the energy rises toward the energy of the saddle points, the dynamics becomes increasingly chaotic. For this reason, the Hénon–Heiles system has become a paradigmatic example in which to study mixed regular-chaotic Hamiltonian dynamics, both the bounded motion below the saddle point energy (e.g., see Refs. 100 and 101) and the chaotic scattering above [102–104]. It is therefore to be expected that the complications of chaotic dynamics remain present if the system is subject to a time-dependent driving force.

Following Ref. 40, we study trajectories that approach the central region from the (arbitrarily chosen) reactant channel. They will be identified by their initial conditions at time $t_0 = -4.2$, just before the onset of the driving pulse. Initial conditions will be sampled on the surface

$$\sqrt{3}x - y = 3 \tag{82}$$

and at an initial energy of $E_0 = 0.3$ (which is conserved before the driving sets in). Some of these trajectories will be deflected by the potential barrier that separates the channel from the central well. They will return into the reactant channel without ever entering the intermediate region. Other trajectories will overcome the barrier and enter the intermediate region. They will participate in the complicated internal dynamics of the activated complex and finally leave across one of the three barriers into the adjacent channel. Figure 7 shows the final channels that the trajectories reach as a function of initial conditions. There are several different regions, or "islands," for every final channel. They intertwine with the islands of the other channels in a complicated manner.

The complex island structure in Fig. 7 is a consequence of the complicated dynamics of the activated complex. When a trajectory approaches a barrier, it can either escape or be deflected by the barrier. In the latter case, it will return into the well and approach one of the barriers again later, until it finally escapes. If this interpretation is correct, the boundaries of the islands should be given by the separatrices between escaping and nonescaping trajectories, that is, by the time-dependent invariant manifolds described in the previous section. To test this hypothesis, Kawai et al. [40] calculated those separatrices in the vicinity of each saddle point through a normal form expansion. Whenever a trajectory approaches a barrier, the value of the reactive-mode action I_1 is calculated. If the trajectory escapes, it is assigned this value of the action as its "escape action".

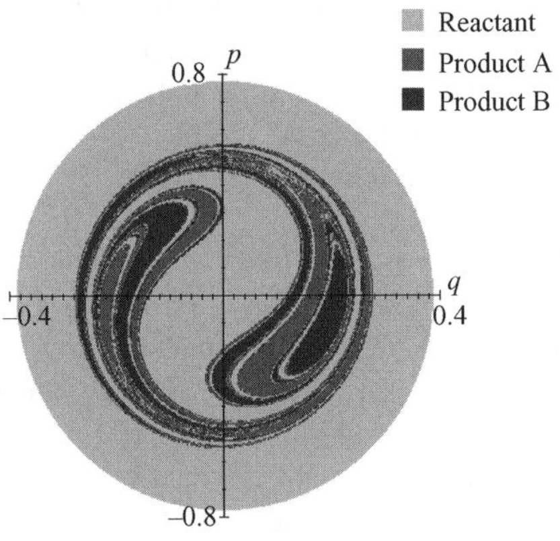

Figure 7. Initial conditions in the surface (82) that lead to reactions into channel A, channel B, or a return into the reactant channel, respectively. (From Ref. 40.) (*See color insert.*)

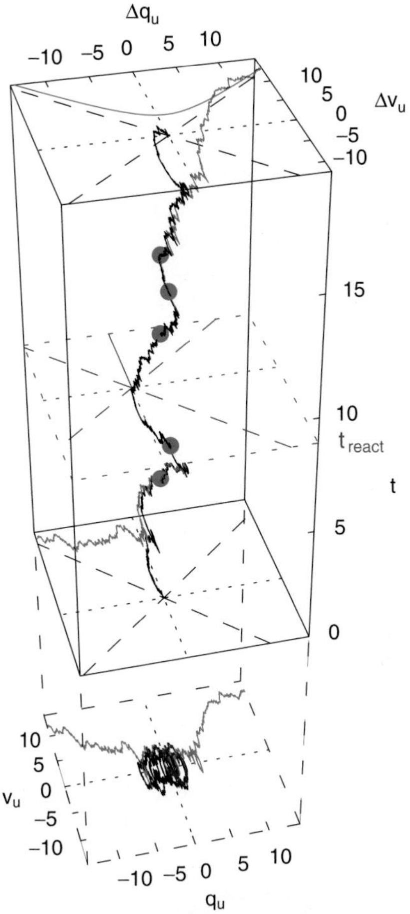

Figure 4. A random instance of the TS trajectory (black) and a reactive trajectory (red) under the influence of the same noise in a system with $N = 2$ degrees of freedom, projected onto the reactive degree of freedom. See text for a detailed description. (From Ref. 38.)

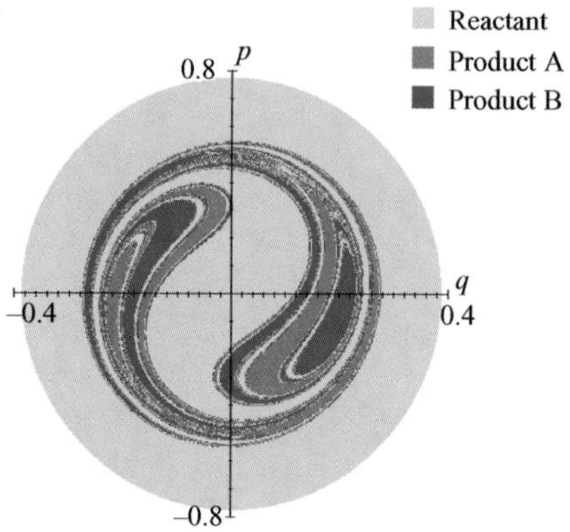

Figure 7. Initial conditions in the surface (82) that lead to reactions into channel A, channel B, or a return into the reactant channel, respectively. (From Ref. 40.)

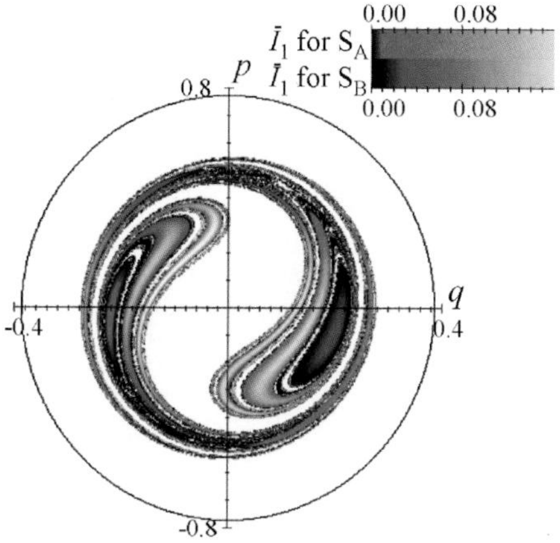

Figure 8. The escape actions for trajectories that lead into channel A or channel B. On the island boundaries, the escape actions tend to zero. (From Ref. 40.)

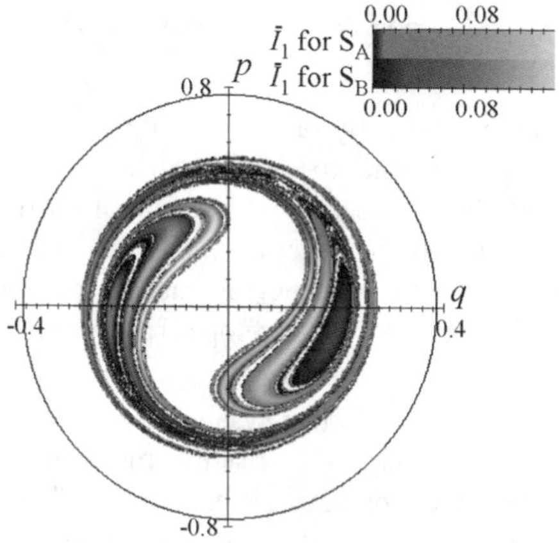

Figure 8. The escape actions for trajectories that lead into channel A or channel B. On the island boundaries, the escape actions tend to zero. (From Ref. 40.) (*See color insert.*)

Figure 8 displays the escape actions thus obtained for trajectories that react into channel A or B. It confirms, first of all, that all escape actions are positive. Furthermore, they take a maximum in the interior of each reactive island and decrease to zero as the boundaries of the islands are approached. These boundaries therefore coincide with the invariant manifolds that are characterized by $\bar{I}_1 = 0$. A more detailed study of the island structure [40] reveals in addition that the time-dependent normal form approach is necessary to describe the islands correctly. Neither the harmonic approximation of Section IVB1 nor the earlier autonomous TST described in Section II yield the correct island boundaries.

V. OUTLOOK: OPEN PROBLEMS

We have outlined how the conceptual tools provided by geometric TST can be generalized to deterministically or stochastically driven systems. The centerpiece of the construction is the TS trajectory, which plays the role of the saddle point in the autonomous setting. It carries invariant manifolds and a TST dividing surface, which thus become time-dependent themselves. Nevertheless, their functions remain the same as in autonomous TST: there is a TST dividing surface that is locally free of recrossings and thus satisfies the fundamental requirement of TST. In addition, invariant manifolds separate reactive from nonreactive trajectories, and their knowledge enables one to predict the fate of a trajectory a priori.

Because the concepts of time-dependent TST have been developed only very recently, their investigation is still in its infancy, and many problems remain open. Exemplarily, some possible directions of further research are the following:

- Whereas in deterministically driven systems a normal form procedure is available that allows one to calculate the time-dependent geometric objects explicitly in the presence of both nonlinearities and configuration-dependent coupling to the external fields, no analogous scheme is yet available for stochastic driving. Because stochastic driving implies damping, it will have to be developed in a framework of dissipative normal form theory [105, 106].
- For time-dependent Hamiltonian systems we chose in Section IVB to use a normal form that decouples the reactive mode from the bath modes, but does not attempt a decoupling of the bath modes. This procedure is always safe, but in many cases it will be overly cautious. If it is relaxed, the dynamics within the center manifold is also transformed into a (suitably defined) normal form. This opens the possibility to study the dynamics within the TS itself, as has been done in the autonomous case, for example in Ref. 107. One can then try to identify structures in the TS that promote or inhibit the transport from the reactant to the product side.
- Once the cases of deterministic and stochastic driving have been incorporated into a framework of time-dependent TST, one can proceed to combine the two. This situation presents special challenges because the deterministic external force will drive the reactive system away from thermal equilibrium with the heat bath. For the description of non-equilibrium reactions, no general theoretical framework is yet available. Such a framework would have to answer two fundamental questions [9]: on the one hand, the statistical distribution of reactants is in general unknown; on the other hand, the phase-space structures that mediate the transition between different asymptotic states have to be identified. The latter problem, which is properly speaking dynamical in nature, can be tackled using the time-dependent TST. A first step in this direction was taken by Lehmann et al. [108–110], who identified a "basin boundary" periodic orbit on a periodically oscillating barrier as the moving separatrix between the two adjacent wells. The results of Section III yield a TS trajectory that can be regarded as a stochastic version of this periodic basin boundary.
- Many solvents do not possess the simple structure that allows their effects to be modeled by the Langevin equation or generalized Langevin equation used earlier to calculate the TS trajectory [58, 111, 112]. Instead, they must be described in atomistic detail if their effects on the effective free energies (i.e., the time-independent properties) and the solvent response (i.e., the nonequilibrium or time-dependent properties) associated with the

chemical reaction are to be modeled accurately. An important extension of the time-dependent TST outlined here lies in its application to such complex solvents.

- If the system is modeled within the Langevin formalism, the fluctuating force exerted by the solvent is assumed to be independent of the dynamics of the reactants and can thus also be assumed to be known a priori. This knowledge allows the construction of a TS trajectory that depends on the future as well as the past behavior of the driving force. As soon as the influence of the solute on the solvent has been taken into account, such complete information will not be readily accessible since extracting an entire sample of the external driving force from a molecular simulation will quickly become prohibitively expensive. Because the diagnostic power of the TS trajectory depends critically on its ability to look ahead to future noise (see Section III), one might not be able to obtain an exact recrossing-free dividing surface under these conditions. However, an approximate solution at short to intermediate times will still be available. The construction of such an approximation is the key to an extension of time-dependent TST to the dynamics of reactions in complex solvents.
- The ultimate aim of TST is the calculation of reaction rates. It remains to be shown how the moving dividing surface can be used to compute a rate in a manner that is analogous to a traditional TST rate calculation [113].

The authors hope that this chapter will inspire further investigation of these and other open questions that will lead us to fully understand the potential of time-dependent transition state theory.

Acknowledgment

This work was partly supported by the U.S. National Science Foundation and by the Alexander von Humboldt Foundation. The computational facilities at the CCMST have been supported under NSF grant CHE 04-43564.

References

1. H. Eyring, The activated complex in chemical reactions, *J. Chem. Phys.* **3**, 107 (1935).
2. L. S. Kassel, Statistical mechanical treatment of the activated complex in chemical reactions, *J. Chem. Phys.* **3**, 399 (1935).
3. E. P. Wigner, Calculation of the rate of elementary association reactions, *J. Chem. Phys.* **5**, 720 (1937).
4. *Trans. Faraday Soc.* **34**, 1 (1938). Herein is a transcript of a general discussion "on the theoretical methods of treating activation energy and reaction velocity" among whose contributors are H. Eyring, M. G. Evans, M. Polanyi, E. Wigner, N. B. Slater, and C. N. Hinshelwood.
5. J. T. Hynes, The theory of reactions in solution, in *Theory of Chemical Reaction Dynamics*, Vol. 4, M. Baer, (ed.), CRC, Boca Raton, FL, 1985, pp. 171–234.

6. P. Hänggi, P. Talkner, and M. Borkovec, Reaction-rate theory: fifty years after Kramers, *Rev. Mod. Phys.* **62**, 251 (1990).
7. D. G. Truhlar, B. C. Garrett, and S. J. Klippenstein, Current status of transition-state theory, *J. Phys. Chem.* **100**, 12771 (1996).
8. W. H. Miller, Spiers memorial lecture quantum and semiclassical theory of chemical reaction rates, *Faraday Disc. Chem. Soc.* **110**, 1 (1998).
9. E. Pollak and P. Talkner, Reaction rate theory: What it was, where it is today, and where is it going? *Chaos* **15**, 026116 (2005).
10. C. Jaffé, D. Farrelly, and T. Uzer, Transition state in atomic physics, *Phys. Rev. A* **60**, 3833 (1999).
11. C. Jaffé, D. Farrelly, and T. Uzer, Transition state theory without time-reversal symmetry: chaotic ionization of the hydrogen atom, *Phys. Rev. Lett.* **84**, 610 (2000).
12. T. Komatsuzaki and R. S. Berry, Regularity in chaotic reaction paths. I. Ar_6, *J. Chem. Phys.* **110**, 9160 (1999).
13. B. Eckhardt, Transition state theory for ballistic electrons, *J. Phys. A* **28**, 3469 (1995).
14. C. Jaffé, S. D. Ross, M. W. Lo, J. Marsden, D. Farrelly, and T. Uzer, Statistical theory of asteroid escape rates, *Phys. Rev. Lett.* **89**, 011101 (2002).
15. H. P. de Oliveira, A. M. Ozorio de Almeida, I. Damiõ Soares, and E. V. Tonini, Homoclinic chaos in the dynamics of a general Bianchi type-IX model, *Phys. Rev. D* **65**, 083511 (2002).
16. J. C. Keck, Variational theory of reaction rates, *Adv. Chem. Phys.* **13**, 85 (1967).
17. D. G. Truhlar and B. C. Garrett, Variational transition state theory, *Annu. Rev. Phys. Chem.* **35**, 159 (1984).
18. J. T. Hynes, Chemical reaction dynamics in solution, *Annu. Rev. Phys. Chem.* **36**, 573 (1985).
19. S. C. Tucker, Variational transition state theory in condensed phases, in *New Trends in Kramers' Reaction Rate Theory*, P. Hänggi and P. Talkner (eds.), Kluwer Academic, The Netherlands, 1995, pp. 5–46.
20. P. Pechukas and F. J. McLafferty, On transition state theory and the classical mechanics of collinear collisions, *J. Chem. Phys.* **58**, 1622 (1973).
21. P. Pechukas and E. Pollak, Classical transition state theory is exact if the transition state is unique, *J. Chem. Phys.* **71**, 2062 (1979).
22. E. Pollak, Periodic orbits and the theory of reactive scattering, in *Theory of Chemical Reaction Dynamics*, Vol. 3, M. Baer (ed.), CRC, Boca Raton, FL, 1985.
23. R. Hernandez, A combined use of perturbation theory and diagonalization: application to bound energy levels and semiclassical rate theory, *J. Chem. Phys.* **101**, 9534 (1994).
24. T. Komatsuzaki and M. Nagaoka, Study on "regularity" of barrier recrossing motion, *J. Chem. Phys.* **105**, 10838 (1996).
25. T. Komatsuzaki and M. Nagaoka, A dividing surface free from a barrier recrossing motion in many-body systems, *Chem. Phys. Lett.* **265**, 91 (1997).
26. T. Komatsuzaki and R. S. Berry, Regularity in chaotic reaction paths. II. Ar_6. Energy dependence and visualization of the reaction bottleneck, *Phys. Chem. Chem. Phys.* **1**, 1387 (1999).
27. T. Komatsuzaki and R. S. Berry, Local regularity and non-recrossing path in transition state – a new strategy in chemical reaction theories, *J. Mol. Struct.* (*Theochem*) **506**, 55 (2000).
28. T. Komatsuzaki and R. S. Berry, Regularity in chaotic reaction paths III: Ar_6 local invariances at the reaction bottleneck, *J. Chem. Phys.* **115**, 4105 (2001).

29. T. Komatsuzaki and R. S. Berry, Dynamical hierarchy in transition states: Why and how does a system climb over the mountain? *Proc. Natl. Acad. Sci. USA* **98**, 7666 (2001).

30. T. Komatsuzaki and R. S. Berry, Chemical reaction dynamics: many-body chaos and regularity, *Adv. Chem. Phys.* **123**, 79 (2002).

31. C.-B. Li, Y. Matsunaga, M. Toda, and T. Komatsuzaki, Phase-space reaction network on a multisaddle energy landscape: HCN isomerization, *J. Chem. Phys.* **123**, 184301 (2005).

32. C. B. Li, A. Shojiguchi, M. Toda, and T. Komatsuzaki, Dynamical hierarchy in transition states in reactions, *Few-Body Systems* **38**, 173 (2006).

33. S. Wiggins, L. Wiesenfeld, C. Jaffé, and T. Uzer, Impenetrable barriers in phase-space, *Phys. Rev. Lett.* **86**, 5478 (2001).

34. T. Uzer, C. Jaffé, J. Palacian, P. Yanguas, and S. Wiggins, The geometry of reaction dynamics, *Nonlinearity* **15**, 957 (2002).

35. C. Jaffé, S. Kawai, J. Palacián, P. Yanguas, and T. Uzer, A new look at the transition state: Wigner's dynamical perspective revisited, *Adv. Chem. Phys.* **130A**, 171 (2005).

36. E. Pollak, Classical and quantum rate theory for condensed phases, in *Theoretical Methods in Condensed Phase Chemistry*, S. D. Schwartz (ed.), Kluwer Academic Publishers, Dordrecht, 2000, pp. 1–46.

37. T. Bartsch, R. Hernandez, and T. Uzer, Transition state in a noisy environment, *Phys. Rev. Lett.* **95**, 058301 (2005).

38. T. Bartsch, T. Uzer, and R. Hernandez, Stochastic transition states: reaction geometry amidst noise, *J. Chem. Phys.* **123**, 204102 (2005).

39. T. Bartsch, T. Uzer, J. M. Moix, and R. Hernandez, Identifying reactive trajectories using a moving transition state, *J. Chem. Phys.* **124**, 244310 (2006).

40. S. Kawai, A. D. Bandrauk, C. Jaffé, T. Bartsch, J. Palacián, and T. Uzer, Transition state theory for laser-driven reactions, *J. Chem. Phys.* **126**, 164306 (2007).

41. G. W. Ford, M. Kac, and P. Mazur, Statistical mechanics of assemblies of coupled oscillators, *J. Math. Phys.* **6**, 504 (1965).

42. H. Mori, Transport, collective motion, and Brownian motion, *Prog. Theor. Phys.* **33**, 423 (1965).

43. R. Zwanzig, Ensemble method in the theory of irreversibility, *Phys. Rev.* **124**, 983 (1961).

44. R. Zwanzig, Nonlinear generalized Langevin equation, *J. Stat. Phys.* **9**, 215 (1973).

45. B. Carmeli and A. Nitzan, Theory of activated rate processes: position dependent friction, *Chem. Phys. Lett.* **102**, 517 (1983).

46. K. Lindenberg, K. E. Shuler, V. Seshadri, and B. J. West, Langevin equations with multiplicative noise: theory and applications to physical processes, in *Probabilistic Analysis and Related Topics*, Vol. 3, A. T. Bharucha-Reid (ed.), Academic Press, San Diego, 1983, pp. 81–125.

47. J. P. Hansen and I. R. McDonald, *Theory of Simple Liquids*, Academic Press, San Diego, 1986.

48. R. Zwanzig, *Nonequilibrium Statistical Mechanics*, Oxford University Press, London, 2001.

49. A. D. Bandrauk (ed.), *Molecules in Laser Fields,* Marcel Dekker, New York, 1994.

50. S. A. Rice and M. Zhao, *Optical Control of Molecular Dynamics*, Wiley, Hoboken, NJ, 2000.

51. T. Brabec and F. Krausz, Intense few-cycle laser fields: frontiers of nonlinear optics, *Rev. Mod. Phys.* **72**, 545 (2000).

52. M. Shapiro and P. Brumer, Quantum control of bound and continuum state dynamics, *Phys. Rep.* **425**, 195 (2006).

53. P. Agostini and L. F. DiMauro, The physics of attosecond light pulses, *Rep. Prog. Phys.* **67**, 813 (2004).
54. A. D. Bandrauk, E.-W. S. Sedik, and C. F. Matta, Effect of absolute laser phase on reaction paths in laser-induced chemical reactions, *J. Chem. Phys.* **121**, 7764 (2004).
55. H. Rabitz, R. de Vivie-Riedle, M. Motzkus, and K. Kompa, Chemistry — whither the future of controlling quantum phenomena? *Science* **288**, 824 (2000).
56. H. A. Kramers, Brownian motion in a field of force and the diffusional model of chemical reactions, *Physica (Utrecht)* **7**, 284 (1940).
57. E. Pollak and P. Talkner, Transition-state recrossing dynamics in activated rate processes, *Phys. Rev. E* **51**, 1868 (1995).
58. P. G. Bolhuis, D. Chandler, C. Dellago, and P. Geissler, Transition path sampling: throwing ropes over mountain passes, in the dark, *Annu. Rev. Phys. Chem.* **53**, 291 (2002).
59. C. Dellago, P. G. Bolhuis, and P. Geissler, Transition path sampling, *Adv. Chem. Phys.* **123**, 1 (2002).
60. A. O. Caldeira and A. J. Leggett, Influence of dissipation on quantum tunneling in macroscopic systems, *Phys. Rev. Lett.* **46**, 211 (1981). *Ann. Phys. (N.Y.)* **149**, 374 (1983).
61. E. Pollak, Theory of activated rate processes: a new derivation of Kramers' expression, *J. Chem. Phys.* **85**, 865 (1986).
62. V. I. Mel'nikov and S. V. Meshkov, Theory of activated rate processes: exact solution of the Kramers problem, *J. Chem. Phys.* **85**, 1018 (1986).
63. E. Pollak, H. Grabert, and P. Hänggi, Theory of activated rate processes for arbitrary frequency dependent friction: solution of the turnover problem, *J. Chem. Phys.* **91**, 4073 (1989).
64. R. Graham, Macroscopic theory of activated decay of metastable states, *J. Stat. Phys.* **60**, 675 (1990).
65. E. Hershkovitz and E. Pollak, Multidimensional generalization of the Pollak–Grabert–Hänggi turnover theory for activated rate processes, *J. Chem. Phys.* **106**, 7678 (1997).
66. C. C. Martens, Qualitative dynamics of generalized Langevin equations and the theory of chemical reaction rates, *J. Chem. Phys.* **116**, 2516 (2002).
67. J. M. A. Danby, Stability of the triangular points in the elliptic restricted problem of three bodies, *Astron. J.* **69**, 165 (1964).
68. À. Jorba and J. Villanueva, On the persistence of lower dimensional invariant tori under quasi-periodic perturbations, *J. Nonlinear Sci.* **7**, 427 (1997).
69. F. Gabern and À. Jorba, Generalizing the restricted three-body problem: the bianular and tricircular coherent problems, *Astron. Astrophys.* **420**, 751 (2004).
70. W. H. Miller, Semi-classical theory for non-separable systems: construction of "good" action-angle variables for reaction rate constants, *Faraday Disc. Chem. Soc.* **62**, 40 (1977).
71. R. Hernandez and W. H. Miller, Semiclassical transition state theory: a new perspective, *Chem. Phys. Lett.* **214**, 129 (1993).
72. S. C. Creagh, Classical transition states in quantum theory, *Nonlinearity* **17**, 1261 (2004).
73. S. C. Creagh, Semiclassical transmission across transition states, *Nonlinearity* **18**, 2089 (2005).
74. C. S. Drew, S. C. Creagh, and R. H. Tew, Uniform approximation of barrier penetration in phase space, *Phys. Rev. A* **72**, 062501 (2005).
75. R. Schubert, H. Waalkens, and S. Wiggins, Efficient computation of transition state resonances and reaction rates from a quantum normal form, *Phys. Rev. Lett.* **96**, 218302 (2006).
76. A. Deprit, Canonical transformations depending on a small parameter, *Cel. Mech.* **1**, 12 (1969).

77. A. Deprit, J. Henrard, J. F. Price, and A. Rom, Birkhoff's normalization, *Cel. Mech. Dyn. Astron.* **1**, 222 (1969).
78. A. J. Dragt and J. M. Finn, Lie series and invariant functions for analytic symplectic maps, *J. Math. Phys.* **17**, 2215 (1976).
79. A. M. Ozorio de Almeida, N. de Leon, M. A. Mehta, and C. C. Marston, Geometry and dynamics of stable and unstable cylinders in Hamiltonian systems, *Physica D* **46**, 265 (1990).
80. N. De Leon, M. A. Mehta, and R. Q. Topper, Cylindrical manifolds in phase space as mediators of chemical reaction dynamics and kinetics. I. Theory, *J. Chem. Phys.* **94**, 8310 (1991).
81. N. De Leon, M. A. Mehta, and R. Q. Topper, Cylindrical manifolds in phase space as mediators of chemical reaction dynamics and kinetics. II. Numerical considerations and applications to models with two degrees of freedom, *J. Chem. Phys.* **94**, 8329 (1991).
82. H. Waalkens, A. Burbanks, and S. Wiggins, A computational procedure to detect a new type of high-dimensional chaotic saddle and its application to the 3D Hill's problem, *J. Phys. A* **37**, L257 (2004).
83. H. Waalkens, A. Burbanks, and S. Wiggins, Phase space conduits for reaction in multidimensional systems: HCN isomerization in three dimensions, *J. Chem. Phys.* **121**, 6207 (2004).
84. H. Waalkens, A. Burbanks, and S. Wiggins, Efficient procedure to compute the microcanonical volume of initial conditions that lead to escape trajectories from a multidimensional potential well, *Phys. Rev. Lett.* **95**, 084301 (2005).
85. H. Waalkens, A. Burbanks, and S. Wiggins, A formula to compute the microcanonical volume of reactive initial conditions in transition state theory, *J. Phys. A* **38**, L759 (2005).
86. F. Gabern, W. S. Koon, J. E. Marsden, and S. D. Ross, Theory and computation of non-RRKM lifetime distributions and rates in chemical systems with three or more degrees of freedom, *Physica D* **211**, 391 (2005).
87. F. Gabern, W. S. Koon, J. E. Marsden, S. D. Ross, and T. Yanao, Application of tube dynamics to non-statistical reaction processes, *Few-Body Systems* **38**, 167 (2006).
88. S. Wiggins, *Normally Hyperbolic Invariant Manifolds in Dynamical Systems*, Springer, New York, 1994.
89. P. Pechukas, Statistical approximations in collision theory, in *Modern Theoretical Chemistry*, Vol. 2, W. H. Miller (ed.), Plenum, New York, 1976, pp. 269–322.
90. R. Kubo, The fluctuation-dissipation theorem, *Rep. Prog. Phys.* **29**, 255 (1966).
91. R. F. Fox, Gaussian stochastic processes in physics, *Phys. Rep.* **48**, 180 (1978).
92. L. Arnold, *Random Dynamical Systems*, Springer, Berlin, 1998.
93. R. Zwanzig, Statistical mechanics of irreversibility, in *Lectures in Theoretical Physics (Boulder)*, vol. 3 W. E. Brittin, B. W. Downs, and J. Downs (eds), Wiley-Interscience, Hoboken, NJ, 1961, pp. 106–141.
94. I. Prigogine and P. Résibois, On the kinetics of the approach to equilibrium, *Physica* **27**, 629 (1961).
95. R. F. Grote and J. T. Hynes, The stable states picture of chemical reactions. II. Rate constants for condensed and gas phase reaction models, *J. Chem. Phys.* **73**, 2715 (1980).
96. D. Chandler, Statistical mechanics of isomerization dynamics in liquids and the transition state approximation, *J. Chem. Phys.* **68**, 2959 (1978).
97. M. Abramowitz and I. A. Stegun, *Pocketbook of Mathematical Functions*, Verlag Harri Deutsch, Frankfurt, 1984.
98. H. Goldstein, *Classical Mechanics*, Addison-Wesley, Reading, MA, 1965.

99. A. J. Dragt and J. M. Finn, Normal form for mirror machine Hamiltonians, *J. Math. Phys.* **20**, 2649 (1979).

100. M. Hénon and C. Heiles, The applicability of the third integral of motion: some numerical experiments, *Astron. J.* **69**, 73 (1964).

101. A. J. Lichtenberg and M. A. Liebermann, *Regular and Stochastic Motion*, Springer, New York, 1982.

102. J. Kaidel, P. Winkler, and M. Brack, Periodic orbit theory for the Hénon–Heiles system in the continuum region, *Phys. Rev. E* **70**, 066208 (2004).

103. J. Aguirre, J. C. Vallejo, and M. A. F. Sanjuán, Wada basins and chaotic invariant sets in the Hénon–Heiles system, *Phys. Rev. E* **64**, 066208 (2001).

104. J. Aguirre and M. A. F. Sanjuán, Limit of small exits in open Hamiltonian systems, *Phys. Rev. E* **67**, 056201 (2003).

105. V. I. Arnold, *Geometrical Methods in the Theory of Ordinary Differential Equations*, Springer, New York, 1988.

106. J. Murdock, *Normal Forms and Unfoldings for Local Dynamical Systems*, Springer, New York, 2002.

107. À. Jorba, A methodology for the numerical computation of normal forms, centre manifolds and first integrals of Hamiltonian systems, *Exp. Math.* **8**, 155 (1999).

108. J. Lehmann, P. Reimann, and P. Hänggi, Surmounting oscillating barriers, *Phys. Rev. Lett.* **84**, 1639 (2000).

109. J. Lehmann, P. Reimann, and P. Hänggi, Surmounting oscillating barriers: path-integral approach for weak noise, *Phys. Rev. E* **62**, 6282 (2000).

110. J. Lehmann, P. Reimann, and P. Hänggi, Activated escape over oscillating barriers: the case of many dimensions, *Phys. Status Solidi B* **237**, 53 (2003).

111. D. Frenkel and B. Smit, *Understanding Molecular Simulation: From Algorithms to Application*, Academic Press, San Diego, 1996.

112. K. Lum, D. Chandler, and J. D. Weeks, Hydrophobicity at small and large length scales, *J. Phys. Chem. B* **103**, 4570 (1999).

113. T. Bartsch, J. M. Moix, R. Hernandez, and T. Uzer, Reaction rate calculation using a moving transition state, *J. Phys. Chem. B* **112**, 206 (2008).

ELECTRONIC STRUCTURE REFERENCE CALCULATIONS FOR DESIGNING AND INTERPRETING P AND T VIOLATION EXPERIMENTS

MALAYA K. NAYAK AND RAJAT K. CHAUDHURI

Indian Institute of Astrophysics, Bangalore 560034, India

CONTENTS

I. INTRODUCTION

It was believed for a long time that the fundamental laws of nature are invariant under space inversion, and hence the conservation of space inversion symmetry (P) is a universally accepted principle. The nonconservation of this symmetry was discovered experimentally by Wu and co-workers in the β decay of ^{60}Co in

Advances in Chemical Physics, Volume 140, edited by Stuart A. Rice
Copyright © 2008 John Wiley & Sons, Inc.

1957 [1]. After the discovery of P violation, the combined operation of charge conjugation C and space inversion P (CP) was thought to be a good symmetry. In 1964, the experiment of Christenson et al. [2] provided evidence of a small violation of CP symmetry in the decay of neutral K_L^0 mesons. This, in conjunction with the so called CPT theorem [3, 4], implies the violation of time reversal symmetry (T). Apart from this indirect evidence of T violation, there is no other instance where such an effect has been observed. Lee and Yang [5] as well as Landau [6] showed that a nonzero permanent electric dipole moment (EDM) of any nondegenerate quantum mechanical system is a signature of the non-conservation of space inversion and time reversal symmetries. Thus the experimental observation of a permanent EDM of an elementary particle, an atom, or a molecule will be direct evidence of the violation of time reversal symmetry (T).

On the other hand, the permanent EDM of an elementary particle vanishes when the discrete symmetries of space inversion (P) and time reversal (T) are both violated. This naturally makes the EDM small in fundamental particles of ordinary matter. For instance, in the standard model (SM) of elementary particle physics, the expected value of the electron EDM d_e is less than 10^{-38} e.cm [7] (which is effectively zero), where e is the charge of the electron. Some popular extensions of the SM, on the other hand, predict the value of the electron EDM in the range 10^{-26}–10^{-28} e.cm. (see Ref. 8 for further details). The search for a nonzero electron EDM is therefore a search for physics beyond the SM and particularly it is a search for T violation. This is, at present, an important and active field of research because the prospects of discovering new physics seems possible.

It is well recognized that heavy atoms and heavy polar diatomic molecules are very promising candidates in the experimental search for permanent EDMs arising from the violation of P and T. The search for nonzero P,T-odd effects in these systems with the presently accessible level of experimental sensitivity would indicate the presence of new physics beyond the SM of electroweak and strong interactions [9], which is certainly of fundamental importance. Despite the well known drawbacks and unresolved problems of the SM, there are no experimental data available that would be in direct contradiction with this theory. In turn, some popular extensions of the SM, which allow one to overcome its disadvantages, are not yet confirmed experimentally [8, 10].

A crucial feature of the search for P,T-odd effects in atoms and molecules is that in order to interpret the measured data in terms of fundamental constants of these interaction, one must calculate specific properties of the systems to establish a connection between the measured data and studied fundamental constants. These properties are described by operators that are prominent in the nuclear region; they cannot be measured, and their theoretical study is a non-trivial task.

During the last several years, there has been growing recognition of the significance of (and requirement for) ab initio calculations of electronic structure providing a high level of reliability and accuracy in accounting for both relativistic and correlation effects associated with these properties. In this chapter, we discuss one of the P,T-odd interaction constants, the so-called W_d, which is a measure of the effective electric field at the unpaired electron in the ground state of heavy polar molecules.

II. SEARCH FOR EDM ARISING FROM P,T-ODD INTERACTIONS

After the discovery of the combined charge and space symmetry violation, or CP violation, in the decay of neutral K_L^0 mesons [2], the search for the EDMs of elementary particles has become one of the fundamental problems in physics. A permanent EDM is induced by the super-weak interactions that violate both space inversion symmetry and time reversal invariance [11]. Considerable experimental efforts have been invested in probing for atomic EDMs (d_a) induced by EDMs of the proton, neutron, and electron, and by the P,T-odd interactions between them. The best available limit for the electron EDM, d_e, was obtained from atomic Tl experiments [12], which established an upper limit of $|d_e| < 1.6 \times 10^{-27}\,e{\cdot}\text{cm}$. The benchmark upper limit on a nuclear EDM is obtained from the atomic EDM experiment on ^{199}Hg [13] as $|d_{\text{Hg}}| < 2.1 \times 10^{-28}\,e{\cdot}\text{cm}$, from which the best restriction on the proton EDM, $|d_p| < 5.4 \times 10^{-24}\,e{\cdot}\text{cm}$, was also obtained by Dmitriev and Senkov [14]. The previous upper limit on the proton EDM was estimated from the molecular TlF experiments by Hinds and co-workers [15].

Since 1967, when Sandars suggested the use of polar heavy-atom molecules for the experimental search for the proton EDM [16], molecules have been considered the most promising candidates for such experiments. Sandars also noticed earlier [17] that the P-odd and P,T-odd effects are strongly enhanced in heavy atoms due to relativistic and other effects. For example, in paramagnetic atoms the enhancement factor for an electron EDM, d_a/d_e, is roughly proportional to $\alpha^2 Z^3 \alpha_D$, where α is the fine structure constant, Z is the nuclear charge, and α_D is the atomic polarizabilities. The enhancement can be of the order of 10^2 or greater for highly polarizable heavy atoms ($Z \geq 50$). Furthermore, the effective intermolecular electric field acting on electrons in polar molecules can be five or more orders of magnitude greater than the maximal electric field accessible in a laboratory. The first molecular EDM experiment was performed on TlF by Hinds et al. [15]; it was interpreted as a search for the proton EDM and other nuclear P,T-odd effects. In the last series of the ^{205}TlF experiments by Cho et al. [18] in 1991, the restriction on the proton EDM, $d_p = (-4 \pm 6) \times 10^{-23}\,e{\cdot}\text{cm}$ was obtained. This was recalculated in 2002 by Petrov et al. [19], who obtained the restriction as $d_p = (-1.7 \pm 2.8) \times 10^{-23}\,e{\cdot}\text{cm}$.

The experimental investigation of the electron EDM and parity nonconservation (PNC) effects was stimulated in 1978 by Labzovsky [20], Gorshkow et al. [21], and Sushkov et al. [22], who clarified the possibilities of additional enhancement of these effects in diatomic radicals like BiS and PbF due to the closeness of energy levels of opposite parity in Ω-doublets having a $^2\Pi_{1/2}$ ground state. Sushkov et al. [23] as well as Flambaum and Khriplovich [24] suggested the use of Ω-doubling in diatomic radicals with a $^2\Sigma_{1/2}$ ground state for such experiments. Soon after that semiempirical calculations were performed for the HgF, HgH, and BaF molecules by Kozlov [25], and at the same time ab initio calculations of PNC effects in PbF initiated by Labzowsky were completed by Titov et al. [26]. A few years later, Hinds launched an experimental search for the electron EDM in the YbF molecule for which his group obtained the first result in 2002 [27], $d_e = (-0.2 \pm 3.2) \times 10^{-26}\, e{\cdot}\text{cm}$. Though that restriction is worse than the best current d_e datum (from the Tl atom experiments), nevertheless, it is only due to limitations on counting statistics, as Hudson et al. [27] pointed out later.

At present, a series of electron EDM experiments on YbF [28] and PbO (DeMille's group at Yale, USA) are in progress. The unique suitability of PbO for searching for the elusive d_e is demonstrated by the very high projected statistical sensitivity of the Yale experiment to the electron EDM, which has the prospect of allowing one to detect d_e on the order of 10^{-29}–10^{-31} $e{\cdot}$cm [29], two to four orders of magnitude lower than the current limit quoted earlier. Some other candidates for the EDM experiments, in particular, HgH, HgF, TeO^*, and HI^+, are being discussed, and an experiment on PbF is planned (Oklahoma University, USA)

III. PARTICLE PHYSICS IMPLICATIONS FOR THE EXISTENCE OF AN ELECTRON EDM

As mentioned in the Introduction, the observation of a nonzero EDM of an electron would be a signature of behavior beyond that described by the standard model (SM) of physics [9]. It would be a more sensitive probe of the SM than the neutron EDM, which could have nonzero EDM due to CP violation in the QCD sector of the SM.

Table I presents the estimate of the electron EDM predicted by different particle physics models [8, 9]. As can be seen from this table, the value of the electron EDM in the SM is 10–12 orders of magnitude smaller than in the other models. This is due to the fact that the first nonvanishing contribution to this quantity arises from three-loop diagrams [30]. There are strong cancellations between diagrams at the one-loop as well as two-loop levels. It is indeed significant that the electron EDM is sensitive to a variety of extensions of the SM including supersymmetry (SUSY), multi-Higgs, left–right symmetry, lepton

TABLE I
Prediction for the Electron EDM, $|d_e|$, in Popular Theoretical Models

Model	$\lvert d_e\rvert$ (in $e{\cdot}$cm)
Standard model	$< 10^{-38}$
Left–right symmetry	10^{-28}–10^{-26}
Lepton flavor-changing	10^{-29}–10^{-26}
Multi-Higgs	10^{-28}–10^{-27}
Supersymmetric	$\leq 10^{-25}$
Experimental limit [12]	$< 1.6 \times 10^{-27}$

flavor-changing, and so on [10]. This is particularly true for the minimal ("naive") SUSY model, which predicts an electron EDM already at the level of 10^{-25} $e{\cdot}$cm. However, the best experimental estimate on the electron EDM, 1.6×10^{-27} $e{\cdot}$cm, from the experiment for the Tl atom [12], is almost two orders of magnitude smaller. More sophisticated SUSY models have many desirable features such as their ability to explain the "gauge hierarchy problem" and to solve the problem of dark matter in astrophysics. Interestingly, they predict the electron EDM at the level of 10^{-27} $e{\cdot}$cm or somewhat smaller. It is certainly remarkable that studies of the electron EDM can shed light on supersymmetry, which is one of the most profound ideas in contemporary physics.

IV. PARITY AND TIME REVERSAL SYMMETRY

The space inversion transformation is $\mathbf{x} \rightarrow -\mathbf{x}$ and the corresponding operator on state vector space is called the *parity* operator (P). The parity operator reverses the sign of the position (Q) and the linear momentum (L) operator,

$$\mathrm{PQP}^{-1} = -\mathrm{Q} \tag{1}$$

$$\mathrm{PLP}^{-1} = -\mathrm{L} \tag{2}$$

Under the action of the parity operator P, the position and momentum commutator [Q,L] = $i\hbar$, becomes

$$\mathrm{PQP}^{-1}\mathrm{PLP}^{-1} - \mathrm{PLP}^{-1}P\mathrm{QP}^{-1} = \mathrm{P}i\hbar\mathrm{P}^{-1} \tag{3}$$

By the use of Eqs.(1) and (2), this becomes

$$\mathrm{QL} - \mathrm{LQ} = \mathrm{P}i\hbar\mathrm{P}^{-1} \tag{4}$$

which is compatible with the original commutation relation provided $\mathrm{P}i\mathrm{P}^{-1} = i$. Therefore the parity operator is *linear* and *unitary*. Since the two consecutive space inversions produce no change at all, it follows that the state described by $|\Psi\rangle$ and by $P^2|\Psi\rangle$ must be the same. Thus the operator P^2 can differ from the identity operator by at most a phase factor. It is convenient to choose this phase factor to be unity and hence we have

$$P = P^{-1} = P^{\dagger} \tag{5}$$

The effect of time reversal operator T is to reverse the linear momentum (L) and the angular momentum (J), leaving the position operator unchanged. Thus, by definition,

$$TQT^{-1} = Q \tag{6}$$

$$TLT^{-1} = -L, \quad TJT^{-1} = -J \tag{7}$$

Under the action of time reversal operator T the position and momentum commutator $[Q,L] = i\hbar$ becomes

$$TQT^{-1}TLT^{-1} - TLT^{-1}TQT^{-1} = Ti\hbar T^{-1} \tag{8}$$

According to Eqs.(6) and (7), this becomes $-(QL - LQ) = Ti\hbar T^{-1}$, which when compared with the original commutation relation yields $TiT^{-1} = -i$. Therefore the time reversal operator is *anti-linear.* It can also be shown that the time reversal operator T is *anti-unitary.*

V. GENERAL PRINCIPLE FOR THE MEASUREMENT OF EDMS

The basic physics governing the measurement of the EDM in all types of electrically neutral systems is almost the same as discussed in this section. If the system under consideration has a magnetic moment $\boldsymbol{\mu}$ and is exposed to a magnetic field $\boldsymbol{B}$, then the interaction Hamiltonian can be written

$$H_{\text{mag}} = -\boldsymbol{\mu} \cdot \boldsymbol{B} \tag{9}$$

From a classical point of view, the magnetic field exerts a torque on the system so that the magnetic moment $\boldsymbol{\mu}$ and hence the angular momentum $\boldsymbol{J}$ begin to precess about the magnetic field $\boldsymbol{B}$. The precession frequency for a system having $J = \frac{1}{2}$ corresponds to the energy separation of $2\mu B$ between $m = \pm\frac{1}{2}$ states, given by

$$\delta\omega = 2\mu B/\hbar \tag{10}$$

If the system under consideration also possesses an electric dipole moment $\boldsymbol{d}$ and is exposed to an electric field $\boldsymbol{E}$, then the interaction Hamiltonian can be written

$$H_{\text{ele}} = -\boldsymbol{d} \cdot \boldsymbol{E} \tag{11}$$

As a result of the projection theorem [31], the expectation value of the EDM operator $\boldsymbol{d}$, which is a vector operator, is proportional to the expectation value of $\boldsymbol{J}$ in the angular momentum eigenstate. This fact, in conjunction with Eq. (9), implies that the electric field modifies the precession frequency of the system because of the additional torque experienced by the system due to the interaction between the electric field and the EDM. It can readily be shown that the modified precession frequency is

$$\Delta\omega_{+} = \frac{2\mu B + 2dE}{\hbar} \tag{12}$$

when the applied electric field $\boldsymbol{E}$ is parallel to the magnetic field $\boldsymbol{B}$, and

$$\Delta\omega_{-} = \frac{2\mu B - 2dE}{\hbar} \tag{13}$$

when they are antiparallel. The aim of the experiment is to measure the change in the precession frequency due to the flipping of the electric field with respect to the magnetic field. From the above two expressions, it is obvious that the change in the precession frequency is

$$\Delta\omega = \Delta\omega_{+} - \Delta\omega_{-} = 4dE/\hbar \tag{14}$$

Equation (14) can be used to determine the value of d, the permanent EDM of the system. The most recent and the best limit available so far for the intrinsic EDM of the electron was set to be $|d_e| = 1.6 \times 10^{-27}\, e\cdot\text{cm}$ by Regan et al. [12] from atomic EDM experiments for the Tl atom. It may be worth mentioning that the above principle was used in the electron EDM experiment by Regan et al. [12].

VI. MECHANISMS GIVING RISE TO PERMANENT EDMS

A permanent EDM of a stable atomic or molecular state can arise only when both P and T invariance are broken (see Fig.1). It is often said that polar molecules possess "permanent" EDMs and exhibit a linear Stark effect. However, the Stark effect exhibited by the polar molecule is not *really* linear for sufficiently small $\boldsymbol{E}$ at zero temperature, and moreover, it violates neither P nor T symmetry [10]. We emphasize that a permanent EDM that exhibits a linear Stark effect even for an infinitesimally weak $\boldsymbol{E}$ is a genuine signature of P and T violation or CP violation

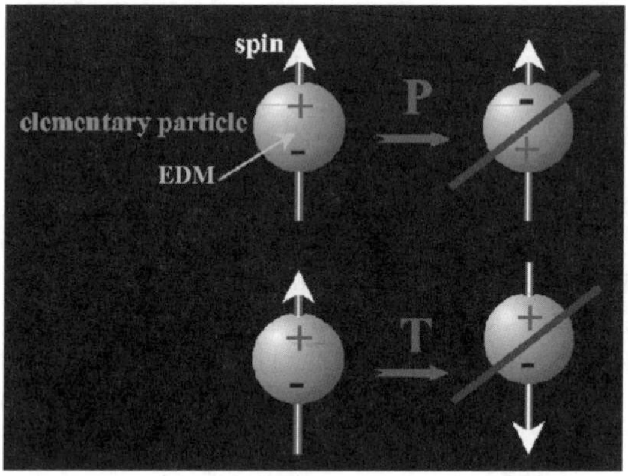

Figure 1. Parity (P) and time (T) reversal symmetry violation.

in conjunction with the CPT theorem. The mechanisms that can give rise to such P,T-odd effects and hence permanent EDMs are the following:

1. Intrinsic electric dipole moments of electrons
2. Intrinsic electric dipole moments of nucleons
3. P and T violating electron–nucleon interactions
4. P and T violating electron–electron interactions
5. P and T violating nucleon–nucleon interactions

The nonrelativistic electronic Hamiltonian for N electrons in the field of M point charges (nuclei) under the Born–Oppenheimer approximation is given by

$$\mathcal{H} = \sum_{i=1}^{N} \left(\frac{p_i^2}{2m} - \sum_{A=1}^{M} \frac{Z_A e}{r_{iA}} + \sum_{i \neq j} \frac{e^2}{r_{ij}} \right) \tag{15}$$

where Z is the atomic number, m is the electron mass, r_{iA} is the distance between the ith electron and the Ath nucleus, and r_{ij} is the distance between the ith and jth electron. Let us first consider the effects of the intrinsic EDM of an electron in an atom or a molecule from the nonrelativistic point of view. Let us also assume that the value of the intrinsic EDM of an electron is d_e. The EDM operator corresponding to the ith electron in the atom/molecule is given by [8]

$$d_i = d_e \boldsymbol{\sigma}_i \tag{16}$$

where $\boldsymbol{\sigma}_i$ are the Pauli spin matrices for the ith electron.

Therefore in the presence of an external electric field of strength E_z in the z-direction, the total nonrelativistic molecular Hamiltonian is

$$\mathrm{H} = \mathcal{H} + \mathrm{H_I} + \mathrm{H_{ext}} \tag{17}$$

where

$$\mathrm{H_I} = -d_e \sum_i \boldsymbol{\sigma}_i \cdot \boldsymbol{E}_i^{\mathrm{int}} \tag{18}$$

and

$$\mathrm{H_{ext}} = -eE_z \sum_i \left(z_i + \frac{d_e}{e} \sigma_{zi} \right) \tag{19}$$

$\boldsymbol{E}_i^{\mathrm{int}}$ in Eq. (18) is the internal electric field that the ith electron experiences due to the other electrons and the nuclei of the molecule. Clearly, we can write

$$e\boldsymbol{E}_i^{\mathrm{int}} = -\boldsymbol{\nabla}_i \left(\sum_{j \neq i} \frac{e^2}{r_{ij}} - \sum_{A=1}^{M} \frac{Z_A e}{r_{iA}} \right) \tag{20}$$

so that we get

$$\mathrm{H_I} = \frac{d_e}{e} \sum_i \boldsymbol{\sigma}_i \cdot \boldsymbol{\nabla}_i \left(\sum_{j \neq i} \frac{e^2}{r_{ij}} - \sum_{A=1}^{M} \frac{Z_A e}{r_{iA}} \right) \tag{21}$$

But it turns out that the above Hamiltonian leads to a vanishing value of the atomic or molecular EDM due to Schiff's theorem [32]. This result is most surprising because it implies that even though the individual electrons in the molecule have nonzero EDMs, the molecular EDM is still zero. Later, Sandars [17] demonstrated that the relativistic treatment (presented later) of electrons in atoms or molecules leads to a nonzero electron EDM.

The relativistic treatment of electron EDM begins by replacing the nonrelativistic Hamiltonian $\mathcal{H}$ and the interaction Hamiltonian $\mathrm{H_I}$ by their relativistic counterparts

$$\mathcal{H}_{\mathrm{DF}} = \sum_i \left(c\boldsymbol{\alpha}_i \cdot \boldsymbol{p}_i + \beta mc^2 - \sum_{A=1}^{M} \frac{Z_A e}{r_{iA}} + \sum_{j \neq i} \frac{e^2}{r_{ij}} \right) \tag{22}$$

and

$$\mathrm{H}_{\mathrm{EDM}} = -d_e \sum_i \beta_i (\boldsymbol{\Sigma}_i \cdot \boldsymbol{E}_i + i\boldsymbol{\alpha}_i \cdot \boldsymbol{B}_i) \tag{23}$$

where

$$\boldsymbol{\Sigma} = \begin{pmatrix} 0 & \hat{\boldsymbol{\sigma}} \\ \hat{\boldsymbol{\sigma}} & 0 \end{pmatrix}, \quad \beta = \begin{pmatrix} I & 0 \\ 0 & -I \end{pmatrix} \tag{24}$$

and $\boldsymbol{E}$ and $\boldsymbol{B}$ are the electric and magnetic fields, respectively. Since the second term of Eq. (23) is of the order of $O(v^2/c^2)$, we can neglect this term and rewrite Eq. (23) as

$$\mathrm{H}_{\mathrm{EDM}} \approx -d_e \sum_i \beta_i \boldsymbol{\Sigma}_i \cdot \boldsymbol{E}_i \tag{25}$$

If the atom as a whole is also exposed to an external electric field E_z in the z-direction, then the total relativistic molecular many-body Hamiltonian can be written

$$\mathrm{H} = \mathcal{H}_{\mathrm{DF}} + \mathrm{H}' \tag{26}$$

where

$$\mathrm{H}' = -d_e \sum_i \beta_i \boldsymbol{\Sigma}_i \cdot \boldsymbol{E}_i^{\mathrm{int}} - eE_z \sum_i \left(z_i + \frac{d_e}{e} \beta_i \boldsymbol{\Sigma}_{zi} \right) \tag{27}$$

Since $\beta = 1$ in the nonrelativistic limit where the contribution vanishes, we can replace β by $(\beta - 1)$ in the expression for H'. In this case the residual EDM interaction of an electron with the internal electric field reduces to

$$\mathrm{H}_d = -d_e(\beta - 1)\boldsymbol{\Sigma} \cdot \boldsymbol{E}^{\mathrm{int}} \tag{28}$$

With the aid of these above two equations, the linear Stark splitting can be evaluated as

$$\Delta E = -2d_e e E_z \sum_{n \neq 0} \frac{\langle \psi_0 | z | \psi_n \rangle \langle \psi_n | (\beta - 1)\boldsymbol{\Sigma} \cdot \boldsymbol{E}^{\mathrm{int}} | \psi_0 \rangle}{E_0 - E_n} - d_e E_z \langle \psi_0 | (\beta - 1) \boldsymbol{\Sigma}_z | \psi_0 \rangle \tag{29}$$

The ψ_0 and ψ_n in Eq. (29) are eigenfunctions of $\mathcal{H}_{DF}$ corresponding to the energy eigenvalues E_0 and E_n, respectively [8, 33].

The high value of the electron density at the nucleus leads to the enhancement of the electron EDM in heavy atoms. The other possible source of the enhancement is the presence of small energy denominators in the sum over states in the first term of Eq.(29). In particular, this takes place when $(E_0 - E_n)$ is of the order of the molecular rotational constant. (It is imperative that a nonperturbative treatment be invoked when the Stark matrix element $eE_z\langle\psi_0|z|\psi_n\rangle$ is comparable to the energy denominator $(E_0 - E_n)$ [33].) Neglecting the second term of the right-hand side of Eq.(29), which does not contain this enhancement factor [8, 27], we get

$$\Delta E = \langle\psi_0|\mathrm{H}_d|\psi_0\rangle = -d_e\langle\psi_0|(\beta - 1)\boldsymbol{\Sigma}\cdot\boldsymbol{E}^{\mathrm{int}}|\psi_0\rangle \tag{30}$$

With the aid of this, the P,T-odd constant W_d can be expressed as [34]

$$W_\mathrm{d} = \frac{2}{d_e}\langle\mathbf{X}^2\boldsymbol{\Sigma}_{1/2}|\mathrm{H}_d|\mathbf{X}^2\boldsymbol{\Sigma}_{1/2}\rangle \tag{31}$$

It can readily be understood from the above expressions that W_d is a measure of the effective electric field at the unpaired electrons.

VII. COMPUTATIONAL PROCEDURE

A. Internal Electric Field of a Diatomic Molecule

In case of a molecule, the internal electric field experienced by an electron can be written

$$\boldsymbol{E}^{\mathrm{int}} = \boldsymbol{E}_i^{\mathrm{mol}} = \sum_m \boldsymbol{E}_i^m + \sum_{j\neq i}\boldsymbol{E}_{ij} \tag{32}$$

where $\boldsymbol{E}_i^m$ is the field due to the mth nucleus at the site of the ith electron and $\sum_{j\neq i}\boldsymbol{E}_{i,j}$ is the electric field due to the jth electron at the site of ith electron. For diatomic molecule, the above equation reduces to

$$\boldsymbol{E}_i^{\mathrm{mol}} = \boldsymbol{E}_i^{\mathrm{M}} + \boldsymbol{E}_i^{\mathrm{F}} + \sum_{j\neq i}\boldsymbol{E}_{ij} \tag{33}$$

For spherically symmetric nuclear charge distribution (*Gaussian*, *Fermi*, or point nucleus), the electric field at a point $\boldsymbol{r}$ outside the nucleus can be evaluated from Gauss' law as

$$E(\boldsymbol{r}) = \frac{1}{4\pi\epsilon_0}\frac{Q}{r^2}\hat{\boldsymbol{r}} \quad \text{(in SI units)} \tag{34}$$

where Q is the total charge inside the nucleus (proportional to the atomic number Z of the atom) and $\boldsymbol{r}$ is the position vector of the point under consideration relative to the center of the nucleus. (Note that although the electric field outside the nucleus is the same for all the above-mentioned charge distributions, the electric field inside the nucleus may be different for different charge distributions.)

By use of Eqs.(33) and (34), the nuclear electric field of a diatomic molecule MF can be written

$$\boldsymbol{E}_i^{\mathrm{mol}} = \frac{1}{4\pi\epsilon_0}\left[\frac{Z_{\mathrm{M}}}{r_{\mathrm{M}}^2}\hat{\boldsymbol{r}}_{\mathrm{M}} + \frac{Z_{\mathrm{F}}}{r_{\mathrm{F}}^2}\hat{\boldsymbol{r}}_{\mathrm{F}}\right] + \sum_{j\neq i}\boldsymbol{E}_{i,j} \tag{35}$$

The last term of the above equation is quite small compared to the other two terms and hence can be neglected. So, neglecting the last term, we get

$$\boldsymbol{E}_i^{\mathrm{mol}} = \frac{1}{4\pi\epsilon_0}\left[\frac{Z_{\mathrm{M}}}{r_{\mathrm{M}}^2}\hat{\boldsymbol{r}}_{\mathrm{M}} + \frac{Z_{\mathrm{F}}}{r_{\mathrm{F}}^2}\hat{\boldsymbol{r}}_{\mathrm{F}}\right] \tag{36}$$

B. Relativistic Molecular Orbitals

The relativistic orbitals are assumed to be of the form

$$\frac{1}{r}\begin{pmatrix} P_{nk}(r)\Omega_{km}(\hat{\boldsymbol{r}}) \\ iQ_{nk}(r)\Omega_{-km}(\hat{\boldsymbol{r}}) \end{pmatrix} \tag{37}$$

where $P_{nk}(r)$ $(Q_{nk}(r))$ are the radial parts of the large (small) component of the basis functions and

$$\Omega_{km}(\hat{\boldsymbol{r}}) = \sum_{\sigma=\pm\frac{1}{2}} C^{jm}_{lm-\sigma\frac{1}{2}\sigma} Y_l^{m-\sigma}(\hat{\boldsymbol{r}})\eta_\sigma \tag{38}$$

is a spinor spherical harmonic, $C^{j_3m_3}_{j_1m_1j_2m_2}$ is a Clebsh–Gordon coefficient, $Y_l^m(\hat{\boldsymbol{r}})$ is a spherical harmonic, and η_σ is a spin basis function

$$\eta_{+\frac{1}{2}} = \begin{pmatrix} 1 \\ 0 \end{pmatrix}, \quad \eta_{-\frac{1}{2}} = \begin{pmatrix} 0 \\ 1 \end{pmatrix} \tag{39}$$

The molecular orbitals used in these calculations are generated using the linear combination of atomic orbitals (LCAO) method, in which an orbital is

$$\phi_{\mathrm{I}}(\boldsymbol{r}) = \sum_{p=1}^{n}\begin{pmatrix} cX^P_{\mathrm{I}p}\chi^P_p(\boldsymbol{r}_p) \\ iX^Q_{p\mathrm{I}}\chi^Q_p(\boldsymbol{r}_p) \end{pmatrix} \tag{40}$$

where ϕ_I are the four-component MOs and X_{Ip}^P are the corresponding MO coefficients. The $\chi_p^P(\boldsymbol{r}_p)$ $(\chi_p^Q(\boldsymbol{r}_p))$ in Eq.(40) are the large (small) two-component basis spinors and $r_p = r - R$, where R is the "center" of the basis function p [35, 36].

C. Matrix Elements of the P,T-Odd Interaction Operator H_d

Using the standard representation of the Dirac matrices β and $\boldsymbol{\Sigma}$, the residual P,T-odd interaction operator H_d can be written

$$\mathrm{H}_d = 2d_e \begin{pmatrix} 0 & 0 \\ 0 & \hat{\boldsymbol{\sigma}} \end{pmatrix} \cdot \boldsymbol{E}^{\mathrm{int}} \tag{41}$$

Neglecting the electric field due to the electrons and replacing $\boldsymbol{E}^{\mathrm{int}}$ by $\boldsymbol{E}_i^{\mathrm{mol}}$ (defined earlier), we can express the P,T-odd interaction operator H_d for the ith electron as

$$\mathrm{H}_d^I = 2d_e \begin{pmatrix} 0 & 0 \\ 0 & \hat{\boldsymbol{\sigma}} \cdot \hat{\boldsymbol{r}}_{\mathrm{M}} \end{pmatrix} \frac{1}{4\pi\varepsilon_0} \frac{Z_{\mathrm{M}}}{r_{\mathrm{M}}^2} + 2d_e \begin{pmatrix} 0 & 0 \\ 0 & \hat{\boldsymbol{\sigma}} \cdot \hat{\boldsymbol{r}}_{\mathrm{F}} \end{pmatrix} \frac{1}{4\pi\varepsilon_0} \frac{Z_{\mathrm{F}}}{r_{\mathrm{F}}^2} \tag{42}$$

It is evident that the operator H_d couples only the small components of the relativistic molecular wavefunctions. Since the small components as well as the nuclear electric fields are prominent in and around the nuclear regions, the dominant contribution to the matrix elements of H_d comes from that region. It should be noted that the absence of the screening term $\boldsymbol{E}_{i,j}$ in Eq.(42) will overestimate the H_{d} matrix element. However, the amount of overestimation is expected to be small.

The value of the matrix element of the operator in Eq.(42) is determined principally by contributions from the regions in and around the nuclei, where both the electric field and the small component (relativistic effect) of the wavefunctions are largest. In the absence of screening ($\boldsymbol{E}_{i,j}$), the nuclear electric field diminishes with the square of the distance from the center of a nucleus; screening further accelerates the decline of the electric field with distance. The electrons of each "constituent atom" have completely screened "their" nuclei at the location of any other nucleus, for which reason, and to a very good approximation, the problem is "uncoupled" for the various nuclear regions.

The evaluation of the integrals using spherical polar coordinates is relatively simple. With slight mathematical manipulation, the matrix element of the operator H_d can be written

$$\langle \mathrm{H}_d^I \rangle_{II} = -2d_e \sum_{\mathrm{X=M,F}} \sum_{k_{\mathrm{X}},m_{\mathrm{X}}} \delta_{-k_{\mathrm{X}},k'_{\mathrm{X}}} \delta_{m_{\mathrm{X}},m'_{\mathrm{X}}} \int_r \chi^Q_{I,k_{\mathrm{X}},m_{\mathrm{X}}}(r) E_{\mathrm{X}}(r) \chi^Q_{I,k'_{\mathrm{X}},m'_{\mathrm{X}}}(r) dr \tag{43}$$

Here, we have removed the factor $1/4\pi\epsilon_0$ because in atomic units (a.u.) its value is unity. We have also used the short-hand notation $E_X(r) = Z_X/r_X{}^2$ for the nuclear electric field. Furthermore, we use the following two identities to arrive at the above expression:

$$(\hat{\boldsymbol{\sigma}} \cdot \hat{\boldsymbol{r}})\Omega_{km} = -\Omega_{-km} \tag{44}$$

and

$$\int \Omega^{\dagger}_{km}\Omega_{k'm'}d\Omega = \delta_{k,k'}\delta_{m,m'} \tag{45}$$

The approximate molecular wavefunction Ψ is a Slater determinant of the single particle orbitals ϕ_I

$$|\Psi\rangle = \frac{1}{\sqrt{N!}}\text{Det}|\phi_1\phi_2\cdots\phi_N| \tag{46}$$

The expectation value of the operator H_d in a specific molecular state Ψ is given by

$$\langle\Psi|H_d^N|\Psi\rangle = \sum_I \langle\phi_I|H_d^I|\phi_I\rangle \tag{47}$$

where

$$H_d^N = \sum_I H_d^I \tag{48}$$

Thus one has to first evaluate the matrix elements $\langle\phi_I|H_d^I|\phi_I\rangle$ at the single particle level to compute the expectation value of the operator H_d^N in a specific molecular state.

It should be noted that the expression for the nuclear electric field $E \propto 1/r^2$ is reasonable if the dimensions of the nuclei are assumed to be negligible (i.e., point charge approximation). Since it is well known that nuclei are of finite size, the given expression overestimates the electric field inside the nuclear region, which can affect the accuracy of the computed W_d. In the case of a uniformly charged spherical nucleus of finite dimension, the appropriate expression for the electric field inside(outside) the nuclear region $E \propto r(E \propto 1/r^2)$ should be used. Some other form of the nuclear charge distribution such as the "Gaussian nucleus" or "Fermi nucleus" can also be used.

The procedure for evaluating the matrix elements of the P,T-odd interaction operator H_d with Cartesian Gaussian spinors is a bit complicated and different from those evaluated from spherical Gaussian spinors as described in Appendices A and B.

VIII. AB INITIO CALCULATION OF P,T-ODD EFFECTS

As mentioned earlier, heavy polar diatomic molecules, such as BaF, YbF, TlF, and PbO, are the prime experimental probes for the search of the violation of space inversion symmetry (P) and time reversal invariance (T). The experimental detection of these effects has important consequences [37, 38] for the theory of fundamental interactions or for physics beyond the standard model [39, 40]. For instance, a series of experiments on TlF [41] have already been reported, which provide the tightest limit available on the tensor coupling constant C_T, proton electric dipole moment (EDM) d_p, and so on. Experiments on the YbF and BaF molecules are also of fundamental significance for the study of symmetry violation in nature, as these experiments have the potential to detect effects due to the electron EDM d_e. Accurate theoretical calculations are also absolutely necessary to interpret these ongoing (and perhaps forthcoming) experimental outcomes. For example, knowledge of the effective electric field E (characterized by W_{d}) on the unpaired electron is required to link the experimentally determined P,T-odd frequency shift with the electron's EDM d_e in the ground ($\mathrm{X}^2\Sigma_{1/2}$) state of YbF and BaF.

The twin facts that heavy-atom compounds like BaF, TlF, and YbF contain many electrons and that the behavior of these electrons must be treated relativistically introduce severe impediments to theoretical treatments, that is, to the inclusion of sufficient electron correlation in this kind of molecule. Due to this computational complexity, calculations of P,T-odd interaction constants have been carried out with "relativistic matching" of nonrelativistic wavefunctions (approximate relativistic spinors) [42], relativistic effective core potentials (RECP) [43, 34], or at the all-electron Dirac–Fock (DF) level [35, 44]. For example, the first calculation of P,T-odd interactions in TlF was carried out in 1980 by Hinds and Sandars [42] using approximate relativistic wavefunctions generated from nonrelativistic single particle orbitals.

The P,T-odd interaction constant W_{d} in YbF was first calculated by Titov et al. [43] using generalized RECP (GRECP), as this procedure provides reasonable accuracy with small computational cost. Titov and co-workers also reported W_{d} computed using a restricted active space self-consistent field (RASSCF) scheme [43, 45] with the GRECP. Assuming that contributions from valence–valence electron correlation are negligible, Parpia [35] in 1998 estimated W_{d} from the all-electron unrestricted DF method (UDF). In the same year, Quiney et al. [44] reported the P,T-odd interaction constant W_{d} computed at the core-polarization level with all-electron DF orbitals. Although pair correlation and higher order contributions to W_{d} are nonnegligible, these terms are not included in Quiney et al.'s calculations. The calculations cited above predict the P,T-odd interaction constant to lie in the rather large range $[-0.62,-1.5] \times 10^{25}$ Hz/$e \cdot$cm. Therefore more precise estimation of W_d is necessary to set reliable limits on the electron EDM d_e.

The first calculation of the P,T-odd interaction constant W_d of BaF was carried out by Kozlov et al. [46] using generalized relativistic effective core potentials (GRECP) and self-consistent field (SCF) and restricted active space SCF (RASSCF) methods. They have also reported W_d computed using the effective-operator (EO) technique at the SCF-EO and RASSCF-EO levels. The computed W_d by Kozlov et al. [46] is quite consistent with the semiempirical result of Kozlov and Labzovsky [26] estimated from the experimental hyperfine structure constants measured by Knight et al. [47]. Although the RASSCF-EO result [46] is close to the semiempirical estimate of Kozlov and Labzovsky, it is worthwhile to compute this constant more accurately using correlated many-body methods like configuration interaction (CI), coupled cluster (CC), and so on.

In this section, we describe calculations of the P,T-odd interaction constant W_d for the ground $(X^2\Sigma_{1/2})$ states of YbF and BaF molecules using all-electron DF orbitals and a restricted active space (RAS) configuration interaction (CI) treatment.

The active space used for both systems in these calculations is sufficiently large to incorporate important core–core, core–valence, and valence–valence electron correlation, and hence should be capable of providing a reliable estimate of W_d. In addition to the P,T-odd interaction constant W_d, we also compute ground to excited state transition energies, the ionization potential, dipole moment (μ_e), ground state equilibrium bond length (R_e), and vibrational frequency (ω_e) for the YbF and μ_e for the BaF molecule.

A. YbF Molecule

The ground and excited state properties of YbF are calculated at the optimized geometry using the restricted active space (RAS) configuration interaction (CI) method. We employ 27s27p12d8f and 15s10p uncontracted Gaussian functions for Yb and F, respectively. The basis employed here is almost the same as that used by Parpia [35] in his YbF calculation. Since the computation of ground and excited state properties of YbF using an all active configuration interaction (CI) method is computationally too expensive, we employ the computationally facile RASCI method, which is capable of providing reasonable accuracy at a reduced computational cost.

There are 39 doubly and one singly occupied orbitals in YbF of which the 25th occupied orbital of YbF corresponds to the 5s occupied spin orbital of Yb. As the contribution of the 5s and 5p orbitals of Yb to W_d is quite significant [43, 48, 49], these orbitals are included in the CI space. The occupied orbitals above the 25th are also included in the active space based on energy considerations. (Note that

TABLE II
P,T-Odd Interaction Constant W_d and Dipole Moment μ_e for the Ground $^2\Sigma$ State of YbF Molecule

Methods or Experiment	$W_d(10^{25}$ Hz/$e\cdot$cm)	μ_e (debyes)
Experiment [38]		3.91(4)
Semiempirical [49]	−1.50	
GRECP/RASSCF [43]	−0.91	
Semiempirical [48]	−1.26	
DHF [44]	−0.31	
DHF + CP [44]	−0.61	
UDF (unpaired electron) [35]	−0.962	
UDF (all electrons) [35]	−1.203	4.00
GRECP/RASSCF-EO [45]	−1.206	
DF (Nayak et al.) [50]	−0.963	3.98
RASCI (Nayak et al.) [50]	−1.088	3.91
MRCI (pseudopotential) [51]		3.55

the 4f orbitals of Yb and the 2p orbitals of F in YbF are energetically quite close (see Table 12 of Ref. 35).) Thus altogether 31 active electrons (16α and 15β) are included in the CI active space.

The W_d estimated from the RASCI is compared with other theoretical calculations [34, 35, 43, 44] in Table II. As can be seen in Table II, the present DF estimate of W_d agrees with the Dirac–Fock (DF) value reported by Parpia [35] but differs by $\sim 7\%$ from Titov et al.'s [43] estimate of W_d. At this juncture we emphasize that the DF estimate of W_d reported by Quiney et al. [44] differs by a factor of *three* from ours (as well as from those of Parpia [35] and Titov et al. [43]) because a single combination of symmetry type is considered in their [44] calculations. While these calculations deviate negligibly in the estimated P,T-odd interaction constant at the Dirac–Fock level, the deviation is quite significant at the post Dirac–Fock level. For example, Quiney *et al.* [44] show that the contribution of first-order *core polarization* is almost 100%. On the other hand, Parpia's unrestricted Dirac–Fock (UDF) calculation indicates that the correlation contribution is $\sim 25\%$. Note that although the core-polarization contribution is quite important, the pair correlation and higher order terms are nonnegligible. We also emphasize that the inclusion of electron correlation through the UDF is *generally* not recommended as the UDF theory suffers from spin contamination.

The variations of W_d and μ_e with respect to the size of the CI space are depicted in Figs. 2 and 3, respectively. With the exception of a very small increase between the final two calculations, the parameter W_d has essentially stabilized. Figures 2 and 3 indicate that the contribution to W_d and μ_e from orbitals 60–75 (CSFs 2–3$\times 10^6$) is most significant compared to other unoccupied (at DF level) active orbitals.

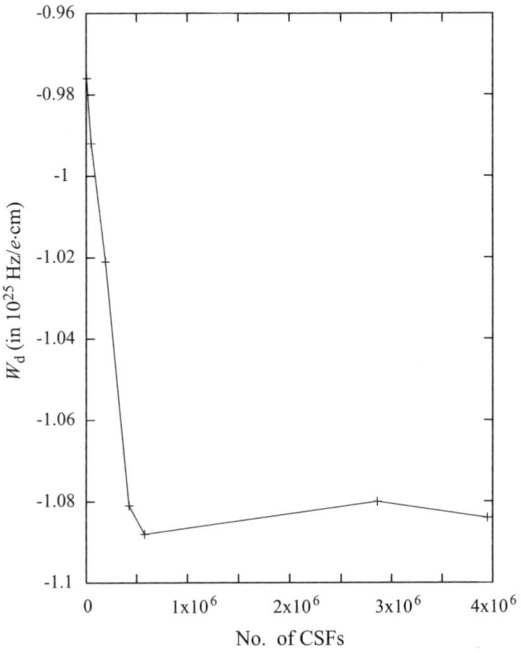

Figure 2. The P,T-odd interaction constant W_d versus the number of configuration state functions (CSFs) for the YbF molecule.

The first ionization potential and low lying ground $(X^2\Sigma_{1/2})$ to excited $(^2\Sigma_{1/2}$, $^2\Pi_{1/2}$, and $^2\Pi_{3/2})$ state transition energies of YbF are compared with experiment [52] in Table III. The RASCI transition energies from our largest CI space are in excellent agreement with experiment. The maximum deviation in the estimated excitation energy is only 419 cm^{-1} (or 2.3%) for the $A^2\Pi_{1/2}$ state. The RASCI method also provides a fairly accurate estimate of the $A^2\Pi_{1/2} - A^2\Pi_{3/2}$ energy gap, which deviates by 3% from experiment.

The equilibrium bond length (R_e) and ground state vibrational frequencies (ω_e) computed at the DF and RASCI levels, are compared with experiment and with other calculations in Table IV. It is evident from Table IV that the RASCI method offers a more accurate estimate of R_e than the DF approximation, while the later method yields a more accurate estimate of the vibrational frequency ω_e. However, the minuscule error in ω_e (at the DF level) is perhaps fortuitous given the larger error in R_e of 2.8%.

B. BaF Molecule

The P,T-odd constant W_d and dipole moment μ_e for the ground state of the BaF molecule are calculated using the restricted active space (RAS) configuration

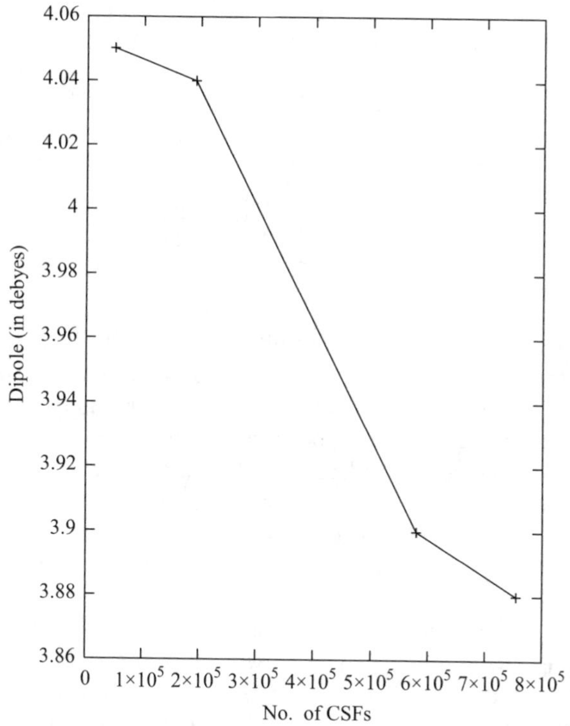

Figure 3. The molecular dipole moment μ_e versus the number of configuration state functions (CSFs) for the YbF molecule.

interaction (CI) method at the experimental geometry $R_e = 2.16$ Å [52]. The ground state properties of BaF are computed using "kinetically balanced"[53] 27s27p12d8f even-tempered Gaussian type of orbitals (GTOs) for the Ba atom and 15s10p GTOs for the F atom.

There are 32 doubly and one singly occupied orbitals in BaF of which the 25th occupied orbital of BaF corresponds to the 5s occupied spin orbital of Ba.

TABLE III
Vertical Ionization Potential and Transition Energies of the YbF Molecule (in cm^{-1}), Computed Using the RASCI Method

State	RASCI [50]	Experiment [52]
IP	48537	
$X^2\Sigma_{1/2}$	0	0
$A^2\Pi_{1/2}$	18509	18090
$A^2\Pi_{3/2}$	19838	19460
$B^2\Sigma_{1/2}$	21505	21067

TABLE IV
Ground State Spectroscopic Constants of YbF Molecule

Spectroscopic Constant [51]	Nayak et al. [50] DF	Nayak et al. [50] RASCI	Others	Experiment [52]
R_e (in Å)	2.073	2.051	2.074 [35] 2.045 [51]	2.016
ω_e (in cm^{-1})	504	529	492 [51]	502

As the contribution of the 5s and 5p orbitals of Ba to W_d is quite significant [43] for the BaF molecule, we have included these orbitals of Ba in our CI active space for the calculation of W_d and μ_e for the ground state of the BaF molecule. The occupied orbitals above the 25th are also included in the RASCI space from energy consideration. Thus altogether 17 active electrons (9α and 8β) are included in the CI space. The present calculations for W_d consider nine sets of RASCI space, which are constructed from 17 active electrons and 16, 21, 26, 31, 36, 46, 56, 66, and 76 active orbitals to analyze the convergence of W_d.

The W_d estimated from the RASCI is compared with other theoretical calculations [46, 54] in Table V. The present DF estimate of W_d deviates by $\sim 21\%(29\%)$ from the SCF (RASSCF) estimate of Kozlov et al. [46] and by $\sim 17\%$ from the semiempirical result of Kozlov and Labzovsky [54], while our RASCI result departs by $\sim 6\%(3\%)$ from the SCF-EO (RASSCF-EO) treatment of Kozlov et al. [46] and is in good agreement with the semi-empirical result of Kozlov and Labzovsky. At this juncture, we emphasize that our computed ground state dipole moment of BaF ($\mu_e = 3.203$ debyes) is also reasonably close to experiment $\mu_e = 3.2$ debyes (see Table 5 of Ref. 38).

Figure 4 plots W_d against the dimension of the CI space used in the calculations. It is evident from Fig. 4 that the W_d decreases with increasing size of the CI space

TABLE V
P,T-Odd Interaction Constant W_d for the Ground $X^2\Sigma$ State of the BaF Molecule

Method	$W_d 10^{25}$ Hz/$e \cdot$cm)
SCF [46]	−0.230
RASSCF [46]	−0.224
SCF-EO [46]	−0.375
RASSCF-EO [46]	−0.364
Semiempirical [54]	−0.35
DF (Nayak et al.) [55]	−0.293
RASCI (Nayak et al.) [55]	−0.352

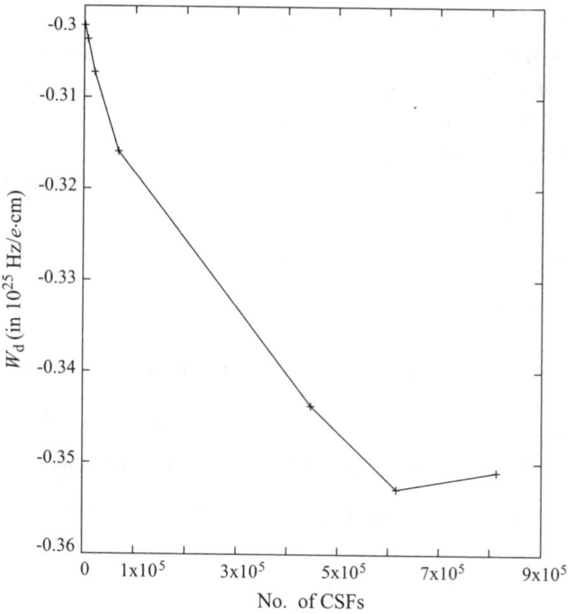

Figure 4. Plot of the P,T-odd interaction constant W_d versus number of CSFs for the BaF molecule.

until it reaches -0.352×10^{25} Hz/$e \cdot$cm. With the exception of a very small increase at the final calculation, the parameter has essentially stabilized. Figure 4 further indicates that contribution to W_d from orbitals 70–100 (CSFs 1–6 $\times$ 10^5) is most significant compared to other unoccupied (at DF level) active orbitals.

At this juncture, we note that experiments on the BaF molecule to detect P,T-symmetry violation has not yet been performed but are being planned. Hence the present estimate for W_d as well as the previous calculations [46, 54] will be useful for experimentalists. As the experiments progress, calculations with much higher accuracies will be performed by including effects that are neglected/omitted in the present calculation such as electric field $\boldsymbol{E}_{i,j}$ in Eq.(42). Therefore the accuracy of the present calculation is sufficient at this moment.

APPENDIX A: CARTESIAN GAUSSIAN SPINORS AND BASIS FUNCTIONS

In terms of Cartesian Gaussian spinors, the basis functions can be defined as a linear combination of the following Gaussian spinors [57]:

$$\phi_I = N_{k,\alpha} \prod_{i=1,3} (x_i - a_i)^{k_i} \exp\left[-\alpha(x_i - a_i)^2\right]\chi_{\pm} \tag{49}$$

where $N_{k,\alpha}$ is a normalization constant, $k = (k_1, k_2, k_3)$, and $\chi_\pm$ is a two-component spinor (i.e., α or β spin function). The Gaussian is centered at a and the orbital angular momentum quantum number is defined as $L = \sum_i k_i$.

Assuming the Gaussian to be centered at the origin of our coordinate system, we can define the s, p, and d type functions as

$$
\begin{aligned}
s &= N_{k_{000},\alpha_s} \exp\left[-\alpha_s r^2\right]\chi_\pm \\
p_x &= N_{k_{100},\alpha_p}(x) \exp\left[-\alpha_p r^2\right]\chi_\pm, \quad p_y = \cdots, \quad p_z = \cdots \\
d_{xy} &= N_{k_{110},\alpha_d}(xy) \exp\left[-\alpha_d r^2\right]\chi_\pm, \quad d_{xz} = \cdots
\end{aligned}
$$

The subscripts s, p, and d denote the different exponents for different kinds of functions.

A.1 Difficulty in Evaluating the Matrix Elements of H_d

The P,T-odd interaction operator H_d used in the calculations of W_d is of the form

$$
\mathrm{H}_d \propto \frac{\hat{\boldsymbol{\sigma}} \cdot \hat{\boldsymbol{r}}}{r^2} \tag{50}
$$

We can write $r^2 = x^2 + y^2 + z^2$ and $\hat{\boldsymbol{\sigma}} \cdot \hat{\boldsymbol{r}}$ as

$$
\hat{\boldsymbol{\sigma}} \cdot \hat{\boldsymbol{r}} = \frac{\vec{\boldsymbol{\sigma}} \cdot \vec{\boldsymbol{r}}}{|\sigma||r|}
$$

where $|r| = (x^2 + y^2 + z^2)^{1/2}$ and $|\sigma| = ({\sigma_x}^2 + {\sigma_y}^2 + {\sigma_z}^2)^{1/2}$. Likewise, $\vec{\boldsymbol{\sigma}} \cdot \vec{\boldsymbol{r}}$ can be written

$$
\vec{\boldsymbol{\sigma}} \cdot \vec{\boldsymbol{r}} = \sigma_x x + \sigma_y y + \sigma_z z
$$

which means

$$
\mathrm{H}_d \propto \frac{\sigma_x x + \sigma_y y + \sigma_z z}{(x^2 + y^2 + z^2)^{3/2}} \tag{51}
$$

Now let us evaluate the matrix element of Eq. (51) between s and p (e.g., p_z) type basis functions. Excluding the constant terms and the spin part, the integral involving the spatial part is of the form

$$
\int_{-\infty}^{+\infty} \int_{-\infty}^{+\infty} \int_{-\infty}^{+\infty} \frac{z^2 \exp[-(\alpha_s + \alpha_p)(x^2 + y^2 + z^2)]}{(x^2 + y^2 + z^2)^{3/2}} \, dx \, dy \, dz \tag{52}
$$

This integral is quite difficult to integrate and so we need to find some other alternative procedure.

APPENDIX B: EXPANSION OF CARTESIAN GAUSSIAN BASIS FUNCTIONS USING SPHERICAL HARMONICS

The problem of using the Cartesian Gaussian can be resolved by converting the Cartesian Gaussian into spherical polar coordinates and then expressing the angular parts in terms of spherical harmonics Y_l^m. The necessity of expressing the angular part in terms of spherical harmonics is just to enable the integration of the angular part. Using this procedure, we express x, y, z and 1 as follows:

$$\begin{aligned}
x &= r\sin\theta\cos\phi = -r\sqrt{\frac{2\pi}{3}}(Y_1^1 - Y_1^{-1}) \\
y &= r\sin\theta\sin\phi = \frac{-r}{i}\sqrt{\frac{2\pi}{3}}(Y_1^1 + Y_1^{-1}) \\
z &= r\cos\theta = r\sqrt{\frac{4\pi}{3}}(Y_1^0) \\
1 &= \sqrt{4\pi}(Y_0^0)
\end{aligned}$$

where Y_l^m are standard spherical harmonics.

Omitting the normalization constant as well as the spin part, we can write the spatial part of s, p, and d type functions as

$$\begin{aligned}
s &= \sqrt{4\pi}\exp(-\alpha_s r^2)[Y_0^0] \\
p_x &= -r\sqrt{\frac{2\pi}{3}}\exp(-\alpha_p r^2)[Y_1^1 - Y_1^{-1}] \\
d_{xy} &= \frac{r^2}{i}\sqrt{\frac{2\pi}{15}}\exp(-\alpha_d r^2)[Y_2^2 - Y_2^{-2}]
\end{aligned}$$

B.1 The P,T-Odd Operator H_d in Spherical Polar Coordinates

Adopting the same procedure, we can also express the operator H_d in spherical polar coordinate. The first step is to write $\vec{\sigma}\cdot\vec{r}$ as

$$\vec{\sigma}\cdot\vec{r} = \sigma_x r\sin\theta\cos\phi + \sigma_y r\sin\theta\sin\phi + \sigma_z r\cos\theta \tag{53}$$

so that the angular part $\hat{\sigma}\cdot\hat{r}$ of the operator H_d can be written

$$\hat{\sigma}\cdot\hat{r} = \frac{\vec{\sigma}\cdot\vec{r}}{|\sigma||r|} = \frac{\sigma_x\sin\theta\cos\phi + \sigma_y\sin\theta\sin\phi + \sigma_z\cos\theta}{|\sigma|} \tag{54}$$

where $|\sigma|$ is a constant. The operators σ_x, σ_y, and σ_z will act on the spin part of the basis functions. So the spin-dependent part can be separated from the spatial part of the integral. Here we concentrate only on the spatial part of the integral, where the spatial part of the P,T-odd interaction operator H_d is of the form

$$\mathrm{H}_d \propto \frac{\sin\theta\cos\phi + \sin\theta\sin\phi + \cos\theta}{r^2} \tag{55}$$

B.2 Matrix Elements of the P,T-Odd Operator H_d

First consider the s and p type functions. The only nonzero matrix element of the operator $\cos\theta/r^2$ (last term of the above expression) is between s and p_z. Matrix elements of H_d between functions of the same parity are zero. Excluding the normalization constants and the spin-dependent part, the matrix element of the operator H_d between s and p_z is

$$\left\langle s\left|\frac{\cos\theta}{r^2}\right|p_z\right\rangle = \frac{4\pi}{\sqrt{3}}\int_0^\infty r\exp[-(\alpha_s+\alpha_p)r^2]dr\int_\Omega Y_0^{0*}\cos\theta Y_1^0 d\Omega \tag{56}$$

Similarly, the nonzero matrix elements of the same operator $(\cos\theta/r^2)$ between p and d type functions can be obtained using the same procedure as

$$\begin{aligned}
\left\langle p_x\left|\frac{\cos\theta}{r^2}\right|d_{xz}\right\rangle &= \frac{2\pi}{3\sqrt{5}}\int_0^\infty R(r)dr\int_\Omega (Y_1^1 - Y_1^{-1})^*\cos\theta(Y_2^1 - Y_2^{-1})d\Omega\\
\left\langle p_y\left|\frac{\cos\theta}{r^2}\right|d_{yz}\right\rangle &= \frac{2\pi}{3\sqrt{5}}\int_0^\infty R(r)dr\int_\Omega (Y_1^1 + Y_1^{-1})^*\cos\theta(Y_2^1 + Y_2^{-1})d\Omega\\
\left\langle p_z\left|\frac{\cos\theta}{r^2}\right|d_{x^2}\right\rangle &= \frac{2\sqrt{2}\pi}{3\sqrt{5}}\int_0^\infty R(r)dr\int_\Omega Y_1^{0*}\cos\theta(Y_2^2 + Y_2^{-2} - Y_2^0 + \sqrt{5}Y_0^0)d\Omega\\
\left\langle p_z\left|\frac{\cos\theta}{r^2}\right|d_{y^2}\right\rangle &= -\frac{2\sqrt{2}\pi}{3\sqrt{5}}\int_0^\infty R(r)dr\int_\Omega Y_1^{0*}\cos\theta(Y_2^2 + Y_2^{-2} + Y_2^0 - \sqrt{5}Y_0^0)d\Omega\\
\left\langle p_z\left|\frac{\cos\theta}{r^2}\right|d_{z^2}\right\rangle &= \frac{8\pi}{3\sqrt{15}}\int_0^\infty R(r)dr\int_\Omega Y_1^{0*}\cos\theta(Y_2^0 + \sqrt{5/4}Y_0^0)d\Omega
\end{aligned} \tag{57}$$

Here, we have used the abbreviated form for the radial part of the integral as $R(r) = r^3\exp[-(\alpha_p+\alpha_d)r^2]$.

Now we can evaluate the radial and angular parts of the integrals separately. The radial part is evaluated using the standard formula

$$\int_0^\infty r^{2n+1}e^{-ar^2}dr = \frac{n!}{2a^{n+1}}$$

where n is a positive integer.

The angular part of the integral is evaluated using the standard expression given by Arfken and Weber [58],

$$\int_{\Omega} Y_{L_1}^{M_1 *} \cos\theta Y_L^M d\Omega = \left[\frac{(L-M+1)(L+M+1)}{(2L+1)(2L+3)}\right]^{1/2} \delta_{M_1,M}\delta_{L_1,L+1} + \left[\frac{(L-M)(L+M)}{(2L-1)(2L+1)}\right]^{1/2} \delta_{M_1,M}\delta_{L_1,L-1} \tag{58}$$

Using these above expressions, we obtain the required integrals (i.e., final matrix elements of the operator $(\cos\theta)/r^2$) as

$$\left\langle s\left|\frac{\cos\theta}{r^2}\right|p_z\right\rangle = \frac{2\pi}{3}\left(\frac{1}{\alpha_s+\alpha_p}\right) \tag{59}$$

$$\left\langle p_x\left|\frac{\cos\theta}{r^2}\right|d_{xz}\right\rangle = \frac{2\pi}{15}\left(\frac{1}{\alpha_p+\alpha_d}\right)^2 \tag{60}$$

$$\left\langle p_y\left|\frac{\cos\theta}{r^2}\right|d_{yz}\right\rangle = \frac{2\pi}{15}\left(\frac{1}{\alpha_p+\alpha_d}\right)^2 \tag{61}$$

$$\left\langle p_z\left|\frac{\cos\theta}{r^2}\right|d_{x^2}\right\rangle = \frac{2\pi}{15}\left(\frac{1}{\alpha_p+\alpha_d}\right)^2 \tag{62}$$

$$\left\langle p_z\left|\frac{\cos\theta}{r^2}\right|d_{y^2}\right\rangle = \frac{2\pi}{15}\left(\frac{1}{\alpha_p+\alpha_d}\right)^2 \tag{63}$$

$$\left\langle p_z\left|\frac{\cos\theta}{r^2}\right|d_{z^2}\right\rangle = \frac{2\pi}{5}\left(\frac{1}{\alpha_p+\alpha_d}\right)^2 \tag{64}$$

Furthermore, we adopt the same procedure to evaluate the matrix elements of the other terms in the expression of the P,T-odd interaction operator H_d (i.e., $(\sin\theta\cos\phi)/r^2$ and $(\sin\theta\sin\phi)/r^2$) as well as between other types of functions with higher l (i.e., between d, f and f, g) and so on. Some other methods also exist to resolve the above-mentioned problem.

Acknowledgments

We thank Prof. Hans Joergen Jensen and his group for providing their DIRAC04 [56] code, which is adapted with our codes. The authors are deeply indebted and grateful to Prof. Karl Freed and Prof. S. L. N. G. Krishnamachari for their valuable comments and suggestions.

This research is supported, in part, by the Department of Science and Technology, India (Grant SR/S1/PC-32/2005).

References

1. C. S. Wu, E. Ambler, R. W. Hayward, and R. P. Hudson, *Phys. Rev* **105**, 1413 (1957).
2. J. H. Christenson, J. W. Cronin, V. L. Fitch, and R. Turlay, *Phys. Rev. Lett.* **13**, 138 (1964).
3. W. Pauli, *Neils Bohr and the Development of Physics*, McGraw-Hill, New York, 1955.
4. G. Luders, *Ann. Phys.* **2**, 1 (1957).
5. T. D. Lee and C. N. Yang, Brookhaven National Lab, *Report No.* **BNL443**, 791 (1957).
6. L. D. Landau, *Nucl. Phys.* **3**, 127 (1957).
7. M. Pospelov and I. B. Khriplovich, *Sov. J. Nucl. Phys.* **53**, 638 (1991).
8. E. D. Commins, *Adv. At. Mol. Opt. Phys.* **40**, 1 (1999).
9. A. V. Titov, N. S. Mosyagin, T. A. Isaev, and D. DeMille, *Prog. Theor. Chem. Phys. B* **15**, 253 (2006).
10. W. Bernreuther and M. Suzuki, *Rev. Mod. Phys.* **63**, 313 (1991).
11. L. D. Landau, *Sov. Phys. JETP* **5**, 336 (1957).
12. B. C. Regan, E. D. Commins, C. J. Schmidt, and D. DeMille, *Phys. Rev. Lett.* **88**, 071805 (2002).
13. M. V. Romalis, W. C. Griffith, and E. N. Fortson, *Phys. Rev. Lett.* **86**, 2505 (2001).
14. V. F. Dmitriev and R. A. Senkov, *Phys. Rev. Lett.* **91**, 212303 (2003).
15. E. A. Hinds, C. E. Loving, and P. G. H Sandars, *Phys. Lett. B* **62**, 97 (1976).
16. P. G. H. Sandars, *Phys. Rev. Lett.* **19**, 1396 (1967).
17. P. G. H. Sandars, *Phys. Lett.* **14**, 194 (1965).
18. D. Cho, K. Sangster, and E. A. Hinds, *Phys. Rev. A* **44**, 2783 (1991).
19. A. N. Petrov, N. S. Mosyagin, T. A. Isaev, A. V. Titov, V. F. Ezov, E. Eliva, and U. Kaldor, *Phys. Rev. Lett.* **88**, 073001 (2002).
20. L. N. Labzovsky, *Sov. Phys. JETP* **48**, 434 (1978).
21. V. G. Gorshkow, L. N. Labzovsky, and A. N. Moskalyov, *Sov. Phys. JETP* **49**, 209 (1979).
22. O. P. Sushkov and V. V. Flambaum, *Sov. Phys. JETP* **48**, 608 (1978).
23. O. P. Sushkov, V. V. Flambaum, and I. B. Khriplovich, *Sov. Phys. JETP* **87**, 1521 (1984).
24. V. V. Flambaum and I. B. Khriplovich, *Phys. Lett. A* **110**, 121 (1985).
25. M. G. Kozlov, *Sov. Phys. JETP* **62**, 1114 (1985).
26. M. G. Kozlov, V. I. Fomichev, Y. Y. Dmitriev, L. N. Labzovsky, and A. V. Titov, *J. Phys. B* **20**, 4939 (1987).
27. J. J. Hudson, B. E. Sauer, M. R. Tarbutt, and E. A. Hinds, *Phys. Rev. Lett.* **89**, 023003 (2002).
28. B. E. Sauer et al., *At. Phys.* **20**, 44 (2006); *XX International Conference on Atomic Physics (ICAP 2006)*, C. Roos, H. Haeffner, and R. Blatt (eds.), American Institute of Physics, Melville, NY, 2006.
29. D. DeMille, F. Bay, S. Bickman, D. Kawall, Jr., D. Krause, S. E. Maxwell, and L. R. Hunter, *Phys. Rev. A* **61**, 052507 (2000).
30. F. Hoogeveen, *Nucl. Phys. B* **341**, 322 (1990).
31. J. J. Sakurai, *Modern Quantum Mechanics*, Addison Wesley, Boston, 1985, p. 241.
32. L. I. Schiff, *Phys. Rev.* **132**, 2194 (1963).
33. M. G. Kozlov and L. N. Labzovsky, *J. Phys. B* **28**, 1933 (1961).
34. N. S. Mosyagin, M. G. Kozlov, and A. V. Titov, *J. Phys. B* **31**, L763 (1998).

35. F. A. Parpia, *J. Phys. B* **31**, 1409 (1998).
36. F. A. Parpia and A. K. Mohanty, *Phys. Rev. A* **52**, 962 (1995).
37. B. E. Sauer, J. Wang, and E. A. Hinds, *Phys. Rev. Lett.* **74**, 1554 (1995).
38. B. E. Sauer, J. Wang, and E. A. Hinds, *J. Chem. Phys.* **105**, 7412 (1996).
39. S. M. Barr, *Phys. Rev. D* **45**, 4148 (1992).
40. S. M. Barr, *Int. J. Mod. Phys. A* **8**, 209 (1993).
41. B. N. Ashkinadzi et al., Petersburg Nuclear Physics Institute, St. Petersburg, *Report No.* **1801**, (1992).
42. E. A. Hinds, and P. G. H. Sandars, *Phys. Rev. A* **21**, 471 (1980).
43. A. V. Titov, N. S. Mosyagin, and V. F. Ezhov, *Phys. Rev. Lett.* **77**, 5346 (1996).
44. H. M. Quiney, H. Skaane, and I. P. Grant, *J. Phys. B* **31**, L85 (1998).
45. N. S. Mosyagin, M. G. Kozlov, and A. V. Titov, *J. Phys. B* **31**, L763 (1998).
46. M. G. Kozlov, A. V. Titov, N. S. Mosyagin, and P. V. Souchkov, *Phys. Rev. A* **56**, R3326 (1997).
47. L. B. Knight, W. C. Easley, W. Weltner, and M. Wilson, *J. Chem. Phys.* **54**, 322 (1971).
48. M. G. Kozlov and V. F. Ezhov, *Phys. Rev. A* **49**, 4502 (1994).
49. M. G. Kozlov, *J. Phys. B* **30**, L607 (1997).
50. M. K. Nayak and R. K. Chaudhuri, *Chem. Phys. Lett.* **419**, 191 (2006).
51. M. Dolg, H. Stoll, and H. Preuss, *Chem. Phys.* **165**, 21 (1992).
52. K. P. Huber and G. Herzberg, *Molecular Spectra and Molecular Structure,Constants of Diatomic Molecules*, Van Nostrand, New York, 1979, Vol. 4.
53. R. E. Stanton and S. Havriliak, *J. Chem. Phys.* **81**, 1910 (1984).
54. M. G. Kozlov and L. N. Labzovsky, *J. Phys. B* **28**, 1933 (1995).
55. M. K. Nayak and R. K. Chaudhuri, *J. Phys. B* **39**, 1231 (2006).
56. http://dirac.chem.sdu.dk/obtain/Dirac-site-license.shtml: *Dirac, a relativistic ab-initio electronic structure program, Release DIARC04 (2004)*, written by H. J. Aa. Jensen, T. Saue, and L. Visscher with contributions from V. Bakken, E. Eliav, T. Enevoldsen, T. Fleig, O. Fossgaard, T. Helgaker, J. Laerdahl, C. V. Larsen, P. Norman, J. Olsen, M. Pernpointner, J. K. Pedersen, K. Ruud, P. Salek, J. N. P. Van Stralen, J. Thyssen, O. Visser, and T. Winther (http://dirac.chem.sdu.dk).
57. P. J. C. Aerts and W. C. Nieuwpoort, *Chem. Phys. Lett.* **113**, 165 (1985).
58. G. B. Arfken and H. J. Weber, *Mathematical Methods for Physicists*, Elsevier, New Delhi, India, 2001, p. 804.

AUTHOR INDEX

Numbers in parentheses are reference numbers and indicate that the author's work is referred to although his name is not mentioned in the text. Numbers in *italic* show the page on which the complete references are listed.

Advances in Chemical Physics, Volume 140, edited by Stuart A. Rice
Copyright © 2008 John Wiley & Sons, Inc.

SUBJECT INDEX

Advances in Chemical Physics, Volume 140, edited by Stuart A. Rice
Copyright © 2008 John Wiley & Sons, Inc.